D0205231

FUNDAMENTALS OF

PLANT PATHOLOGY

FUNDAMENTALS OF PLANT PATHOLOGY

Daniel A. Roberts
UNIVERSITY OF FLORIDA

Carl W. Boothroyd
CORNELL UNIVERSITY

W. H. FREEMAN AND COMPANY
San Francisco

Printed in the United States of America

International Standard Book Number: 0-7167-0822-1
Library of Congress Catalog Card Number: 77-169737

4 5 6 7 8 9

to
Ruth Remsen Roberts
and
Loretta Ranney Boothroyd

Contents

Preface

Fully 90 percent of the 4000-odd students who take college courses in plant pathology every year are undergraduates studying the subject for the first and last time. We have written this book for that vast, sometimes silent majority, as well as for those who decide to make a career in this field.

We believe the beginning student is often bewildered by the details usually set forth in the textbooks of our science. Consequently, we have spared the minutiae, but have tried to present concisely the principles of plant pathology, illustrated by selected examples of important plant diseases. It is important that the student learn a few basic concepts and accumulate a few specific facts about plant diseases. However, it is just as important that he gain an appreciation of the relation of plant pathology to the day-to-day activities around him and that his study of this science in some way be useful and productive, whatever his ultimate profession may be.

Our experience during the past twenty years has determined the organization of the contents. The basic laws of plant pathology constitute Part I, which was drafted earlier by the senior author, supported by a Guggenheim Fellowship and a sabbatical leave from Cornell University. Examples of the laws of plant pathology make up Part II. We teach by example, and present to our students the contents of both parts simultaneously. For example, postharvest diseases (Chapter 10) are studied first and are used as specific examples of the principles defined in Chapter 1. Throughout our courses of instruction, these principles are discussed in the light of the specific diseases they exemplify.

The technical language of this textbook is based on the terminology of the late H. H. Whetzel, whose perception and insistence on precision stimulated us to make this effort. Also, Whetzel's view that "the diseased plant plays the central role in the phytopathological drama" is the thesis of this book. Unlike many introductory texts on plant pathology, this book logically places emphasis upon consideration of the diseased plant, rather than upon the specific causes of disease.

Plant disease is continuous dysfunction, so it is logical to analyze plant diseases in terms of impaired physiology. We have classified plant diseases for what they are, and we follow the scheme proposed more than twenty years ago by G. L. McNew. This classification, which is based on the physiological activities of the plant that are adversely affected by the disease in question, digresses from Whetzel's, which arranged diseases according to the kinds of symptoms produced. McNew's classification properly emphasizes the physiological aspects of plant disease, but it also gives the diseased plant the leading role.

We owe special thanks to our teaching assistants, whose enthusiasm, industry, and patience in the laboratory often go unrewarded. We are also indebted to all of our students, whose struggles with our subject have clearly defined the needs we have tried to meet with this book. To the extent that we have succeeded in meeting those needs, we owe much to the inspiration and criticism of our own teachers and colleagues. For the help they gave, we are particularly grateful to the teaching staff at Cornell University, and to R. E. Stall, J. W. Kimbrough, and G. F. Weber of the University of Florida.

Administrative matters were handled with the cooperation of the heads of the Department of Plant Pathology at the University of Florida, Gainesville, and the Department of Plant Pathology at Cornell University, Ithaca, New York. We thank Phares Decker, G. C. Kent, L. H. Purdy, and D. F. Bateman for giving us the time and secretarial help we needed to prepare the manuscript, and we deeply appreciate the assistance of H. H. Lyon of Cornell University, who prepared most of the halftones for publication.

Finally, we submit our effort to students beginning their study of the science of plant pathology with the hope that it serves them well.

April 1972

Daniel A. Roberts
Carl W. Boothroyd

I

THE THEORY OF PLANT PATHOLOGY

1

Disease in Plants and the Science of Plant Pathology

Every crop plant is in jeopardy from the moment its seed is sown: if it is to yield full measure, a plant must endure the buffeting of the elements, competition from weeds, plagues of insects, and the ravages of disease. Plant diseases alone exacted an annual toll of some $3 billion in the United States during the mid–twentieth century—this despite disease-control efforts based on science and technology that are considered to be highly advanced. Crop losses due to plant diseases are discussed in Box 1-1.

Three of the natural enemies of cultivated plants, weather, weeds, and insects, are dealt with, respectively, in the disciplines of climatology, agronomy and horticulture, and entomology; a fourth is dealt with in the science of **plant pathology**, the study of the nature, causes, and prevention of plant diseases. Plant pathologists have two objectives: first, to minimize crop losses through application of the principles of plant-disease prevention; and second, to develop a deeper understanding of biological science through knowledge of the principles of plant disease. The relation of plant pathology to other scientific disciplines is indicated in Box 1-2.

BOX 1-1
CROP LOSSES DUE TO PLANT DISEASES

We believe that estimates of financial losses resulting from plant diseases are seldom accurate, and that, when they do approximate actual values, they do so only by chance. Estimation of crop losses is based on two very different principles: (1) measuring yield loss and (2) predicting what that lost yield might have represented in terms of farm income.

The first principle is difficult to express in quantitative terms, the second, impossible. The income from a crop yielding only 80% of its potential may exceed the income expected from 100% yield; the increased supply could depress the unit price by more than enough to offset the value of the increased yield. And there is yet no way to predict market economy. All the estimator can do is to equate the value of the lost yield with that actually obtained. This is about as accurate as expressing the national economy in terms of "full employment."

Scientists have made progress in the measurement of crop losses due to diseases, but too little is known about just how the continuous dysfunction of diseased plants results in decreased crop yields. One would expect a yield loss to be directly related to the number of missing plants in a crop suffering from a lethal disease. But this is not necessarily true, particularly in root and tuber crops. An isolated "hill" of potatoes, for example, will yield about half again as much as one between two others in a row. Such a dramatic increase would probably not occur, however, in an isolated plant of a crop whose marketable product is borne above the ground. More information is needed on the "missing-hill effect" as it applies to different species of crop plants, particularly those affected by seedling blights, root and stem rots, vascular wilts, and other lethal diseases.

Two major problems are associated with measuring losses due to plant diseases in which foliage is spotted or blighted. The first—to develop a method of determining the amount of actual necrosis by visual inspection—has been generally overcome. Visual grades progress in logarithmic steps, so amounts of necrosis can be expressed on a scale $\log [a/(100 - a)]$, where a is the percentage of the leaf or leaflet that is necrotic at the time of observation. The values for a can be determined visually only by comparing diseased foliage with standard diagrams prepared in advance for each crop species and each disease likely to be encountered. Details of the techniques of measuring plant diseases have been given by Large (1966).

The second problem in measuring losses due to foliage diseases is determining the effect of different measured amounts of necrosis on the yield and quality of the marketable product. The question of how much disease can be tolerated without significant loss of yield is most difficult to answer. The time that a given amount of necrosis occurs and the part of the plant in which it develops determine the ultimate loss in yield. Moreover, necrosis in the product itself, although it does not always reduce the yield, may well lower the quality so that it may be drastically down-graded or even culled.

In the late-blight disease of potatoes, 50–100% of the crop will be lost if 75% of the leaflets have at least one lesion within two months of planting. If foliage is so blighted three months after planting, approximately 20% of the crop will be lost, but if the disease does not affect 75% of the leaflets until within two to four weeks of harvest, only about 5% will be lost.

Little is known of the role played by the different parts of the plant in contributing to the dry weight of the farm product. In small grains, however, it is known that photosynthates from the flag leaf, the upper stem, and flower parts are responsible for the bulk of the grain yield. Lower leaves contribute only 15–20% of the yield. A wheat plant, therefore, can tolerate a foliage infection so extensive that the lowest leaves are completely necrotic and still yield almost full measure, provided its upper parts are unaffected.

Despite the errors inherent in the methods of estimating crop losses, we believe that the figures listed below, based on the report by LeClerg (1964) are reasonable estimates of annual crop losses due to infectious plant diseases in the United States.

Barley	14%	Ornamental Plants	15%
Beans	22%	Potatoes	23%
Corn	15%	Sugarcane	14%
Cotton	14%	Tomatoes	23%
Fruit and Nuts	12–30%	Wheat	28%
Oats	21%		

(*continued*)

BOX 1-1 (*continued*)

On the basis of these figures, it would perhaps be safe to generalize that plant diseases, on the average, reduce the potential crop yields in the United States by 15–20%; but the Georgia farmer who lost 100% of his corn crop in the 1970 epidemic of southern leaf blight takes no comfort in the fact that the average loss to that disease was only 15% on a nationwide basis.

More important than average crop losses is the potential loss that could occur if plant-disease control measures were not practiced. The potential loss is difficult to estimate, but parasitic nematodes alone take at least 60% of the potential yield of potatoes in some countries in which agricultural technology is not nearly so far advanced as, for example, in the United States.

Obviously, much more research is needed on the problems of estimating losses from infectious plant diseases. Perhaps the most significant contributions in the immediate future will come from studies that seek to find out how much plant disease we can afford. We need desperately to know the exact relations between yield reductions and specific intensities of individual plant diseases—intensities determined at stated intervals during the cropping season.

A CONCEPT OF DISEASE IN PLANTS

Disease is continuous impairment of metabolism, but the complex interactions of biological phenomena that together constitute plant disease make further definition difficult. Most published definitions are inadequate—they are descriptive but not simultaneously exclusive. Whetzel (1935) proposed a precise definition of the concept of disease, paraphrased below, that serves as a logical basis for study of plant pathology: *Disease in a plant consists of a series of harmful physiological processes caused by continuous irritation of the plant by a primary agent.* It is exhibited through morbific cellular activity, and is expressed by morphological and histological conditions called symptoms.

Disease itself is not a condition, but a process or, rather, an interrelated group of harmful processes. It harms green plants by affecting

BOX 1-2
THE RELATION OF PLANT PATHOLOGY TO OTHER SCIENCES

The objective of the plant pathologist is to prevent plant-disease epidemics or to delay their onset until after harvest time, so that serious crop losses cannot occur. Knowledge of basic biological and physical sciences, as well as comprehension of environmental and social sciences, are the foundation stones upon which the science of plant pathology rests.

The plant pathologist, to be effective, must focus the knowledge of these basic sciences, through the lens of plant pathology, upon the objective of preventing disease epidemics in crop plants. The interrelations of the basic and applied aspects of plant pathology are illustrated below.

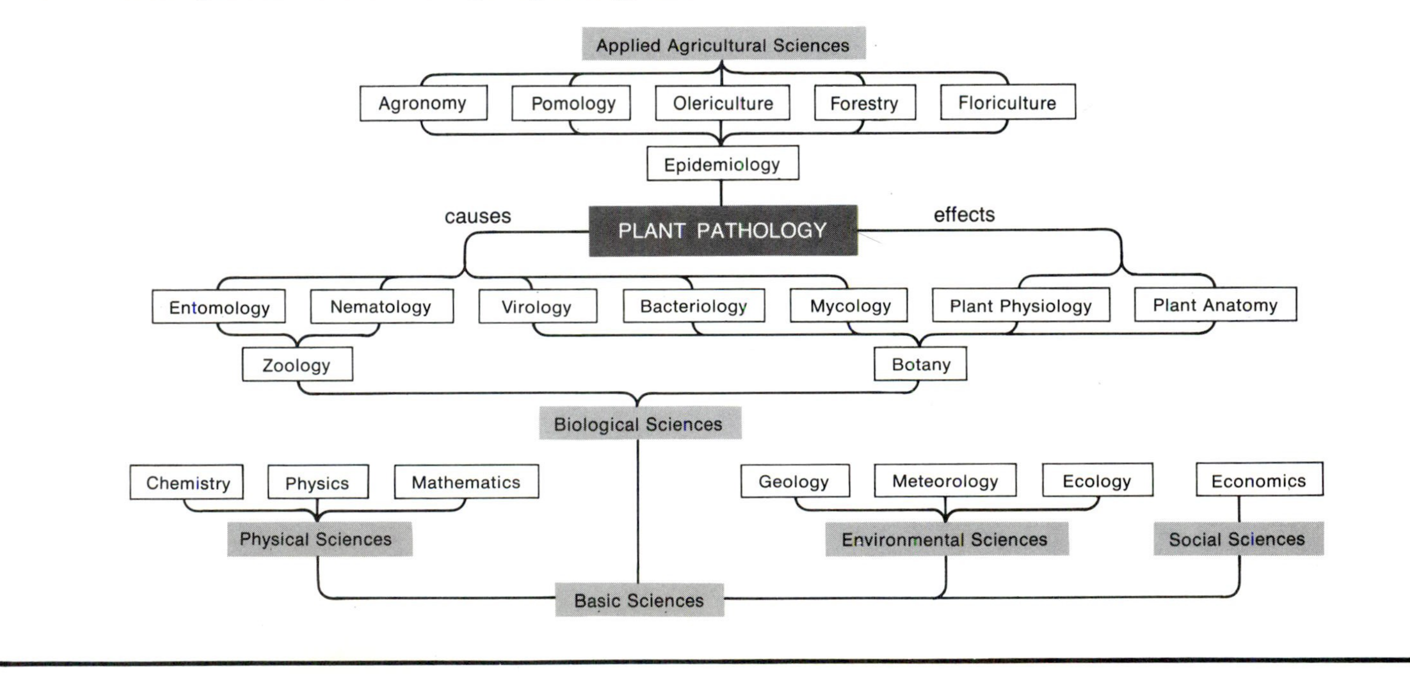

functions normally involved in their growth and reproduction (Duggar, 1909; Sorauer, 1914). In essence, it is physiological (Blackman, 1924; Duggar, 1909; Kühn, 1858; Morstatt, 1922) and processive or dynamic (Whetzel, 1929). Injuries, whether mechanical or induced by certain insects or other agents, are also harmful, and may impair vital functions. Disease differs from injury in one important way, however: the cause of disease exerts its irritation continuously, whereas an agent of injury affects the plant intermittently, its irritation being discontinuous, temporary, or transient.

Plant disease was once thought to result from supernatural causes, and was later attributed to bad weather. It is true that the environment, especially the weather, can determine the severity of a plant disease, but it is causally involved at a secondary level: there can be no disease in the absence of the primary agent. If this primary agent, the **pathogen,** is transmissible from diseased to healthy plants in which disease may subsequently develop, the agent (a fungus or bacterium, for example) is said to be **infectious,** and the term **infectious disease** is commonly used when such an agent is the causative factor. Nontransmissible agents, such as deficiencies or excesses of mineral nutrients, are termed **noninfectious;** similarly, plant diseases resulting from such causes are referred to as **noninfectious diseases.** In this book we are concerned both with diseases of plants caused by infectious agents and those caused by noninfectious agents. Because both kinds of agents cause diseases, both may be termed "pathogens." In common usage, however, that term denotes only transmissible agents of disease, such as bacteria, fungi, and viruses. To avoid confusion, therefore, the term will be restricted to transmissible agents in the following chapters.

Cellular irritability is the basis of the science of plant pathology (Duggar, 1911), and the continuous irritation of a plant by an exogenous causal agent is manifested in morbific or abnormal activity at the cellular level. Abnormal cellular activity of autogenic origin sometimes results in teratoses, or "misdevelopments" of organs (Whetzel, 1935), such as the "doubling" of rose blossoms. In contrast to plant disease, teratoses rarely affect the physiological process of affected plants in a harmful way.

Acceptance of a sound concept of disease is prerequisite to a proper study of plant pathology as a science. "Science is knowledge reduced to law and embodied in system" (Funk & Wagnalls Unabridged Dictionary, 1923), and it seeks to establish causal connections among facts and to expound laws that correlate facts (Duggar, 1909). Our discussion of the fundamentals of plant pathology from this point on through Chapter 22 is based on principles illustrated by infectious diseases. We discuss certain noninfectious diseases in Chapter 23.

THE CLASSIFICATION OF INFECTIOUS PLANT DISEASES

Science demands that the laws or principles of a discipline be embodied in a logical system, and various schemes of classifying plant diseases have been presented to meet this demand. Classifications of infectious plant diseases have been based on (1) crops affected, (2) organs attacked, (3) taxonomy of pathogens, (4) symptoms, (5) source of inoculum, and (6) physiology of diseased plants. Classification by crops affected or organs attacked may be convenient approaches to the study of individual diseases, but they have contributed little to the synthesis of the unified body of knowledge that constitutes the science of plant pathology. Classification of disease by the taxonomy of pathogens is illogical; a disease consists of a complex of harmful physiological processes, and these processes are ignored when only the primary agent of the complex, the pathogen, is the basis for classification. Moreover, there is usually little or no correlation between the systematics of pathogens and the nature of the diseases caused by them (McNew, 1960). Classification on the basis of symptoms properly directs attention to studies of the diseased plant. Such a classification, however, is actually a classification of the conditions of disease and not one of disease itself: nor does it take into account that symptoms, however useful as aids to diagnosis, vary with different environments. Classification by the source of inoculum divides the thousands of plant diseases into too few categories to be useful.

McNew (1950, 1960) has proposed a system based upon seven physiological processes: (1) storage of food, (2) hydrolysis and utilization of stored food, (3) absorption and accumulation of water and minerals, (4) growth (meristematic activity), (5) conduction of water, (6) photosynthesis, and (7) translocation of elaborated food materials. Depending upon which of these vital functions is adversely affected, a plant disease would be classified in one of the following groups: (1) soft rots and seed decays, (2) damping-off and seedling blights, (3) root rots, (4) gall diseases and others in which meristematic activity is impaired, (5) vascular wilts, (6) diseases affecting photosynthesis (bacterial and fungal spots and blights, downy and powdery mildews, and rusts), and (7) diseases interfering with translocation (viral diseases and those caused by mycoplasmalike organisms).

Despite its limitations, this classification is the only one thus far proposed that attempts to group infectious plant diseases in a natural system in keeping with the concept of the physiological nature of disease in plants. With this classification, it is possible to demonstrate the complex interactions among pathogens, susceptible plants, and the environment, and to demonstrate how these interactions are related to different kinds of disease and to the principles of plant-disease prevention.

The relative severity of infectious disease may be limited by any one of the following variables: (1) the avoidance, by chance, of agents of disease (often termed "disease escape"), (2) the inherent degree of resistance of the plant, (3) the number of pathogens, (4) the virulence (degree of pathogenicity) of pathogens, (5) the suitability to the pathogen of environmental conditions, and (6) the duration of those conditions. In a given crop season, therefore, a widespread outbreak of disease can be expected if there is an abundance of susceptible plants, if there is a large supply of a virulent pathogen, and if environmental conditions favorable to the pathogen persist over a long period. Disease will not be widespread, however, unless all six variables strongly favor the pathogen. Interrelationships among pathogens, susceptible plants, and the environment are shown in Figure 1-1.

PRINCIPLES OF PLANT PATHOLOGY

A principle is a fundamental truth or basic law; the ultimate objective of every science is to gain knowledge of principles, and principles are discovered by studying facts. The rate of development of any science is determined by the rapidity with which its principles are understood, and plant pathology is no exception. As knowledge of the science has increased, new concepts have arisen and old ones have been modified to conform to new discoveries. Inevitably, the discovery of new facts and the formulation of new concepts have required the use of new terms and the modification of old ones. **Terminology** is the science of the correct application and usage of the names given to concepts, structures, and phenomena (Whetzel, 1929), and precise terminology is indispensable to any science. In this book, where commonly used terms that relate to the principles of plant pathology are first presented, they are set in **boldface type.**

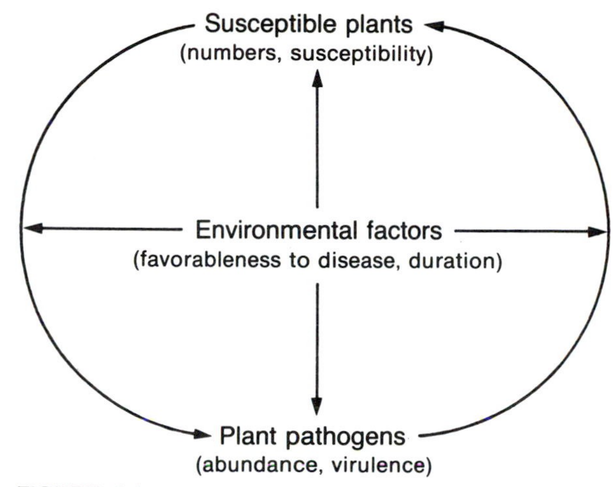

FIGURE 1-1
Interrelations among susceptible plants, plant pathogens, and the environment.

Principles of Plant Pathogenesis

It is axiomatic that a pathogen must become established in an area before it can cause disease in plants growing there. The geographical area in which a plant pathogen is established is its **range.** The transport of a pathogen into an area in which it becomes established for the first time is termed **distribution.** To become established and persist in a region, a plant pathogen must have not only an ability to infect the plants there during their growing season, but also an ability to survive between growing seasons.

The life history of a pathogen already established in an area consists of distinct units of continuous existence termed **cycles.** Each cycle begins with the transfer of the pathogen to a new plant environment and ends with cessation of its activities there. Disease results both from the activities of pathogens and those of the attacked plants, as they are influenced by the environment. **Primary cycles** begin only after a period of rest or seasonal inactivity (dormancy) by the pathogen; **secondary cycles** originate during the growing season, when the pathogen neither rests nor is dormant (Whetzel, 1929).

The pathogen may pass through two phases in a cycle, pathogenesis and survival. In **pathogenesis,** it is associated with the tissues of the attacked plant where it exercises its **pathogenicity,** its ability to cause disease. After passing through its pathogenic phase during the growing season, a pathogen must exist until the next growing season; this phase of its life history is termed **survival.** In the temperate zones, the term **overwintering** is used for the survival phase of the life history of a plant pathogen.

Pathogenesis, in primary and secondary cycles of the pathogen, proceeds in several stages: (1) production and dispersal of inocula, (2) inoculation of susceptible plants with those inocula, (3) penetration of the susceptible plant by the pathogen, and (4) infection and disease in the susceptible plant.

Inoculum (*pl.* inocula) is a pathogen, or any part of a pathogen, that is capable of initiating a pathogenic attack. **Inoculation** is the transfer of inoculum to or into an **infection court,** that part of the susceptible plant from which the pathogen will eventually establish a pathogenic relationship with the plant. The inoculum may remain quiescent in the infection court or begin development immediately: spores of a fungus may germinate, bacterial cells divide. Such activities may be considered **prepenetration phenomena. Penetration** of the plant follows, the pathogen entering through wounds, through natural openings in the plant, or by direct penetration of plant tissues.

Infection is established when the pathogen becomes associated with the cells and tissues of the inoculated susceptible plant. The time for which the association may exist with little or no injury to the plant varies greatly, but the interrelationship eventually does become injurious. **Disease** begins with the first injurious response and does not end until either the plant ceases to respond to the action of the pathogen or dies from its effects, or until the pathogen itself is absent from the plant's tissues.

During infection and disease, two distinct relationships prevail between the pathogen and the diseased plant. One is a food relationship, the other a disease relationship. The diseased plant is the central figure in the phytopathological drama (Whetzel, 1929), and the disease relationship, therefore, is of primary importance to the plant pathologist.

The food relationship is important also; because plant pathogens obtain nourishment from living plants, they are **parasites** as well as pathogens, and must be studied as both. Not all parasites are pathogens, however; parasitic nodule-forming bacteria, for example, do not normally cause disease in the leguminous plants whose roots they parasitize. The living organism that provides nourishment for a parasite is termed a **host,** and the couplet "host-parasite" refers to a relationship between two organisms in which the latter is dependent upon the former for food. The term "host-pathogen" is illogical because each half of the couplet refers to a different relationship—"host" to a food relationship and "pathogen" to a disease relationship. A plant capable of responding pathologically is susceptible to disease, and has consequently been termed a **suscept** (Whetzel, 1937). Thus, the couplet "suscept-pathogen" should be used to designate a relationship that results in disease.

There are different degrees of parasitism: **obligate parasites** obtain food only from living hosts, and cannot grow on nonliving organic matter, as **obligate saprophytes** must; **facultative saprophytes** are parasites that may also grow saprophytically (consequently, they may be grown on nonliving culture media in the laboratory); **facultative parasites** usually grow saprophytically, but can also parasitize plants, obtaining nourishment from host cells that have been killed by toxins or enzymes secreted by the facultative parasites themselves.

Environment and Disease in Plants

Suscept-pathogen interactions in a diseased plant are often strikingly affected by environmental factors; moreover, such noninfectious factors as mineral-nutrient deficiencies may act as primary causal agents of diseases.

The severity of a given plant disease during a growing season is frequently determined by the duration of environmental conditions favorable to initiation and development of disease. The term **epidemic** is used to denote the occurrence of disease in a high percentage of the individuals of a population, and the term **epidemiology** refers to the study of such phenomena. (Strictly speaking, "epidemic" refers to the widespread occurrence of a disease in man, whereas "epiphytotic" and "epizootic" refer to such occurrence of disease in plants and animals, respectively.) **Sporadic diseases** (Walker, 1957) are epidemics that occur occasionally and at irregular intervals. It should be noted that the term "epidemic" does not necessarily imply that the diseased individuals are severely affected.

The adjective **endemic** is used to describe a disease whose range is geographically restricted, or one that is constantly present, to a greater or lesser degree, in a particular place, as distinct from epidemic and sporadic disease; the term carries no connotation of the mildness or severity of disease.

Principles of the Prevention of Plant Disease

The objective of the pure science of plant pathology is to attain knowledge of certain complex biological phenomena, and that of practical or applied plant pathology is to prevent crop losses from plant diseases. Plant diseases are prevented by the application of **control measures**—that is, any profitable means of reducing crop losses that are due to disease (Whetzel, 1929). Disease-control measures are classified in accord with the four basic principles of plant-disease prevention that they exemplify: (1) exclusion, (2) eradication, (3) protection, and (4) development of resistance. The first two principles involve the control of the pathogen; the second two deal primarily with defense of the plant.

Exclusion is prevention of the entrance of a pathogen into an area not yet infested or prevention of its establishment there. When the noninfested area is outside the range of the pathogen, exclusion operates against distribution; when the area is within the range of the pathogen, exclusion prevents dispersal. **Eradication** is the elimination of a pathogen from an area into which it has been introduced. Eradicative measures seldom result in total elimination of the pathogen.

Protection is the prevention of entrance of a pathogen and the establishment of a restraint between it and a susceptible plant by the development of a chemical or physical barrier between the inoculum and the infection court. **Development of resistance** refers to the manipulation of the morphology of a crop plant, or its physiology, or both, in such a way that the pathogen cannot become established in the plant. Such changes usually are of genetic origin and are the result of selective breeding for resistance.

Some generalizations can be made about the interrelationships among the principles of control and those of pathogenesis. Obviously, if a pathogen has been excluded, inoculation cannot occur. When exclusionary measures have not been undertaken or have not been successful, the pathogen, ideally, must be eradicated before suscepts are inoculated. Most protective measures are taken before inoculation occurs, and the pathogen is destroyed in the infection court before its entrance into the plant. When immunity or resistance is a property of the living protoplast, inoculation and entrance by the pathogen may be completed, but neither a parasitic relationship nor a disease can be established. Relationships among stages of pathogenesis and principles of plant disease control are shown in Table 1-1.

The principles of pathogenesis and those of control, along with their

TABLE 1-1
Relations between the prevention and control of plant diseases and the elements of plant pathogenesis

Prevention and control of plant diseases	Plant pathogenesis
Control of the pathogen	
Exclusionary measures	Distribution of the pathogen
Eradicative measures	Survival of the pathogen
Defense of the suscept	
Protective measures	Prepenetration phenomena
Breeding for resistance	Infection and disease

interrelationships, are subject to profound environmental effects. Knowledge of these interactions is the heart of the science of plant pathology.

The Study of Individual Plant Diseases

Every plant disease exemplifies the principles of plant pathology. The study of a single plant disease, therefore, demands comprehension of the ways in which the principles summarized above and discussed in detail in subsequent chapters apply to that disease. A logical approach to the understanding of a specific disease of plants is to study, in sequence, the disease itself, its causality, its epidemiology, and its prevention (Table 1-2).

Every disease should be descriptively named and can be classified on the basis of the physiology of the affected plants. Knowledge of its history, range, economic importance, and symptoms is necessary for a thorough understanding of the disease. Throughout the study, disease should be viewed as the effect, as distinguished from the cause.

Etiology is the science of cause, and includes the study of all factors involved, directly or indirectly, in the initiation and development of disease. Specifically, etiology includes the study of the form and classification of the pathogen, proof of its pathogenicity, and an understanding of the principles of pathogenesis as they are manifested in the cycles of the pathogen.

The principles of survival, dispersal, inoculation, prepenetration, penetration, infection, and disease have already been described. The rules of proof of pathogenicity, however, should be stated here. For practical purposes, the constant association of a particular pathogen with a given symptom of disease is usually accepted as evidence of the causal responsibility of that pathogen. Association does not prove

TABLE 1-2
Topics for the study of a plant disease

- I. Name and classification
- II. Susceptible plants
- III. History and geographical distribution
- IV. Economic importance
  - A. Extent of losses
  - B. Nature of losses
- V. Symptoms
- VI. Etiology
  - A. Morphology, taxonomy, and nomenclature of the pathogen
  - B. Proof of pathogenicity
  - C. The primary cycle of the pathogen
    - 1. Survival of the pathogen
    - 2. Production and dispersal of inoculum
    - 3. Inoculation of the susceptible plant with inoculm
    - 4. Prepenetration phenomena and penetration of the plant by the pathogen
    - 5. Infection and disease in the plant
      - a. Physiology of the diseased plant
      - b. Pathological anatomy
  - D. Secondary cycles of the pathogen
    - 1. Production and dispersal of inoculum
    - 2. Inoculation of the susceptible plant with inoculum
    - 3. Prepenetration phenomena and penetration of the plant by the pathogen
    - 4. Infection and disease in the plant
      - a. Physiology of the diseased plant
      - b. Pathological anatomy
- VII. Epidemiology
- VIII. Prevention and treatment
  - A. Control of the pathogen
    - 1. Exclusion
    - 2. Eradication
  - B. Defense of the plant
    - 1. Protection
    - 2. Development of resistance

pathogenicity, however, and erroneous diagnoses have been made on the assumption that it does. Saprophytic organisms that invade necrotic tissues initially attacked by a primary causal organism may be constantly associated with diseased plants and may grow so luxuriantly as to mask the presence of the pathogen. Three rules for experimental proof of the pathogenicity of an organism were presented in 1883 by the German bacteriologist Robert Koch; a fourth was appended by E. F. Smith (1905). Briefly, these rules are: (1) the suspected causal organism must be constantly associated with the disease, and (2) it must be isolated and grown in pure culture; (3) when a healthy suscept is inoculated with the pathogen from a pure culture, symptoms of the original disease must develop, and (4) the same pathogen must be reisolated from plants infected under experimental conditions. These rules of proof are often referred to as Koch's Postulates.

Environmental and biological factors that influence disease development are properly considered in the study of etiology. The ways in which these factors affect widespread outbreaks of disease, however, are properly in the realm of epidemiology.

Knowledge of the disease itself, of its etiology, and of its epidemiology, should make possible the recommendation of control measures to prevent losses due to the disease under investigation. Information on distribution, survival, dispersal, and infection often suggests the type of control measures—exclusionary, eradicative, protective, or genetic—that will be effective in preventing the plant disease (see Figure 1-1).

Literature Cited

Blackman, V. H., 1924. Physiological aspects of parasitism. *Proc. Brit. Assoc. Sect. K Toronto 1924,* pp. 1–14.

Duggar, B. M., 1909. *Fungous Diseases of Plants.* New York: Ginn.

———, 1911. Physiological plant pathology. *Phytopathology* 1:71–79.

Koch, R., 1883. *Über die Milzbrandimpfung, eine Entgegnung auf den von Pasteur in Genf gehaltenen Vortrag.* Kassel und Berlin: Theodor Fischer.

Kühn, J., 1858. *Die Krankheiten der Kulturgewächse, ihre Urasachen und ihre Verhütung.* Berlin: Boselmann.

Large, E. C., 1966. Measuring plant disease. *Annu. Rev. Phytopathol.* 4:9–28.

LeClerg, E. L., 1964. Crop losses due to plant diseases in the United States. *Phytopathology* 54:1309–1313.

McNew, G. L., 1950. Outline of a new approach in teaching plant pathology. *Plant Dis. Rep.* 34:106–110.

———, 1960. The nature, origin, and evolution of parasitism. *In* J. G. Horsfall and A. E. Dimond, eds., *Plant Pathology—an Advanced Treatise,* vol. 2, pp. 19–69. New York: Academic Press.

Morstatt, H., 1922. Die wissenschaftlichen Grundlagen der Pflazenpathologie. *Angew. Bot.* 4:16–32.

Smith, E. F., 1905. *Bacteria in Relation to Plant Disease,* vol. 1. Washington, D. C.: Carnegie Institution of Washington.

Sorauer, P., 1914. *Manual of Plant Diseases* (3rd ed., transl. from German by Frances Dorrance), vol. 1. Wilkes-Barre, Pa.: The Record Press.

Walker, J. C., 1969. *Plant Pathology* (3rd ed.). New York: McGraw-Hill.

Whetzel, H. H., 1929. The terminology of phytopathology. *Int. Congr. Plant Sci. Proc.* 2:1204–1215.

———, 1935. Nature of disease in plants. Mimeographed paper, Plant Pathology Department, Cornell University.

———, 1937. Etiology. Mimeographed paper, Plant Pathology Department, Cornell University.

2

Morphological Symptoms of Disease in Plants

Symptoms are the conditions of disease, the manifestations of physiological reactions of plants to the harmful activities of the causal agents. For any disease in a given plant, there is characteristic expression of symptoms, usually occurring in a sequential series during the course of the disease. This series of symptoms is called the **syndrome,** or disease picture, and diagnosis of a plant disease in the field is largely dependent upon recognition of its syndrome.

Externally detectable symptoms may be exhibited by whole plants or by any organ of a plant. Such symptoms are termed **morphological symptoms,** and are usually detected visually, although some may be detected through the senses of smell, taste, or touch. Symptoms that can be detected only by microscopic examination of diseased tissue are **histological symptoms,** and their study has been called **pathological** (or **morbid**) **anatomy** (Butler and Jones, 1949). The three kinds of morphological symptoms are necroses, hypoplases, and hyperplases. These symptoms are described below in terminology similar to that of Whetzel (1929, 1935).

NECROSES

Necroses are characterized by degeneration of protoplasts, followed by death (**necrosis**) of cells, tissues, organs, and whole plants. Necrotic symptoms expressed before death of protoplasts are **plesionecrotic** (nearly dead), and those expressed after death of protoplasts are **holonecrotic** (entirely dead).

Plesionecroses

Yellowing, wilting, and hydrosis are commonly occurring plesionecroses. **Yellowing** (**chloranemia** or **icterus**) results from chlorophyll breakdown. Yellowing is distinct from **chlorosis,** a hypoplastic symptom characterized by the failure of chlorophyll to develop fully (Whetzel, 1935), although the latter term is often used to designate any pathological yellow coloration in tissues that are normally green (Heald, 1933).

Wilting is the drooping of plants or parts of plants as a consequence of loss of turgor in the cells of their leaves or stems. Wilting due to water deficiency in the soil is reversible, provided water is supplied before the permanent wilting point has been reached. Pathological wilting is irreversible, however, and is caused by pathogen-induced impairment of water conduction.

Hydrosis is the term used to describe the water-soaked translucent appearance of diseased tissues whose intercellular spaces contain liquids released from cells with damaged plasma membranes. Hydrosis often precedes development of such holonecrotic symptoms as rot, spot, and blight.

Holonecroses

Holonecroses may develop in any part of a diseased plant. As a group, they are probably the most familiar of all symptoms, because they are among the most conspicuous: plant tissues in which necrosis occurs usually turn some shade of brown (Heald, 1933). Holonecrotic symptoms are treated here in three groups, depending upon whether they occur in storage organs, in green tissues, or in woody tissues.

Necroses of Storage Organs. Many diseases of storage organs—such as fruit, seed, bulbs, corms, tubers, and roots—terminate in decomposition or decay, called **rot** (Figure 2-1). Hydrosis often precedes rot,

FIGURE 2-1
Soft rot of potato, caused by *Erwinia carotovora.*

which is then soft and wet: exudation of juices from tissues with soft rot is called **leak.** Some rots of fleshy tissues, however, develop without hydrosis and are hard and dry. In some diseases, especially those of fruit, water is lost rapidly, and the decaying fruit dries and becomes shriveled, wrinkled, and tough, thus showing the symptom **mummification** (Figure 2-2).

Necroses of Green Plant Tissues. **Damping-off** is the sudden wilting and toppling over of seedlings as a consequence of extensive necrosis of the stem near the soil line (Figure 2-3). Pathogens that cause damping-off may also infect seeds or hypocotyls beneath the soil surface and thereby prevent emergence of seedlings, a condition illogically named "pre-emergence damping-off."

Probably the most familiar necrotic symptom in green leaves and in fruit is **spot** (Figure 2-4), a well-defined area of gray, tan, or brown necrotic tissue that may be surrounded by margins of purple or some other dark color. When the necrotic tissue within a leaf spot cracks and falls away from the surrounding green tissue, the symptom is referred to as **shot-hole** (Figure 2-4). Minute spots are sometimes called **flecks** or **specks.** When dark mycelia of a fungal pathogen appear on the surface of a necrotic spot, the symptom is called **blotch.** Spot and

FIGURE 2-2
Mummified fruit of peach infected by the brown-rot fungus, *Monilinia fructicola.*

blotch are examples of so-called restricted symptoms, because they do not spread throughout the part of the plant on which they occur.

Other restricted holonecrotic symptoms include streak and stripe. Both are elongated areas of necrosis: **streak** occurs along stems and veins, and **stripe** in laminar tissue between the veins of grasses. **Net necrosis** is a symptom resulting from an irregular pattern of anastomoses between streaks or stripes.

Extensive unrestricted necroses that involve entire organs include blight, scorch, firing, scald, blast, and shelling. **Blight** is the rapid death of whole organs or parts of leaves, including veins (Figure 2-5). The coalescence of many spots in leaf tissue is often improperly called blight; improperly, because blight results from rapid development of a pathogen from a single point of infection and coalesced spots result from development of a pathogen at many points of infection (spots). **Scorch** resembles blight, but the necrosis occurs in irregular patterns between the veins or along leaf margins. **Firing** is the sudden drying, collapse, and death of whole leaves; the leaves themselves are not invaded by a pathogen, but they express firing in response to the

FIGURE 2-3
Damping-off (seedling blight) of cabbage seedlings infected by *Rhizoctonia solani.* (Courtesy of H. D. Thurston, Cornell University.)

activities of root-rot and wilt pathogens. **Scald** is the blanching of epidermal and adjacent tissues of fruit and, occasionally, of leaves. **Blast** is the sudden death of unopened buds or inflorescences. Extensive necrosis that results in premature dropping of fruit is called **shelling.**

Necroses of Woody Tissues. Extensive necrosis of shoots from the tips backwards is termed **die-back** or **stag-head;** this symptom usually becomes evident during or shortly after dormancy. Restricted necrosis of bark and cortical tissues of stems or roots is termed **canker** (Figure 2-6); in cankers, the necrotic tissue is sharply delimited, usually by callus, from surrounding healthy tissue.

Exudation from injured or diseased woody tissues is called **bleeding** when the exudate is neither gummy nor resinous. Exudation of gummy substances like those from diseased citrus and stone-fruit trees is called **gummosis;** exudation of resin by conifers is termed **resinosis.**

FIGURE 2-4
Leafspot and shot-hole of cherry, caused by *Coccomyces hiemalis.*

FIGURE 2-5
Late blight of potato, caused by *Phytophthora infestans.*

FIGURE 2-6
Fireblight of apple, caused by *Erwinia amylovora: top left,* bacterial cell × 10,000; *bottom left,* fruit canker; *right,* limb canker.

HYPOPLASES

Hypoplasia is failure of plants or organs to develop fully; subnormal size and pale coloration are the most common hypoplastic symptoms.

Dwarfing is the failure of plants to attain full size. **Rosetting** is the condition in which internodes do not elongate and leaves are clustered like the petals of a rose.

Repression of color may occur and it may be complete or partial. When complete, it is called **albication;** when partial, normally green tissues are yellow, and the symptom is termed **chlorosis. Mosaic** is a mottling of yellow and green areas in a leaf.

The complete failure of organs to develop is termed **suppression.** A complex hypoplastic symptom, **etiolation,** which occurs in plants kept in darkness, combines dwarfing, chlorosis, and spindliness.

HYPERPLASES

Hyperplastic symptoms are the result of overdevelopment in size or color of plants or plant organs, or the abnormally early development of plant organs.

Gigantism

An excessive increase in size of a cell, a tissue, an organ, or an entire plant is known as **gigantism.**

Gigantism of Leaves and Fruit. **Curl** is the bending of shoots or the rolling of leaves as a result of overgrowth on one side of the organ. The puckering or crinkling of leaves produced by a differential rate of growth of their parts is termed **savoying.** Although the symptom is usually hyperplastic, as when it results from overdevelopment of laminar tissue, it may be hypoplastic, as when it results from under-development of veins or leaf margins. It may, however, be hypoplastic and hyperplastic at the same time: the essential feature of the sympton is that the cells of adjacent areas or tissues develop at different rates.

Overgrowth of epidermal and underlying tissues of leaves, fruit, stems, or tubers may result in **scab,** a symptom consisting of raised, rough, discrete lesions (Figure 2-7). Cell walls in scab lesions are usually **suberized**—that is, hardened by corky deposits of suberin. **Intumescence,** the diagnostic symptom of oedema, is a blistering; it is

FIGURE 2-7
Scab of 'Temple' orange, caused by *Sphaceloma fawcetti.*

caused by localized swelling of epidermal and subepidermal cells because of an excessive accumulation of water.

Gigantism of Stems and Roots. Excessive accumulation of elaborated food materials in stems above a girdled or constricted area produces a swelling termed **sarcody.** Localized swellings that involve entire organs are **tumefactions** (Figure 2-8). Familiar tumefactions are galls, knots, and clubs. Development of adventitious organs results in **fasciculation,** a clustering of organs around a focal point: "witches' brooms" and hairy roots are examples. When cylindrical organs, such as stems, become broadened and flattened as if several were growing together, they are exhibiting the symptom known as **fasciation. Proliferation** is continued development of an organ after it has reached a stage beyond which it normally does not grow. The overgrowth of tissue in response to wounding is **callus;** it forms around most cankers and often prevents the spread of pathogens into adjacent areas of healthy tissue.

Hyperchromic Symptoms

Hyperchromic symptoms, which consist of an overdevelopment of color, include **virescence,** the development of chlorophyll in tissues

FIGURE 2-8
Crown gall induced experimentally in a tomato stem inoculated with *Agrobacterium tumefaciens.*

normally devoid of it; **anthocyanescence,** the purplish coloration resulting from overdevelopment of anthocyanin pigments; and **bronzing,** a coppery appearance, such as that of potassium-deficient potato leaves.

Metaplastic Symptoms

Metaplasia occurs when tissues change from one form into another. If this results in the development of organs in unexpected locations, it is called **heterotopy;** two examples are **phyllody,** the development of petals and other floral organs into leaflike structures; and **juvenillody,** the development of juvenile leaves, like those of a seedling, on mature plants. **Russeting,** a superficial brown roughening of surfaces of fruit and tubers, results from suberization of epidermal and subepidermal cell walls.

Proleptic Symptoms

Development of tissues earlier than usual is the basic characteristic of proleptic symptoms. **Prolepsis** refers to premature development of a shoot from a bud, usually following die-back or injury to twigs. Premature formation of the abscission layer at the base of petioles results in early falling of leaves, or **proleptic abscission.** Unexpected development of organs that are usually rudimentary is termed **restoration.**

A KEY TO FAMILIAR SYMPTOMS OF DISEASE IN PLANTS

Most of the symptoms described herein can be distinguished in the following key, which is based on one by Whetzel (1935).

1.	Cells dying or dead (necroses)	2
	Cells not dying or dead (hypoplases or hyperplases)	23
2.	Cells in process of dying (plesionecroses)	3
	Cells dead (holonecroses)	5
3.	Chlorophyll destroyed (chloranemia or icterus)	Yellowing
	Chlorophyll not destroyed	4
4.	Tissues flaccid, shoots and leaves drooping	Wilting
	Tissues not flaccid, but with excessive intercellular water	Hydrosis
5.	Storage tissues affected	6
	Other tissues affected	7
6.	Maceration, followed by collapse of affected organ	Rot
	Dry decay, skin becoming hard, dry, wrinkled	Mummification
7.	Green tissues affected	8
	Woody tissues affected	19
8.	Seedling stems attacked, plants toppling	Damping-off
	Mature tissues attacked	9
9.	Necrotic areas sharply delimited	10
	Necrotic areas spreading, perhaps affecting whole organ	14
10.	Symptoms usually in dicotyledonous plants	11
	Symptoms usually in monocotyledonous plants	13
11.	Limited necrotic areas covered with mold	Blotch
	No mold growth on necrotic areas	12

12. Necrotic tissue intact within the lesion . . . . Spot
    Necrotic tissue cracking and falling away . . . . Shot-hole
13. Necroses along veins . . . . Streak
    Necroses between veins . . . . Stripe
14. Leaves affected 15
    Fruit and buds affected 18
15. General necroses resulting from indirect effects . . . . Firing
    Necroses resulting from direct effects of pathogen 16
16. Only epidermal and subepidermal tissues affected . . . . Scald
    Tissues below epidermal layer killed 17
17. Necroses around leaf margins only . . . . Scorch
    Necroses not limited to leaf margins . . . . Blight
18. Buds dropping prematurely, fruit not set . . . . Blast
    Fruit falling prematurely . . . . Shelling
19. Necrotic areas 20
    Exudations 21
20. Necrosis localized, usually surrounded by callus . . . . Canker
    Necrosis extensive, twigs dying from tips . . . . . Die-back
21. Exudate gummy . . . . Gummosis
    Exudate not gummy 22
22. Exudate resinous . . . . Resinosis
    Exudate neither gummy nor resinous . . . . Bleeding
23. Underdevelopment of tissues (hypoplases) 24
    Overdevelopment of tissues (hyperplases) 30
24. Underdevelopment with respect to size (nanistic symptoms) 25
    Underdevelopment with respect to color (hypochromic symptoms) 28
25. Entire plant reduced in size . . . . Dwarfing
    Certain organs failing to attain full size 26
26. Complete failure of an organ to develop . . . . Suppression
    Partial development of organs 27
27. Internodes failing to develop . . . . Rosetting
    Budlike, chlorotic leaves from spindly stems . . . . Etiolation
28. Absence of color . . . . Albication
    Failure of chlorophyll to develop, plants light green 29
29. Yellow color in plants receiving light . . . . Chlorosis
    Yellow color in spindly plants grown in the dark . . . . Etiolation

30. Overdevelopment with respect to size
    (gigantism) 31
    Other overdevelopment 40
31. Symptoms usually in leaves or fruit 32
    Symptoms usually in stems or roots 35
32. Symptoms localized 33
    Symptoms not localized 34
33. Hyperplasia of epidermal cells, suber-
    ization of cell walls . . . . Scab
    Hypertrophy of epidermal cells due to
    water accumulation, blistering (oedema) . . Intumescence
34. Increased growth rate on one side of
    stems or leaves . . . . Curl
    Leaf puckering due to different growth
    rates in adjacent tissues . . . . Savoying
35. Localized overgrowths in response to
    girdling or wounding 36
    Overgrowths not in response to girdling
    or wounding 37
36. Swellings above constricted (girdled)
    areas of stems . . . . Sarcody
    Swollen, smooth overgrowth around
    wounds or cankers . . . . Callus
37. Localized swellings (galls, knots,
    clubbed roots) . . . . Tumefaction
    Overgrowths not involving swelling
    of shoots or roots 38
38. Elongation after shoots or roots
    normally cease to elongate . . . . Proliferation
    Overgrowths not involving swelling
    or elongation 39
39. Fusion of stems resulting in flattened
    shoots . . . . Fasciation
    Adventitious root or shoot development
    from a focal point (witches' broom,
    hairy root) . . . . Fasciculation
40. Overdevelopment of color (hyperchromic
    symptoms) 41
    Other overdevelopment 43
41. Greening of tissues normally devoid
    of chlorophyll . . . . Virescence
    Development of colors other than green 42
42. Development of reddish or purplish
    colors in tisues normally green . . . . Anthocyanescence
    Appearance of coppery hues in leaves
    or fruit . . . . Bronzing

43. Change in kind of tissue or organ (metaplastic symptoms) 44
    Premature development of organs or tissues (proleptic symptoms) 45
44. Production of organs in an unexpected location (phyllody, juvenillody) . . . . . . Heterotopy
    Suberization of epidermal cell walls without hyperplasia of cells . . . . Russeting
45. Unexpected development of organs normally rudimentary . . . . Restoration
    Other premature development of organs or tissues 46
46. Growth of shoots from buds that are normally dormant . . . . Prolepsis
    Premature dropping of leaves . . . . . Proleptic abscission

Literature Cited

Butler, E. J., and S. G. Jones, 1949. *Plant Pathology*. London: Macmillan.

Heald, F. D., 1933. *Manual of Plant Diseases* (2nd ed.). New York: McGraw-Hill.

Whetzel, H. H., 1929. The terminology of phytopathology. *Int. Congr. Plant Sci. Proc.* 2:1204–1215.

———, 1935. Symptomatology. Mimeographed paper, Plant Pathology Department, Cornell University.

3

The Infectious Agents of Disease in Plants

Viruses, bacteria, fungi, and nematodes are by far the most numerous and important infectious agents of plant disease, but algae (Wolf, 1930), and such seed plants as dodder (Yuncker, 1931), mistletoe (Trelease, 1916), and witchweed (Garris and Wells, 1956) cause a few diseases of plants (see Box 3-1). Some insects also cause diseases (Carter, 1939; Felt, 1940)—for example, hopperburn of potato (Ball, 1918) and mealybug wilt of pineapple (Illingworth, 1931). Stahel (1933) obtained evidence that a flagellate may be the pathogen of phloem necrosis of coffee, but his experiments did not eliminate the possibility that the disease is of viral origin. In humid climates, certain lichens may be harmful to plants, but diseases due to the few species of lichens said to be pathogenic are of little economic importance.

The four predominant groups of plant pathogens will be discussed in this chapter.

VIRUSES AND MYCOPLASMALIKE ORGANISMS

Flower breaking of tulips (Figure 3-1), a disease now known to be caused by a virus, was described as early as the sixteenth century, but

BOX 3-1
PLANT-PARASITIC SEED PLANTS

More than a thousand species of seed plants in at least seven families parasitize other seed plants and induce in them harmful physiological processes that we recognize as disease. Little or nothing is known, however, about the economic importance of most plant-parasitic seed plants. Certainly, the mistletoes, dodders, and broomrapes have reduced crop yields and have killed economically important plants; but what of the hundreds of other plant-parasitic seed plants? Do they also reduce the quality and quantity of the marketable quality of their hosts? If so, in what ways and by how much? If not, might they even be beneficial to their hosts? Not only are we ignorant of the extent to which a plant-parasitic seed plant affects its host, we do not yet completely understand the nature of how the host is affected.

The mistletoes (family Loranthaceae) are perhaps the best known plant-parasitic seed plants. More than 900 species parasitize trees, mostly in the tropics. Species of *Viscum,* however, occur in Europe, and species of *Arceuthobium* and *Phoradendron* occur in the United States. *Arceuthobium* spp. parasitize conifers, and *Phoradendron* spp., mainly hardwoods. The members of both genera are dioecious and produce seed that are imbedded in sticky matrices. *Arceuthobium* spp. forcibly eject their seed at maturity; the sessile seeds of *Phoradendron* are disseminated by birds and, sometimes, by wind.

When a seed of either plant lands on the bark of a host, it attaches itself by means of its sticky matrix. In the presence of water, and when the temperature is appropriate, the seed germinates, producing a radicle that contacts the surface of the host. The radicle expands to form a disclike attachment, which later develops a projection (the primary haustorium) that grows into the cortex, apparently progressing by mechanical force and enzymatic digestion of host tissue. The primary haustorium spreads in the cork cambium, but does not itself penetrate further. It later produces toothlike projections (sinkers) that grow into the cambium. Because the sinkers contain no phloem elements—only tracheary elements—the mistletoes rob their hosts of water and minerals, but not of elaborated food materials, which move in the phloem. Mistletoes, therefore, are sometimes referred to as "water parasites."

Other parasitic seed plants sink peglike haustoria into conductive tissues, where they connect with the phloem as well as the xylem. Well-known examples are the dodders (*Cuscuta* spp.), parasitic vines that occur in temperate and tropical climates, and members of the genus *Cassytha,* which are common in the tropics and subtropics. Members of both genera lack chlorophyll and presumably contribute nothing to the welfare of their hosts. Dodder seed, which may contaminate seed of a host plant, such as flax, alfalfa, or clover, germinate with seed of the host and send up spindly stems. If a stem contacts a stem or petiole of the host, it will first twine around it and then attach itself by means of peglike branches. The pegs (haustoria) grow into the stem of the host, penetrating by enzymatic action and mechanical pressure, until connections are made with the phloem and xylem. Feeding upon nutrients provided by the host, the dodder stem elongates and twines around other stems of the host plant and those of adjacent plants. It produces bractlike leaves at its nodes and finally forms clusters of small flowers. At present, dodder is best controlled in crops by planting uncontaminated seed. Dodder seeds are sometimes killed in the field by burning crop residues in the areas previously infested by the parasite.

Species of the genus *Cassytha* (of the family Lauraceae) parasitize their hosts much as the species of dodder; but most of them live in warm climates, where they are able to survive perennially in the vegetative parasitic state. Consequently, they are more difficult to control. In Florida citrus, for example, *Cassytha* plants are removed from parasitized trees by pruning infected branches.

Some parasitic seed plants, notably the broomrapes (*Orobanche* spp.) and the witchweeds (*Striga* spp.) attack the roots of their hosts. Members of both genera may be common in such field-grown crops as tobacco and corn, where they are controlled by eradicative measures, such as burning or the use of herbicides.

Another root parasite, beech-drops (*Epifagus virginiana*), is widespread in stands of beech in the eastern United States. It probably causes little damage to its host, but is frequently seen by hikers during late summer and early autumn in the beech forests of the Adirondack Mountains of northern New York. Its tan or brown stems with bractlike leaves emerge singly or in clusters and grow to a height of 15–45 cm.

Another group of parasitic seed plants, those of the family Santalaceae, bear green leaves and true roots. The roots, however, produce haustorial discs that attach to the roots of their hosts and penetrate to the conductive tissues. Members of the family Santalaceae, of which sandalwood is a well-known example, may thus live either independently or semiparasitically, causing little or no harm.

Finally, some parasitic plants produce their vegetative bodies completely within the stems of their hosts, forming a cylinder between the wood and cortex. Flowers of such parasites later break through the bark and are borne on the stems of the hosts. One species, *Rafflesia arnoldii,* produces a flower one meter in diameter —it is said to be the largest flower in the world. Members of its family (Rafflesiaceae) occur in the tropical and subtropical regions of Asia and America.

FIGURE 3-1
Flower-breaking of tulips.

the first experimental proof of the infectious nature of a plant virus was not obtained until late in the nineteenth century, when Mayer (1886) demonstrated that tobacco-mosaic virus (Figure 3-2) is transmissible. Beijerinck (1898) proved that the virus reproduces only within living plants. Later, Iwanowski (1903) demonstrated that tobacco-mosaic virus (TMV), is so small that it will pass through filters with pores too small to permit passage of the smallest bacteria. Thus, the three intrinsic properties of plant viruses—infectiousness, ability to reproduce only *in vivo,* and small size relative to other pathogens—were demonstrated near the beginning of the twentieth century. Since that time, viruses have been causally implicated in more than 600 plant diseases (Hopkins, 1957; Martyn, 1968).

Although much has been learned about plant viruses, a natural classification is not yet possible because our knowledge of the nature of these entities is incomplete. Scarcity of information on structure and the mechanism of reproduction, the bases of natural classifications, make the grouping of plant viruses largely dependent upon other characteristics, such as the symptoms of the diseases they induce, their

FIGURE 3-2
Mosaic pattern in tobacco systemically infected by tobacco-mosaic virus; *inset*, particles of the virus × 34,000.

insect-vector relationships, their physical and chemical properties, their serological relationships, and their interference phenomena. A useful working classification of viruses has been devised on the basis of knowledge of these properties (Holmes, 1948).

Symptoms of Viral Diseases in Plants

The following discussion of the symptoms of viral diseases in plants is based on a discussion by Holmes (1964).

Morphological symptoms of viral diseases in plants vary from a mild clearing of veins and chlorotic mottling to severe lethal necroses. Usually, the symptoms of a viral disease are systemic or generalized as a result of the spread of the virus throughout the diseased plant, but in certain plants, the symptoms may be localized. The discovery that TMV remains localized in necrotic lesions in the inoculated leaves of *Nicotiana glutinosa* (Holmes, 1929; see Figure 3-3), and that the

FIGURE 3-3
Localized necrotic lesions in leaves of *Nicotina glutinosa* mechanically inoculated with tobacco-mosaic virus.

number of lesions is positively correlated with concentration of active virus in the inoculum, provided the basis for extensive quantitative work with plant viruses.

The universal symptom of viral infections is a reduction of plant size, which results in decreased yield. Diseased shoots may blight, and infected leaves may be chlorotic, or they may show patterns of chlorotic or necrotic spots, ringspots (Figure 3-4), or streaks. Leaf deformation may be no more than slight crinkling (savoying), or may be so drastic that affected leaves are scarcely recognizable; laminar development may be so repressed that a normally broad leaf assumes a shoestring form. Local hyperplastic symptoms often occur in plants that have been stunted: outgrowths or enations from veins are familiar examples. Tumefaction is a diagnostic symptom of several viral diseases, notably of swollen shoot of cacao and wound tumor of clover.

A number of different viral diseases may be characterized by the same or similar syndromes. The severity and the kind of symptom induced by a particular virus vary with such factors as the species and

FIGURE 3-4
Leaves of a tobacco plant infected by tobacco-ringspot virus: *bottom,* inoculated leaves; *top,* systemically infected leaves.

age of the affected plant, the environmental conditions before infection and during the course of disease development, and variability within the virus itself. Variants of a single virus are usually detected by differences in such properties as the ability to infect certain plants, to produce characteristic symptoms in suscepts, and to be transmitted by particular insects (Carsner, 1925; Johnson, 1926; McKinney, 1926). Such variants are called **strains** of the virus.

FIGURE 3-5
Stunted corn plant (*foreground*).

The Relations of Plant Viruses to Arthropod Vectors

This discussion of arthropod transmission of plant viruses is based on a paper by Black (1954).

Most plant viruses—TMV is a noteworthy exception—are disseminated by arthropod **vectors,** the agents of dispersal and inoculation. Aphids, which usually transport mosaic and related viruses, transmit more plant viruses than any other group of insects; leafhoppers, which transmit the so-called yellows viruses, are second to aphids in the number of different viruses transmitted. Other arthropod vectors of plant viruses include thrips, white flies, mealy bugs, and mites. Varying degrees of specificity exist between plant viruses and their insect vectors, and plant viruses can be crudely classified by the vectors that transmit them. Arthropod transmission of plant viruses will be discussed in more detail in Chapter 5 in the section "Agents of Dispersal and Inoculation."

Physical and Chemical Properties of Plant Viruses

Although increasing knowledge of the physical and chemical nature of plant viruses may, in time, lead to a natural classification, the data now available are insufficient for this purpose. Such properties as dilution end point, thermal-inactivation point, and longevity *in vitro* are useful in identifying viruses that cause similar symptoms; knowledge of these properties is of doubtful value, however, in determining relationships among viruses that cause different symptoms (Bawden, 1964). Tests to determine physical properties are often controlled by factors other than those involving intrinsic properties of the virus strain under test. For example, the dilution end point of cucumber-mosaic virus has been reported variously from 1:1000 to 1:100,000, and its thermal-inactivation point has been recorded as low as 55°C and as high as 70°C. These properties vary according to the strain of the virus tested, the environment, the species of suscept involved, and the methods of measurement.

Because of its stability relative to that of other plant viruses, TMV has been the subject of many chemical and physical studies. On the basis of his experiments, Mulvania (1926) concluded that TMV is a nonliving entity that might be proteinaceous, having some properties of enzymes. Vinson and Petre (1929) obtained additional evidence that TMV is a protein. The active principle precipitated in alcohol and in safranin, and could be salted out of aqueous solutions by treatment with ammonium sulfate or magnesium sulfate. Moreover, acetone precipitates from clarified juice of diseased plants contained all the active virus and only 10% of the solids present in the original crude juice. These early experiments were the basis for Stanley's research in which he reported, for the first time, the crystallization of a virus (Stanley, 1935). Virus was salted out with ammonium sulfate; crystals formed when a 20:1 mixture of ammonium sulfate and glacial acetic acid was added, with stirring, to a concentrated solution of partly purified virus. Bawden et al. (1936) also purified TMV, and were the first to prove that the virus contains ribonucleic acid (RNA). They

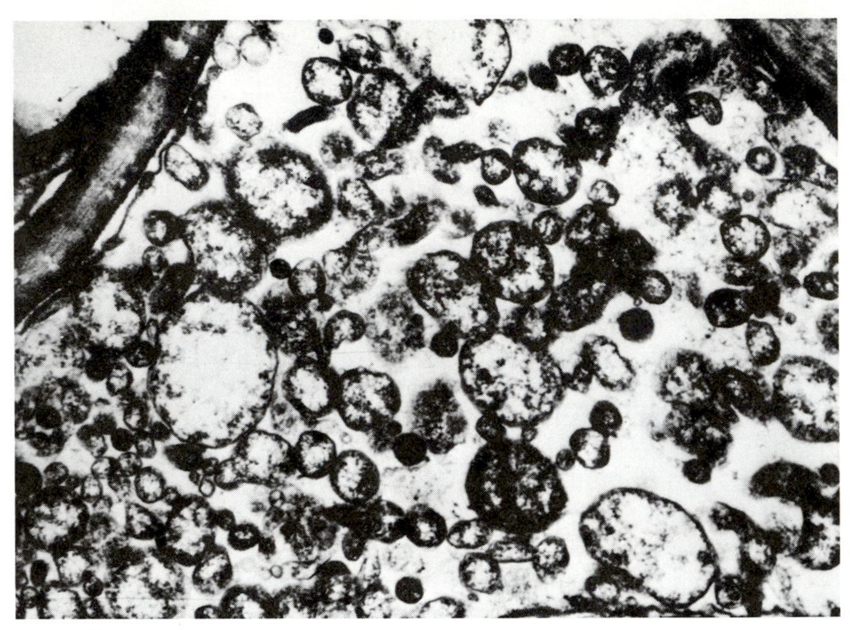

FIGURE 3-6
Mycoplasmalike bodies in phloem cells of a stunted corn plant. (Courtesy of Karl Maramorosch, Boyce Thompson Institute.)

also found that liquid crystalline preparations of TMV show stream double refraction (birefringence, or anisotropy of flow). Birefringence due to the presence of rod-shaped particles is detectable through crossed polaroid lenses when a shearing force is applied to the suspension; spheres or spheroids make a concentrated suspension or solution spontaneously birefringent. Therefore, Bawden et al. (1936) considered TMV to be a rod-shaped nucleoprotein. Later, iron, copper, calcium, and magnesium were found to be associated with purified TMV (Loring and Waritz, 1957). Electron microscope studies have confirmed that TMV is a rigid rod, with dimensions averaging 15×280 mμ. All other plant viruses purified up to now are nucleoproteins; some are rigid or flexuous rods, others are polyhedrals or spheroids. It is probable that most plant viruses in the mosaic group are nucleoproteins that contain RNA, but it must be remembered that only a few of the known plant viruses have been studied in purified form. At least one plant virus, cauliflower-mosaic virus, is a nucleoprotein that contains deoxyribonucleic acid (DNA) (Shepherd et al., 1968). Relatively few of the yellows group of plant pathogens have been purified. A few, such as those of corn stunt (Figure 3-6), aster

BOX 3-2
MYCOPLASMALIKE ORGANISMS AS PLANT PATHOGENS

The discovery of mycoplasmalike bodies (MLBs) in plants thought to be virus-infected has opened up a new frontier in the science of plant pathology. In terms of clinical pictures and transmissibility by insects (usually leafhoppers), mycoplasmalike bodies resemble viruses, but, because of their form, culturability, and their susceptibility to control with certain antibiotics, they resemble bacteria without cell walls. Mycoplasmalike bodies have already been associated with more than forty plant diseases of the "yellows" type, and Koch's rules of proof of pathogenicity have been satisfied for at least one of the diseases they are associated with.

Mycoplasmalike bodies, unlike viruses and like the true mycoplasmas of animals, are pleomorphic (from 100 mμ to 1 μ in diameter), are surrounded by a 10-mμ unit membrane, contain ribosomes and thin-stranded DNA, and are sensitive to antibiotics of the tetracycline group. They are also more sensitive than viruses to *in vivo* heat treatments.

The state of our knowledge of plant-infecting mycoplasmalike bodies has been summarized by Hull (1971) and by Davis and Whitcomb (1971).

yellows, and mulberry dwarf, appear to consist of pleomorphic particles as large as 200 mμ in diameter. These particles (Figure 3-6) resemble the mycoplasmas (Box 3-2) that parasitize man and animals (Doi et al., 1967; Ishiie et al., 1967).

Evidence has been obtained that infectivity of the TMV particle resides in the RNA (Gierer and Schramm, 1956); the structure of the protein moiety, however, seems to be specific for a given strain of the virus. Within attacked cells, virus somehow mobilizes nitrogenous and other compounds of the plant and replicates itself therewith. Possibly, nucleic acids from the cytoplasm—or even from the nucleus—are incorporated directly into the virus. Apparently, however, viral protein is built up from amino acids of attacked cells (Pollard et al., 1958). Because the RNA of TMV is infective, and because it is possible that the RNA and protein components of TMV can be reconstituted *in vitro* (Fraenkel-Conrat and Williams, 1955), it seems likely that, in diseased cells, the viral RNA governs the redirection of cellular synthetic mechanisms toward synthesis of viral nucleoprotein at the expense of normal plant protein synthesis. A few plant viruses apparently exist as naked RNA.

Viruses that attack bacteria (bacteriophages) recombine genetically in infected cells, and strains of the influenza virus recombine in humans; it would be reasonable, therefore, to assume that plant viruses also recombine in their host cells. Although more data are required for proof of genetic recombination of plant viruses, results of experiments conducted by Best (1954, 1956) are not inconsistent with the recombination hypothesis. When tomato plants were inoculated simultaneously with two strains of the spotted-wilt virus, certain symptoms developed that were not characteristic of those produced by either strain. From a leaf showing atypical symptoms, a strain of the virus was isolated that possessed some of the properties of both of the original strains.

Serological Relationships

The discovery that plant viruses are antigenic in animals (Dvorak, 1927; Purdy, 1928) led to extensive research on serological relationships among these viruses. Viruses are specific antigens, and antiserum prepared against one virus will react with the juice of any plant infected by that virus or by one of its strains, provided that the virus content is high enough. Moreover, such antiserum will not react with the juice of a healthy plant or with that of a plant infected by a different virus. Serological tests, therefore, make rapid and accurate methods for determining relationships between certain plant viruses (Ball, 1961). It is important, however, to use purified viral preparations as antigens. Otherwise, the low concentration of antigen (virus) makes the specificity of serological reactions difficult to interpret.

Interference Phenomena

Plants already infected by one strain of a given virus are usually protected against infection by another strain (a serologically related one) of the same virus (Kunkel, 1934; McKinney, 1926, 1929; Salaman, 1933; Thung, 1931), but not against infection by others. The cross-protection test, therefore, provides an excellent method for determining viral relationships. When infection by a known strain prevents infection by an unknown strain, or vice versa, the two strains are presumed to be variants of the same virus. Results of such tests must be interpreted with caution: the metabolism of a plant may be so altered by one virus infection that it cannot subsequently support the replication of an unrelated virus (Bawden and Kassanis, 1945). Results of carefully conducted cross-protection tests, however, provide strong evidence for the relatedness of one plant virus to another.

BACTERIA

Bacteria are probably the simplest nonchlorophyllous plants. Their vegetative bodies consist of single cells grouped in masses or in chains and their reproduction is by binary fission. The size of individual cells varies with species and with environment, but most are approximately 0.5 μ wide and 1–3 μ long. Bacterial cell walls are apparently composed largely of nitrogenous materials (Burkholder, 1948), whereas those of fungi and of higher plants are of complex carbohydrates. Morphologically, bacteria are spherical, rod-shaped, or spiral-shaped, these three kinds being known, respectively, as cocci, bacilli, and spirilla. Only the rod-shaped forms have been demonstrated to cause disease in plants, and only a few of these produce endospores. Flagellation of those plant-pathogenic bacteria that do not form spores may be **monotrichous** (one flagellum at one end), **lophotrichous** (more than one flagellum at one end), or **peritrichous** (flagella attached to points all over the surface of the cell).

Taxonomic relationships among bacteria cannot be determined entirely by morphology; consequently, physiological properties are widely used as bases for classification (Graham, 1964; Stolp, et al., 1965). The *in vitro* physiological characteristics that are used to group bacteria include their Gram-staining reactions, their ability to liquify gelatin, their growth on different culture media, and their production of acid and gas in culture. Additional bases for classifying plant-pathogenic bacteria are their *in vivo* physiological properties, the kinds of diseases they induce in suscepts, and, occasionally, their serological reactions.

Some 180 species of bacteria cause disease in plants: soft rots, vascular wilts, leafspots and blights, galls, and scabs. About one-half of the plant-pathogenic bacteria belong to the genus *Pseudomonas*, about one-third to the genus *Xanthomonas*, and about one-eighth to the genus *Erwinia. Pseudomonas* is responsible for many leafspots, blights, and wilts. *P. phaseolicola,* for example, causes halo blight of bean, and *P. solanacearum* is an agent of wilt in many solanaceous plants.

Burrill (1878, 1880) presented the first proof that a bacterium can cause disease in a plant by showing that fire blight of pear is caused by a bacterium now called *Erwinia amylovora.* At about the same time, Wakker (1883) proved that hyacinth yellows is caused by a bacterium. Although some spots and blights are caused by species of *Erwinia,* which have peritrichous flagella, most are induced by species of *Pseudomonas* or *Xanthomonas. Pseudomonas* has polar flagella and several species produce a greenish, water-soluble, fluorescent pigment in culture; *Xanthomonas* has one polar flagellum and produces yellow, mucoid colonies in culture.

By 1895, when E. F. Smith obtained proof of the bacterial nature of cucurbit wilt (a disease caused by *Erwinia tracheiphila*), most scientists were convinced of the plant-pathogenic nature of certain bacteria. Acceptance of this concept was not universal, however (Fischer, 1899), although Smith (1901) published a detailed account of the available evidence proving the bacterial origin of eight plant diseases.

Aside from a few species of *Erwinia,* some ten species of *Corynebacterium* and a few species of *Xanthomonas* cause vascular wilts. *Corynebacterium* is a taxonomically heterogeneous genus; corynebacteria are Gram-positive pleomorphic rods that are usually wedge-shaped. Ring rot, an important disease of potato, is caused by *Corynebacterium sepedonicum.*

Members of the genus *Xanthomonas* are flagellate and produce yellow colonies on agar.

The soft rot of carrots and other vegetables, caused by *Erwinia carotovora,* was recognized as a bacterial disease by L. R. Jones (1901); by 1907, Smith and Townsend had proved that crown gall of plants is caused by a bacterium, *Agrobacterium tumefaciens.* The three species belonging to the genus *Agrobacterium* induce hyperplasia, and, although they do not fix atmospheric nitrogen, they are related taxonomically to the nitrogen-fixing bacteria of the genus *Rhizobium.*

Scab of potatoes is caused by *Streptomyces scabies;* this organism and its relatives are unlike true bacteria in that their thalli are threadlike and consist of elongated cells in chains. There is some branching, and rudimentary hyphae are formed; exogenous vegetative spores—conidia—may also be formed. Thus, this group of organisms resembles the fungi, and may be intermediate between bacteria and fungi. Indeed, many taxonomists place them in a separate division, the Actinomycetes.

Bacteria and the diseases they cause are subjects of several books or parts of books (Breed, et al., 1957; Dowson, 1957; Elliott, 1951; Stapp, 1961) and of review articles (Burkholder, 1948; Starr, 1959).

FUNGI

Plant diseases caused by fungi have been known since the time of the first recorded history, but scientific proof of the pathogenicity of fungi was not obtained until 1807. In a classical paper, Prévost (1807) confirmed Tillet's (1755) earlier report that the bunt (smut) disease developed in wheat plants grown from seeds that had been dusted with the black masses of smut spores obtained from diseased kernels. Moreover, Prévost provided convincing evidence that the fungus is the cause, not the effect, of the disease. Later, deBary (1853), Kühn (1858), and others clearly demonstrated that certain fungi associated with diseased plants are, in fact, the primary causal agents of disease in those plants. Thus, the science of plant pathology was launched, at first as an offshoot of mycology. Plant pathology now stands as a separate science, but no one can gainsay the importance of mycology to modern plant pathology. Whereas the species of viruses, bacteria, and nematodes that cause plant disease are numbered in the hundreds, plant-pathogenic fungi are numbered in the thousands. Species of fungi cause approximately 100,000 diseases in green plants (McNew, 1966). It is the purpose of this section to review briefly some of the mycological knowledge that is essential to an understanding of plant pathology as a separate science. For thorough considerations of the fungi, standard mycological texts should be consulted (Alexopoulos, 1962; Bessey, 1950; Fitzpatrick, 1930; Gäumann and Dodge, 1928; Wolf and Wolf, 1947).

Fungi are microscopic, nonchlorophyllous plants whose thalli are threadlike or—in some groups—amoeboid, and whose reproductive structures are usually well-defined spores. Knowledge of taxonomic relationships among fungi is indispensable to diagnosis of fungal diseases, but it should be remembered that, with the exception of the downy and powdery mildews and the rusts, the taxonomic positions of plant-pathogenic fungi are not correlated with the kinds of diseases induced by those fungi. For example, one species of *Ceratocystis* causes a rot of sweet potato; a second species of the same genus

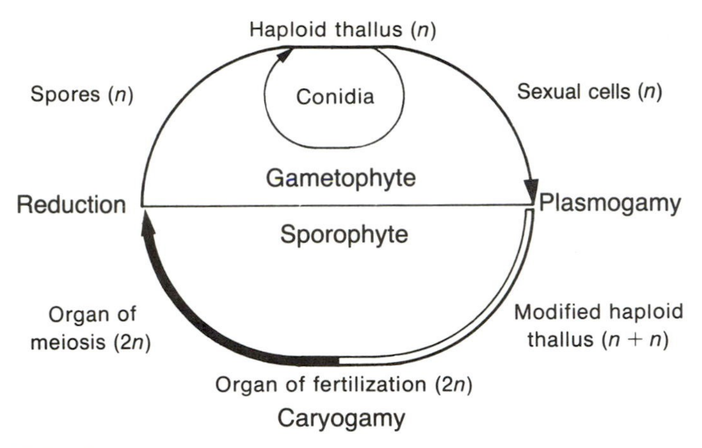

FIGURE 3-7
Generalized nuclear life cycle of fungi.

causes a vascular wilt of elm trees, and an unrelated organism, the fungus *Fusarium oxysporum* f. sp. *lycopersici,* causes a vascular wilt of tomato.

Classification of fungi is based upon their life cycles and the development attained by vegetative and reproductive organs in both portions of their life cycles (Gäumann and Dodge, 1928). If we consider only the nuclear phenomena that occur during these cycles, all fungi are essentially alike; the structures within which these phenomena occur, however, vary widely among the many groups of fungi. Organisms that reproduce sexually contain haploid nuclei during the gametophytic phase of the life cycle and diploid nuclei during the sporophytic phase. Gametes form in gametangia during the haploid phase. This phase ends when two sexual cells fuse, a process termed **plasmogamy,** and the diploid phase begins. Fusion of the two sexual nuclei is termed **caryogamy.** The product of fertilization is the **zygote,** which develops into the organ in which meiosis (reduction division) occurs. The products of meiosis are haploid, and give rise to the vegetative or gametophytic thallus. The nuclear life cycle typical of fungi is diagrammed in Figure 3-7. This generalized life cycle can be adapted to fit all groups of fungi.

Fungi with complete life cycles are either monoecious or dioecious, but, unlike higher plants, any division of the sexes is limited to the haploid phase. Monoecious forms of fungi are termed **homothallic** and are designated with a plus-minus sign ($\pm$); dioecious formed are termed **heterothallic** (Blakeslee, 1904; Gäumann and Dodge, 1928), and are designated either with a plus sign (+) or a minus sign (−).

The structures within which the cytological activities of fungal life cycles occur are diagnostic for the various groups of fungi, and form the basis for their classification. Gäumann and Dodge (1928) recognize four natural classes of fungi, all of which contain plant-pathogenic species. The first two of these classes are characterized by primitive thalli, which are naked in Plasmodiophoromycetes (primitive fungi) and surrounded by a cell wall in Phycomycetes (algal fungi). In both classes, the zygote is the organ of meiosis. In the other two classes, diploid mycelium develops from the zygote, and special reproductive organs of meiosis develop on the diploid mycelium. In one of these classes, the Ascomycetes (sac fungi), sexual spores (ascospores) form within the organs of meiosis (asci); in the other class, Basidiomycetes (club fungi), sexual spores (basidiospores) are produced exogenously on the organs of meiosis (basidia). A fifth class of fungi, the form class Deuteromycetes (Fungi Imperfecti), comprises those fungi for which only the vegetative, or haploid, phase of the life cycle is known.

Recently, the fungi have been reclassified (Alexopoulos, 1962; Martin, 1963; Ainsworth, 1966). Although there is more general agreement on a broad classification of the fungi than is first apparent, the status of taxonomic mycology is understandably unsettled. Here, we shall follow a classification of the fungi (Table 3-1) that might be considered a combination of those of Gäumann and Dodge (1928), Martin (1963), and Ainsworth (1966).

Plasmodiophoromycetes

The taxonomic position of Plasmodiophoromycetes remains in doubt; for example, order Plasmodiophorales of this class is placed in the Phycomycetes by some (Fitzpatrick, 1930) and in the Myxomycetes (slime molds) by others (Wolf and Wolf, 1947). The naked, amoeboid thalli of plasmodiophoromycetes resemble those of the nonpathogenic myxomycetes, but spore formation is similar to that of phycomycetes. The plasmodiophoromycetes are unique among plant-pathogenic fungi, however, in that their reproductive organs are formed by fragmentation of whole thalli, a process termed **holocarpic reproduction** (Gäumann and Dodge, 1928).

The diploid phase of the life cycle of the plasmodiophoromycetes is not nearly so prominent as that of the higher fungi. In most orders, amoeboid zoospores act as sexual cells; after fusion (plasmogamy),

TABLE 3-1
Classification of plant-pathogenic fungi

Division MYXOMYCOTA (primitive fungi)

Class MYXOMYCETES (slime molds)

Class PLASMODIOPHOROMYCETES
Order PLASMODIOPHORALES
Family PLASMODIOPHORACEAE
[example: *Plasmodiophora brassicae* (of club root of crucifers)]

Division EUMYCOTA [Eumycetes]

Class PHYCOMYCETES (algal fungi)
Subclass CHYTRIDIOMYCETES [Chytridiomycetidae]
Order CHYTRIDIALES
Family CHYTRIDIACEAE (chytrids)
[example: *Synchytrium endobioticum* (of potato wart)]
Subclass OOMYCETES [Oomycetidae]
Order PERONOSPORALES
Family PYTHIACEAE
[examples: *Pythium debaryanum* (of damping-off of seedlings) and *Phytophthora infestans* (of late blight of potato and tomato)]
Family PERONOSPORACEAE (downy mildews)
[example: *Peronospora tabacina* (of downy mildew of tobacco)]
Subclass ZYGOMYCETES [Zygomycetidae]
Order MUCORALES
Family MUCORACEAE
[example: *Rhizopus stolonifer* (of soft rot of fruits and vegetables)]

Class ASCOMYCETES (sac fungi)
Subclass HEMIASCOMYCETES [Hemiascomycetidae]
Order ENDOMYCETALES
Family SACCHAROMYCETACEAE (yeasts)
[example: *Nematospora coryli* (of yeast rot of tomato and citrus fruit)]
Order TAPHRINALES
Family TAPHRINACEAE
[example: *Taphrina deformans* (of peach leaf curl)]
Subclass EUASCOMYCETES [Euascomycetidae]
Series PLECTOMYCETES
Order ERYSIPHALES
Family ERYSIPHACEAE (powdery mildews)
[example: *Erysiphe graminis* (of powdery mildew of cereals)]
Series PYRENOMYCETES
Order DOTHIDIALES
Family MYCOSPHAERELLACEAE
[example: *Mycosphaerella fragariae* (of leaf spot of strawberry)]

Order PLEOSPORALES
Family PLEOSPORACEAE
[example: *Venturia inaequalis* (of apple scab)]
Series DISCOMYCETES
Order HELOTIALES
Family SCLEROTINACEAE
[example: *Monilinia fructicola* (of brown rot of stone fruit)]

Class BASIDIOMYCETES (club fungi)
Subclass HETEROBASIDIOMYCETES [Heterobasidiomycetidae]
Order UREDINALES (rusts)
Family UREDINACEAE
[example: *Puccinia graminis* (of stem rust of cereals)]
Order USTILAGINALES (smuts)
Family USTILAGINACEAE
[example: *Ustilago maydis* (of corn smut)]
Family TILLETIACEAE
[example: *Tilletia foetida* (of bunt of wheat)]
Subclass HOMOBASIDIOMYCETES [Homobasidiomycetidae]
Order EXOBASIDIALES
Family EXOBASIDIACEAE
[example: *Exobasidium vaccinii* (of leaf gall of azalea)]
Order POLYPORALES
Family POLYPORACEAE (pore fungi)
[example: *Fomes applanatus* (of heart rot of trees)]
Order AGARICALES
Family AGARICACEAE (gill fungi)
[example: *Armillaria mellea* (of mushroom root rot of trees in the temperate zone)]

Class DEUTEROMYCETES [Fungi Imperfecti]
Order MONILIALES
Family MONILIACEAE
[example: *Botrytis cinerea* (of gray mold of grapes)]
Order MELANCONIALES
Family MELANCONIACEAE
[example: *Colletotrichum lindemuthianum* (of bean anthracnose)]
Order SPHAEROPSIDALES
Family SPHAEROPSIDACEAE
[example: *Diplodia zeae* (of stalk rot of corn)]
Order MYCELIA STERILIA (sterile fungi)
[example: *Sclerotium rolfsii* (of southern blight of herbaceous plants)]

Source: Modified after Gäumann and Dodge (1928), Martin (1963), and Ainsworth (1966).

they produce the plasmodium that serves both as the organ of fertilization (caryogamy) and as the organ of meiosis. Holocarpic reproduction results in the formation of thick-walled resting spores termed **hypnospores.** In general, a hypnospore of a plasmodiophoromycete germinates to produce a single amoeboid flagellate spore, the **zoospore,** or swarmspore. Then, in some species, + and − zoospores fuse; the resulting zygote invades the suscept, and gives rise to the plasmodium, a naked fungal protoplast. Finally, the plasmodium is fragmented into hypnospores, each of which contains one haploid nucleus. In some plasmodiophoromycetes, the life cycle is somewhat more complex. For example, *Plasmodiophora brassicae,* the causal agent of club root of cabbage, produces hypnospores, zoospores, amoeboid thalli within the suscept, secondary zoospores, plasmodia, and, finally, hypnospores again (Figure 3-8). Although the details of their life cycles are not thoroughly understood, it is clear that plasmogamy, caryogamy, and meiosis are not much separated from one another either by time or by space. For this reason, and because of the absence of threadlike thalli, the plasmodiophoromycetes are considered primitive in comparison with other plant pathogenic fungi (Gäumann and Dodge, 1928).

Phycomycetes

The thalli of the phycomycetes, which are surrounded by thin walls, consist of multinucleate threads, which form a **coenocytic mycelium.**

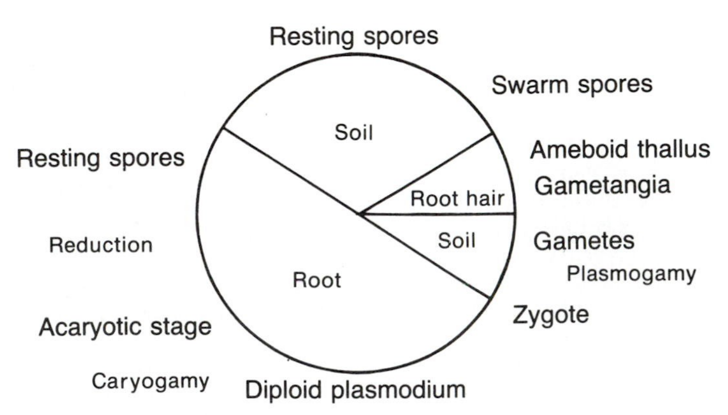

FIGURE 3-8
Sequence of events in the life cycle of *Plasmodiophora brassicae.*

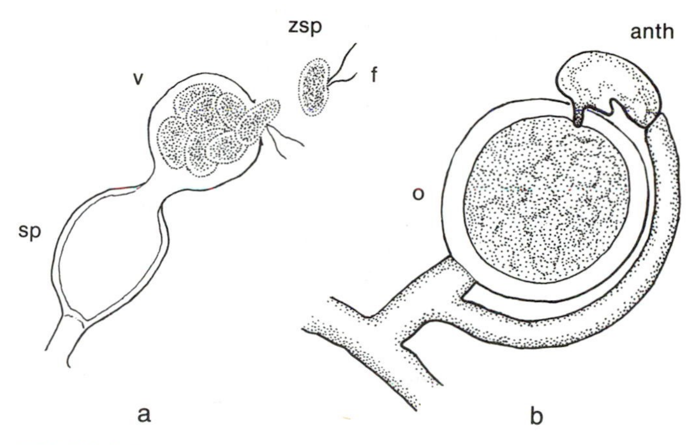

FIGURE 3-9
Vegetative reproduction (*a*) and sexual reproduction (*b*) of *Pythium*, an oomycete: *sp*, sporangium; *v*, vesicle; *zsp*, zoospore; *f*, flagellum; *anth*, antheridium; *o*, oogonium.

In the highest forms, hyphal threads may be abjointed into several cells by cross walls, or **septa.** Vegetative fructifications bear either sporangia or conidia. In the subclass Oomycetes (Figure 3-9), the sexual organs are of different morphology (heterogamous), whereas members of another subclass, Zygomycetes, bear equivalent gametes and are termed isogamous. In the heterogamous forms, the male gamete, a slender projection from a hypha, is termed the **antheridium;** the female gametes (**oogonia**) form as terminal or intercalary spherical cells on hyphae. Sexual spores of most phycomycetes are **oospores** (heterogamous forms) or **zygospores** (isogamous forms); in Chytridiales, an order considered to be primitive by some authors, mycelium is either absent or poorly developed, and hypnospores form after fertilization (Gäumann and Dodge, 1928). Although in the past the chytrids have been placed with the primitive fungi, the current tendency is to consider them in Phycomycetes (Ainsworth, 1966).

The higher oomycetes and zygomycetes are transitional between the phycomycetes and the ascomycetes. It is in the subclass Zygomycetes that plasmogamy and caryogamy are first separated in time and space, making the diploid phase (sporophytic generation) apparent. Further, the sporangia function as conidia, a decreasing number of nuclei function sexually, and zygospores become increasingly specialized as resting spores, often collecting in conspicuous fructifications.

Ascomycetes

In the class Ascomycetes, meiosis occurs immediately after caryogamy, and takes place in characteristic saclike "sporangia" called **asci.** Ascospores are formed endogenously.

The sexual organs of ascomycetes have undergone extensive morphological modification. The male copulation branch, or antheridium, is similar to that of the phycomycetes, and the female organ, **ascogonium,** resembles the oogonium of the phycomycetes. Ascogonia initiate **trichogynes,** receptive branches analogous to the fertilization tubes of the oomycetes. Unlike the plasmodiophoromycetes, and all but the higher phycomycetes, ascomycetes do not undergo caryogamy directly after plasmogamy. Instead, **dicaryons,** paired sexual nuclei that divide conjugately, are formed at plasmogamy. In most ascomycetes, dicaryons occur in **ascogenous hyphae,** which grow from the ascogonium after plasmogamy and give rise to the asci. Asci usually develop by a process known as **crozier formation.** A tip cell of an ascogenous hypha containing but one dicaryon puts forth a lateral branch in which the two nuclei become widely separated. The protuberance bends around to form a crook (crosier), and the nuclei—one in the tip of the crook and one in the stalk of the original branch—divide conjugately. Afterwards, the tip and the stalk are cut off from the crook by septa. The two nuclei in the crook fuse (caryogamy), and the crook develops into an ascus. Within the ascus, the diploid nucleus undergoes meiosis (reduction), and the four resulting haploid nuclei usually undergo mitosis. Thus, eight nuclei appear in the ascus. Eight ascospores, each with one haploid nucleus, are cut out from the cytoplasm within the ascus (Alexopoulos, 1962).

In the simplest ascomycetes, asci are formed free or on an undifferentiated mass of **stroma,** a dark, hard fungal tissue composed of thick-walled, more or less rounded cells. In the higher ascomycetes, however, haploid thalli produce complex fructifications of two general kinds, perithecia and apothecia. **Perithecia** are spherical or flask-shaped, with carbonous exteriors, and are formed on or imbedded in plant tissues that serve as the substrates. Most perithecia are barely visible as black specks; they can be seen easily, however, with the aid of a 10- to 15-power hand lens. **Apothecia** are fleshy, saucer-shaped structures that are usually plainly visible; they often develop on stalks that arise from sclerotia.

Imperfect (asexual) stages are developed most highly in the class Ascomycetes (Gäumann and Dodge, 1928). **Conidia** are asexual spores of diverse morphology, and may be borne singly or in chains on simple or branched conidiophores, or on or within such highly specialized fruiting structures as sporodochia, acervuli, and pycnidia. **Sporodochia** are cushionlike masses of fungal tissue that protrude above the substrate. **Acervuli** are small, saucer-shaped bodies that form beneath the surface of the substrate; the masses of conidia on their conidiophores break through the surface at maturity. **Pycnidia** are tiny flask-shaped bodies whose tips protrude through the surface of the substrate.

Thick-walled resting conidia formed in and on the mycelium are termed **chlamydospores.** Thalli of some fungi form **sclerotia,** masses of hyphae differentiated into a hard, dark rind and a lighter pseudoparenchymatous cortex. Sclerotia are similar to stroma, but, unlike stroma, do not have spore-bearing structures within. Fruiting and surviving structures of ascomycetes and deuteromycetes are illustrated in Figure 3-10.

The taxonomic relationships among the thousands of ascomycetes are still obscure. In the members of Hemiascomycetes, the subclass that includes the yeasts, there is no true dicaryotic phase. The subclass Euascomycetes has been separated into three major groups, depending upon the structure of the sexual fruiting body: asci are produced in **cleistothecia** (nonostiolate perithecia), true perithecia, and apothecia by members of the series Plectomycetes, Pyrenomycetes, and Discomycetes, respectively. Important plant pathogens occur in all of these groups. There is no apparent correlation between the taxonomy of the ascomycetes and the kinds of plant disease caused by them, except occasionally at the level of genus or family.

Deuteromycetes (Fungi Imperfecti)

Deuteromycetes (or Fungi Imperfecti) is an artificial group of some 20,000 species of imperfect fungi (that is, fungi whose sexual, or perfect, stage is unknown). Undoubtedly, perfect stages of many of these fungi will be discovered; many probably do not have a sexual stage, however: either they have never possessed one, or, if they have, they have lost it in the course of evolution.

Many imperfect fungi (in the broad sense of the term) are really ascomycetes for which only the mycelial, conidial, or sclerotial stages are known: those imperfect fungi with the coenocytic mycelia and typical zoosporangia characteristic of the phycomycetes are classified with that group. Many imperfect fungi have clamp connections, a characteristic of the basidiomycetes, and are therefore classified with that group; and the imperfect fructifications of certain rust fungi are sufficiently distinctive to permit them to be grouped in Basidiomycetes with the rust fungi whose sexual stages are known. Many of the imperfect fungi are plant-pathogenic; moreover, in a number of plant-pathogenic ascomycetes, the vegetative or imperfect stage is usually more prominent than the sexual stage during the course of disease.

Important fruiting bodies of the deuteromycetes are shown in Figure 3-10.

Basidiomycetes

In the members of the class Basidiomycetes, sexual spores are borne exogenously on club-shaped cells called **basidia.** Practically all basidiomycetes are heterothallic.

In the smut fungi, which are members of the subclass Heterobasidiomycetes, the nucleus of the organ of meiosis does not move out into the sexual spore as in other basidiomycetes. Instead, it divides mitotically, and a daughter nucleus becomes the spore nucleus. Thus, many spores may be borne by a single, septate organ of meiosis. In the plant-pathogenic rust fungi, which are members of the same subclass, the organ of meiosis is also a resting spore. After reduction, each nucleus becomes a spore nucleus; true basidia are not formed, but analogous structures—septate promycelia—perform the function of basidia, producing haploid spores (sporidia) that are analogous to basidiospores.

Rust fungi do not grow saprophytically in nature, and are, therefore, obligately parasitic pathogens. Two rust fungi, however—the cedar–apple rust (Hotson and Cutter, 1951; Cutter, 1959) and the wheat stem rust (Williams et al., 1966)—have been grown in culture.

All remaining basidiomycetes are members of the subclass Homobasidiomycetes; they all produce clavate, nonseptate basidia that arise from clamp formations on binucleate hyphae. Included here are the orders Exobasidiales, Polyporales, and Agaricales. Meiosis occurs in the developing basidium, and one haploid nucleus moves through a sterigma into each of the four young basidiospores (Figure 3-11). With few exceptions, the basidiospores finally produce mycelia that contain uninucleate haploid "cells." Two, or sometimes as many as four or five, mating types occur. Plasmogamy is accomplished through anastomosis of two haploid hyphae of different types, and binucleate clamp mycelium develops. Caryogamy occurs in the young basidium (Alexopoulos, 1962; Gäumann and Dodge, 1928).

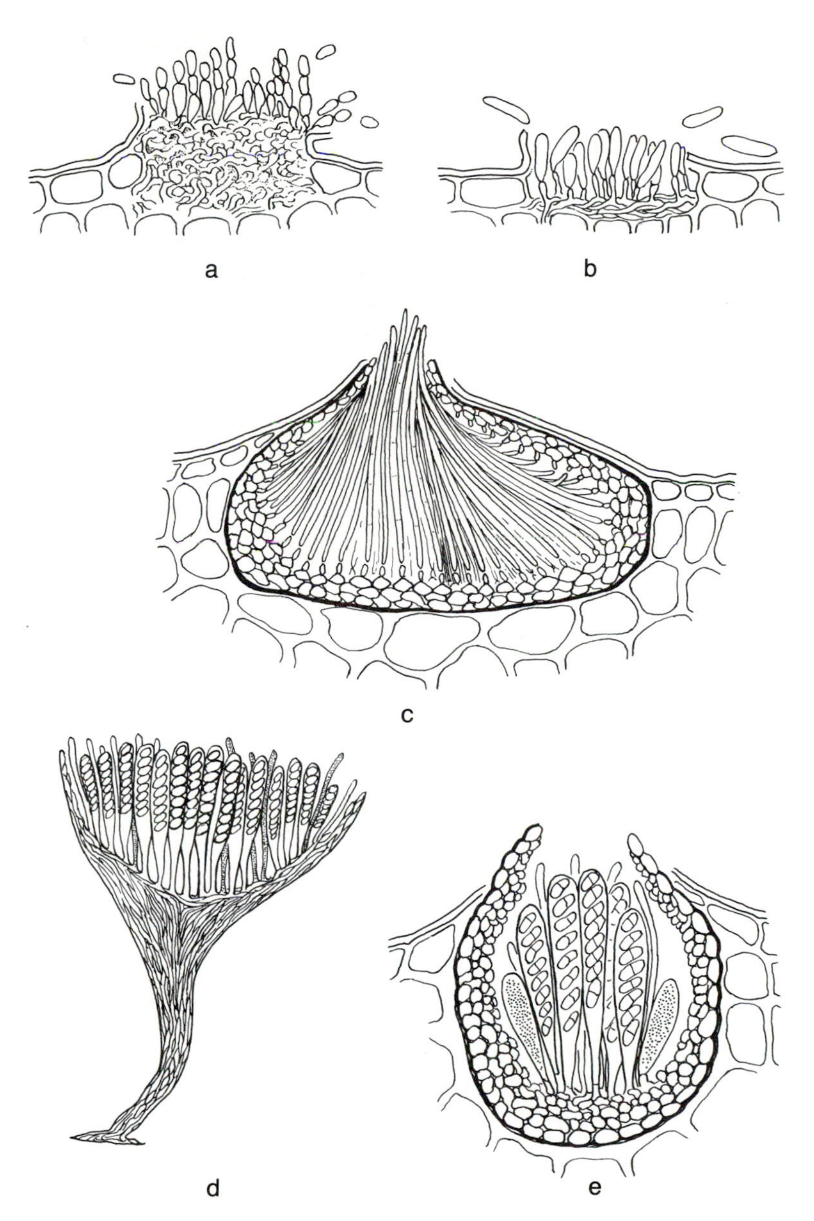

FIGURE 3-10
Some fruiting bodies and structures of ascomycetes and imperfect fungi: sporodochium (*a*), acervulus (*b*), pycnidium (*c*), apothecium (*d*), and perithecium (*e*).

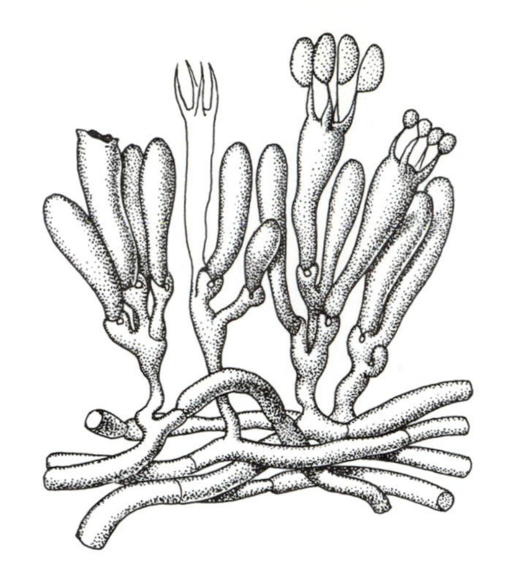

FIGURE 3-11
Basidia and basidiospores.

The haploid mycelium corresponds to the vegetative mycelium of the ascomycetes, but the conidial stage is much reduced relative to that of the ascomycetes. Moreover, haploid basidiomycetous mycelia do not produce functional sexual organs.

Many basidiomycetes produce complex fruiting bodies, such as mushrooms, conks, or punks. The **hymenium** (spore-bearing surface) of fungi in the order Agaricales is in the form of radiating gills or lamellae on the undersurfaces of the caps of mushrooms. Members of the Polyporales have hymenia consisting of pores or irregular wrinkles. Members of both groups are plant-pathogenic, causing root rots and wood decays and, occasionally, leaf diseases. Most homobasidiomycetes, however, are saprophytic. Some root-attacking species, such as *Armillaria mellea,* produce **rhizomorphs,** which are masses of parallel hyphae with dark, tough rinds and lighter-colored cores of pseudoparenchymatous tissue. Rhizomorphs appear as dark cords or strings; they serve both as organs of survival and as attacking organs.

NEMATODES

Nematodes are triploblastic, bilaterally symmetric, nonsegmented invertebrate animals (Chitwood and Chitwood, 1950). Most nematodes, or eelworms, are free-living aquatic, marine, or terrestrial forms that feed on microscopic plants and animals. Many species, however, parasitize man and other animals, and there are almost a thousand species that attack plant organs, including roots, bulbs, stems, leaves, and flowers. In roots, nematodes may cause galls or tumefactions (root knot), rots (root lesions), or they may so damage roots as to prevent their further growth (stubby root). Nematode infections of the roots of crop plants result in dwarfing and decline of the plants and a subsequent reduction in crop yield. Those plant-pathogenic nematodes that complete most of their life cycles within attacked plants are **endoparasitic,** and those that feed on epidermal and cortical tissues, and whose life processes proceed outside their host plants, are **ectoparasitic.**

Goodey (1933) classified nematodes as a class (Nematoda) of the phylum Nemathelminthes, whereas Chitwood and Chitwood (1950) considered the Nematoda a separate phylum. Most plant-pathogenic nematodes are members of the order Tylenchida but several members of the order Dorylaimida are known to attack plants (Thorne, 1961). Table 3-2 outlines the classification of plant-pathogenic nematodes.

Plant-pathogenic nematodes are equipped with **stylets** or spears, specialized feeding structures that are not produced in saprophytic forms. The stylet is a hollow tube in the buccal cavity; it is connected with the oesophagus, has a muscular, bulbous base, and can be pushed out beyond the lip region and retracted.

The anterior end of a nematode begins with an oral opening, or mouth, surrounded by lips that contain sensory openings. The digestive tract includes the mouth, buccal cavity (in which the stylet occurs in plant parasites), oesophagus (often swollen by anterior and posterior oesophageal bulbs), intestine, and rectum, which terminates in an anus, or cloacal opening. An excretory cell is located within the body cavity at the anterior end of the intestine; it is connected to an opening in the cuticle (outer covering) termed the excretory pore. Sexes are usually separate, although many individuals are hermaphroditic (Christie, 1959). In the female (Figure 3-12), ovaries open into a uterus; the uterus is connected with the vagina, which terminates in a vulva. The male testis is connected to the vas deferens, which empties into the cloacal aperture. Some males have a spicule (an accessory organ of attachment) and a bursa (a saclike extension across the cloacal aperture. Nematodes have a primitive nervous system extending back in lateral lines from a ring of nervous tissue that encircles the oesophagus between the anterior and posterior bulbs. A detailed anatomy of the nematodes is given by Chitwood and Chitwood (1950).

Eggs of plant-pathogenic nematodes are deposited, depending on the species, in or on plant tissues or in the soil. Some females, such as those of the sugar-beet nematode and the potato-golden nematode,

TABLE 3-2
Classification of plant-pathogenic nematodes (phylum Nematoda).

Class SECERNENTEA
- Order TYLENCHIDA
  - Superfamily TYLENCHOIDEA
    - Family TYLENCHIDAE
      - Subfamily TYLENCHINAE
        - Genus *Anguina* (seed and gall nemas)
        - Genus *Ditylenchus* (bulb and stem nemas)
      - Subfamily PRATYLENCHINAE
        - Genus *Pratylenchus* (lesion nemas)
      - Subfamily RADOPHOLINAE
        - Genus *Radopholus* (burrowing nemas)
      - Subfamily HOPLOLAIMINAE
        - Genus *Helicotylenchus* (spiral nemas)
    - Family HETERODERIDAE
      - Subfamily HETERODERINAE
        - Genus *Heterodera* (cyst nemas)
        - Genus *Meloidogyne* (root-knot nemas)
    - Family CRICONEMATIDAE
      - Subfamily CRICONEMATINAE
        - Genus *Criconema* (ring nemas)
      - Subfamily HEMICYCLIOPHORINAE
        - Genus *Hemicycliophora* (sheath nemas)
      - Subfamily PARATYLENCHINAE
        - Genus *Paratylenchus* (pin nemas)
  - Superfamily APHELENCHOIDEA
    - Family APHELENCHOIDIDAE
        - Subfamily APHELENCHOIDINAE
        - Genus *Aphelenchoides* (foliar nemas)

Class ADENOPHOREA
- Order DORYLAIMIDA
  - Superfamily DORYLAIMOIDEA
    - Family LONGIDORIDAE
      - Subfamily LONGIDORINAE
        - Genus *Longidorus* (needle nemas)
        - Genus *Xiphinema* (dagger nemas)
  - Superfamily DIPHTHEROPHOROIDEA
    - Family TRICHODORIDAE
      - Subfamily TRICHODORINAE
        - Genus *Trichodorus* (stubby-root nemas)

Source: Modified after Chitwood and Chitwood (1950), Thorne (1961), and Zukerman, Mai, and Rhode (1972).

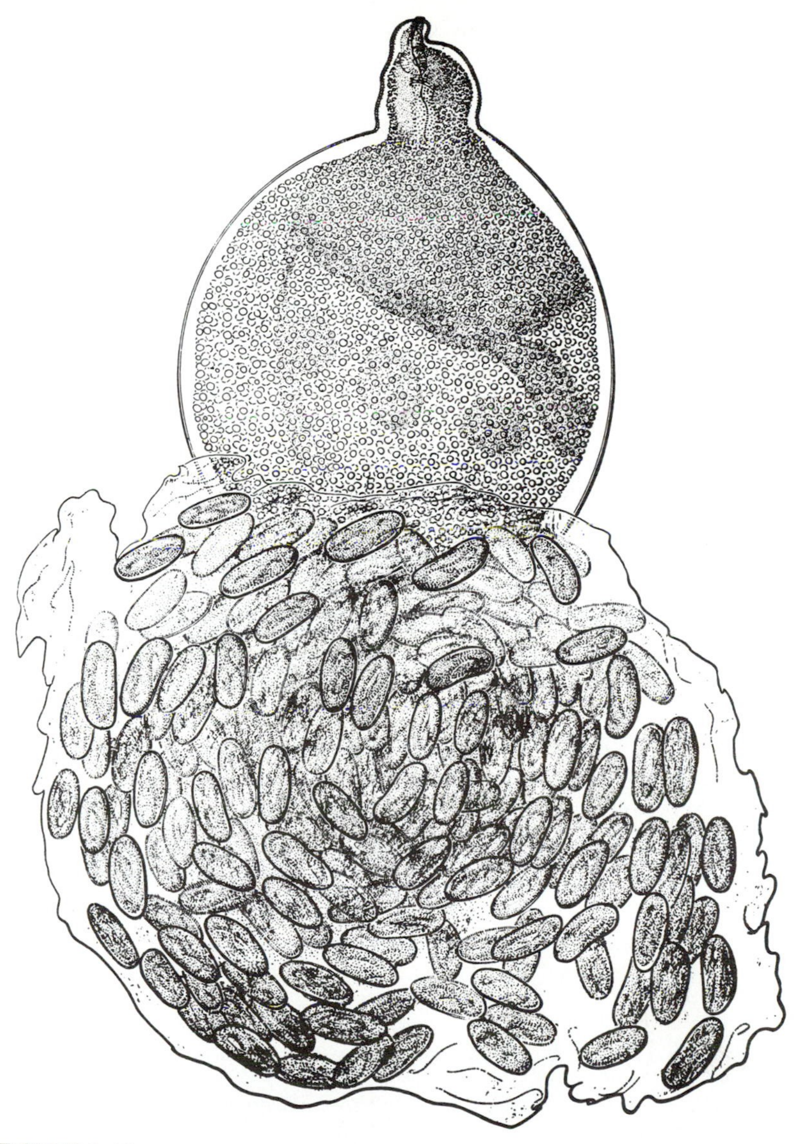

FIGURE 3-12
Egg-laying female of the nematode *Meloidogyne incognita*. (Courtesy of the Soil Science Society of Florida.)

die soon after the formation of eggs, and their bodies develop into thick-walled **cysts** that enclose and protect the eggs during the survival phase of the life cycle. Larvae hatch from the eggs and may undergo several molts before becoming adults.

Literature Cited

MISCELLANEOUS PLANT PATHOGENS

Ball, E. D., 1918. Leaf burn of the potato and its relation to the potato leafhopper. *Science* 48:194.

Carter, W., 1939. Injuries to plants caused by insect toxins. *Bot. Rev.* 5:273–326.

Felt, E. P., 1940. *Plant Galls and Gall Makers.* Ithaca, N. Y.: Comstock.

Garriss, H. R., and J. C. Wells, 1956. Parasitic herbaceous annual associated with corn diseases in North Carolina. *Plant Dis. Rep.* 40:837–839.

Illingworth, J. F., 1931. Preliminary report on evidence that mealy bugs are an important factor in pineapple wilt. *J. Econ. Entomol.* 24:877–889.

Stahel, G., 1933. Zur Kenntniss der Siebrohrenkrankheit (Phloemnekrose) des Kaffebaumes in Surinam, III. *Phytopathol. Z.* 6:335–337.

Trelease, W., 1916. *The Genus Phoradendron, a Monographic Revision.* Urbana, Ill.: University of Illinois Press.

Wolf, F. A. 1930. A parasitic alga, *Cephaleuros virescens* Kunze, on citrus and certain other plants. *J. Elisha Mitchell Sci. Soc.* 45:187–205.

Yuncker, T. G., 1931. *Revision of the North American and West Indian Species of Cuscuta* [Ill. Biol. Monogr. 6]. Urbana, Ill.: University of Illinois Press.

VIRUSES AND MYCOPLASMALIKE ORGANISMS

Ball, E. M., 1961. *Serological Tests for the Identification of Plant Viruses.* Ithaca, N. Y.: American Phytopathological Society Committee on Plant Virology, Plant Pathology Department, Cornell University.

Bawden, F. C., 1964. *Plant Viruses and Virus Diseases* (4th ed.). New York: Ronald Press.

———, and B. Kassanis, 1945. The suppression of one plant virus by another. *Ann. Appl. Biol.* 32:52–57.

———, N. W. Pirie, J. D. Bernal, and S. Fankuchen, 1936. Liquid crystalline substances from virus-infected plants. *Nature* 138:1051.

Beijerinck, M. W., 1898. Über ein *Contagium vivum fluidium* als Ursache der Flekenkrankheit der Tabaksblatter. *Vehr. Akad. Wettensch. Amsterdam* 65(2):3–21. (English transl. by J. Johnson, 1942. *Concerning a contagium vivum fluidium* as the cause of the fleck disease of tobacco leaves. *Phytopathol. Classics* 7:33–52.)

Best, R. J., 1954. Cross protection by strains of tomato spotted wilt virus and a new theory to explain it. *Aust. J. Biol. Sci.* 7:415–424.

———, 1956. Living molecules: a survey of recent advances in our understanding of the nature of viruses. *Aust. J. Sci.* 18:106–120.

Black, L. M., 1954. Arthropod transmission of plant viruses. *Exp. Parasitol.* 3:72–104.

Carsner, E., 1925. Attenuation of the virus of sugar beet curly-top. *Phytopathology* 15:745–757.

Davis, R. E., and R. F. Whitcomb, 1971. Mycoplasmas, rickettsiae, and chlamydiae: possible relation to yellows diseases and other disorders of plants and insects. *Annu. Rev. Phytopathol.* 9:119–154.

Doi, Y., M. Teranaka, K. Yora, and H. Asuyama, 1967. Mycoplasma- or PLT group-like microorganisms found in the phloem elements of plants infected with mulberry dwarf, potato witches' broom, aster yellows, or paulownia witches' broom [in Japanese, English abstract]. *Ann. Phytopathol. Soc. Jap.* 33:259–266.

Dvorak, M., 1927. The effect of mosaic on the globulin of potato. *J. Infect. Dis.* 41:215–221.

Fraenkel-Conrat, H., and R. C. Williams, 1955. Reconstitution of active tobacco mosaic virus from its inactive protein and nucleic acid components. *Proc. Nat. Acad. Sci. U. S. A.* 41:690–698.

Gierer, A., and G. Schramm, 1956. Infectivity of ribonucleic acid from tobacco mosaic virus. *Nature* 177:702–703.

Holmes, F. O., 1929. Local lesions in tobacco mosaic. *Bot. Gaz.* 87:39–55.

———, 1948. Order Virales. The filterable viruses. *In* R. S. Breed, E. G. D. Murray, and A. P. Hitchens, eds., *Bergey's Manual of Determinative Bacteriology* (6th ed.), pp. 1127–1286. Baltimore: Williams & Wilkins.

———, 1964. Symptomatology of viral diseases in plants. *In* M. K. Corbett and H. D. Sisler, eds., *Plant Virology,* pp. 17–38. Gainesville, Fla.: University of Florida Press.

Hopkins, J. C. F., ed., 1957. Common names of plant-virus diseases used in the *Review of Applied Mycology. Rev. Appl. Mycol.* 35(Suppl.).

Hull, R., 1971. Mycoplasma-like organisms in plants. *Rev. Plant Pathol.* 50:121–130.

Ishiie, T., Y. Doi, K. Yora, and H. Asuyama, 1967. Suppressive effects of antibiotics of tetracycline group on symptom development of mulberry dwarf disease [in Japanese, English abstract]. *Ann. Phytopathol. Soc. Jap.* 33:267–275.

Iwanowski, D., 1903. Über der Mosaikkrankheit der Tabakspflanze. *Z. Pflanzenkr.* 13:1–41. (English transl. by J. Johnson, 1942. Concerning the mosaic disease of tobacco plants. *Phytopathol. Classics* 7:27–30.)

Johnson, J., 1926. The attenuation of plant viruses and the inactivating influence of oxygen. *Science* 64:210.

Kunkel, L. O., 1934. Studies on acquired immunity with tobacco and aucuba mosaics. *Phytopathology* 24:437–466.

Loring, A. S., and R. S. Waritz, 1957. Occurrence of iron, copper, calcium and magnesium in tobacco mosaic virus. *Science* 125:646–648.

Martyn, E. B., ed., 1968. An annotated list of names and synonyms of plant viruses and diseases. *Commonw. Mycol. Inst. Phytopathol. Pap.* (9).

Mayer, A., 1886. Über die Mosaikkrankheit des tabaks. *Landwirt. Versuchs-Sta.* 32:451–467. (English transl. by J. Johnson, 1942. On the mosaic disease of tobacco. *Phytopathol. Classics* 7:11–24.)

McKinney, H. H., 1926. Virus mixtures that may not be detected in young tobacco plants. *Phytopathology* 16:893.

———, 1929. Mosaic diseases in the Canary Islands, West Africa, and Gibraltar. *J. Agr. Res.* 39:557–578.

Mulvania, M., 1926. Studies on the nature of the virus of tobacco mosaic. *Phytopathology* 16:853–871.

Pollard, J. K., W. F. Rochow, and F. C. Steward, 1958. The incorporation of certain C^{14}-labelled amino acids into tobacco mosaic virus. *Plant Physiol.* 33:xii–xiii.

Purdy, H. A., 1928. Immunologic reactions with tobacco mosaic virus. *Proc. Soc. Exp. Biol. Med.* 25:702–703.

Salaman, R. N., 1933. Protective inoculation against a plant virus. *Nature* 131:468.

Shepherd, R. J., R. J. Wakeman, and R. R. Romanko, 1968. DNA in cauliflower mosaic virus. *Virology* 36:150–152.

Stanley, W. M., 1935. Isolation of a crystalline protein possessing the properties of tobacco-mosaic virus. *Science* 81:644–645.

Thung, T. H., 1931. Infective principle and plant cell in some virus diseases of the tobacco plant. *Handel. Ned.-Indish. Natuurventensch. Congr.* 1931:450–463.

Vinson, C. J., and A. W. Petre, 1929. Mosaic disease of tobacco, I. Progress in freeing the virus of accompanying solids. *Bot. Gaz.* 87:14–38.

BACTERIA

Breed, R. S., E. G. D. Murray, and N. R. Smith, eds. 1957. *Bergey's Manual of Determinative Bacteriology* (7th ed.). Baltimore: Williams & Wilkins.

Burkholder, W. H., 1948. Bacteria as plant pathogens. *Annu. Rev. Microbiol.* 1:389–412.

Burrill, T. J., 1878. [Remarks in a discussion]. *Trans. Ill. Hort. Soc.* (N.S.) 11:79–80.

———, 1880. Blight of pear and apple trees. *Ill. Agr. Exp. Sta. Rep.* 1880:62–84.

Dowson, W. J., 1957. *Plant Diseases Due to Bacteria* (2nd ed.). Cambridge: Cambridge University Press.

Elliott, C., 1951. *Manual of Bacterial Plant Pathogens* (2nd ed.). Waltham, Mass.: Chronica Botanica.

Fischer, A., 1899. Die Bakterienkrankheiten der Pflanzen. *Zentralbl. Bakteriol.* (2)5:279–287.

Graham, D. C., 1964. Taxonomy of the soft rot coliform bacteria. *Annu. Rev. Phytopathol.* 2:13–42.

Jones, L. R., 1901. A soft rot of carrot and other vegetables. *Vt. Agr. Exp. Sta. Rep.* 13:299–332.

McNew, G. L., 1966. The nature and cause of disease in plants. *Amer. Biol. Teacher* 28:445–461.

Smith, E. F., 1895. *Bacillus tracheiphilus,* sp. nov., die Ursache des Vorwelkens verschiedener Cucurbitaceen. *Zentralbl. Bakteriol.* (2)1: 364–373.

———, 1901. Entgegnung auf Alfred Fischer's "Antwort" in betreff der Existenz voh durch Bakterien verursachten Pflazenkrankheiten. *Zentralbl. Bakteriol.* (2)7:88–100, 128–239, 190–199.

———, and C. O. Townsend, 1907. A plant-tumor of bacterial origin. *Science* 25:671–673.

Stapp. C., 1961. *Bacterial Plant Pathogens.* London: Oxford University Press.

Starr, M. P., 1959. Bacteria as plant pathogens. *Annu. Rev. Microbiol.* 13:211–238.

Stolp, H., M. P. Starr, and N. L. Baigent, 1965. Problems in speciation of phytopathogenic pseudomonads and xanthomonads. *Annu. Rev. Phytopathol.* 3:231–264.

Wakker, J. H., 1883. Vorlaufige Mittheilungen über Hyacinthenkrankheiten. *Bot. Centralbl.* 14:315–317.

FUNGI

Ainsworth, G. C., 1966. A general purpose classification of the fungi. *Bibliogr. Syst. Mycol.* 1966, part I, no. 1–415.

Alexopoulos, C. J., 1962. *Introductory Mycology* (2nd ed.), pp. 3–41. New York: Wiley.

Bessey, E. A., 1950. *Morphology and Taxonomy of Fungi.* Philadelphia: Blakiston.

Blakeslee, A. F., 1904. Sexual reproduction in the Mucorineae. *Proc. Amer. Acad. Arts Sci.* 40:205–315.

Cutter, V. M., 1959. Studies on the isolation and growth of plant rusts in host tissue cultures and upon synthetic media, I. *Gymnosporangium. Mycologia* 51:248–295.

deBary, A., 1853. *Untersuchungen über die Brandpilze und die durch sie verursachten Krankheiten der Pflanzen mit Rücksicht auf das Getreide und andere Nutzpflanzen.* Berlin: Müller.

Fitzpatrick, H. M., 1930. *The Lower Fungi. Phycomycetes.* New York: McGraw-Hill.

Gäumann, E. A., and C. W. Dodge, 1928. *Comparative Morphology of Fungi.* New York: McGraw-Hill.

Hotson, H. H., and V. M. Cutter, Jr., 1951. The isolation and culture of *Gymnosporangium juniperi-virgininae* (*sic.*) Schw. upon artificial media. *Proc. Nat. Acad. Sci. U. S. A.* 37:400–403.

Kühn, J., 1858. *Die Krankheiten der Kulturgewächse, ihre Ursachen und ihre Verhütung.* Berlin: Bosselmann.

Martin, G. W., 1963. Key to the families of fungi. *In* G. C. Ainsworth, *Ainsworth and Bisby's Dictionary of the Fungi* (5th ed.), pp. 411–427. Kew: Commonwealth Mycological Institute.

Prévost, B., 1807. *Memoire sur la cause immediate de la carie ou charbon des blés, et de plusieurs autres maladies des plantes, et sur les preservatifs de la carie.* Paris: Bernard. (English transl. by G. W. Keitt, 1939. Memoir on the immediate cause of bunt or smut of wheat, and of several other diseases of plants, and on prevention of bunt. *Phytopathol. Classics* 6:21–94.)

Tillet, M., 1755. *Dissertation sur la cause qui corrompt et noircit les grains de bled dans les epis; et sur les moyens de prevenir ces accidens.* Bordeaux: Brun. (English transl. by H. B. Humphrey, 1937. Dissertation on the cause of the corruption and smutting of the kernels of wheat in the head and on the means of preventing these untoward circumstances. *Phytopathol. Classics* 5:8–189.)

Williams, P. G., K. J. Scott, and J. L. Kuhl, 1966. Vegetative growth of *Puccinia graminis* f. sp. *tritici* in vitro. *Phytopathology* 56:1418–1419.

Wolf, F. A., and F. T. Wolf, 1947. *The Fungi,* vol. 1. New York: Wiley.

NEMATODES

Chitwood, B. G., and M. B. Chitwood, 1950. *An Introduction to Nematology,* sect. 1 (Anatomy). Baltimore: privately published by the authors.

Christie, J. R., 1959. *Plant Nematodes, their Bionomics and Control.* Gainesville, Fla.: University of Florida Agricultural Experiment Station.

Goodey, T., 1933. *Plant Parasitic Nematodes.* London: Methuen.

Thorne, G., 1961. *Principles of Nematology.* New York: McGraw-Hill.

4

Distribution and Survival of Plant Pathogens

Comprehension of the phenomena associated with initiation and development of disease is the basis for the prevention of plant disease, or the applied science of plant pathology. Of paramount importance to the plant pathologist is an understanding of the geographical distribution of plant pathogens and their survival between crop season, of the production and dispersal of inocula, of the entrance of pathogens into plants, and of infection proper. Distribution and survival of plant pathogens are discussed in this chapter.

GEOGRAPHICAL DISTRIBUTION OF PLANT PATHOGENS

Transportation of Plant Pathogens

Structures of plant pathogens that are well adapted for dispersal to areas previously uninvaded include a number of structures in vital associations with their respective suscepts—such as viral particles, bacterial cells, fungal mycelia, and nematode larvae. They are carried in living plant parts, including seeds, bulbs, tubers, seedlings, and cuttings. Certain other structures are able to endure in nonvital associations—such as the oospores of phycomycetes; the chlamydospores, sclerotia, stromata, mycelium, immature pycnidia, perithecia, and some conidia of ascomycetes and imperfect fungi; smut spores; the teliospores, amphiospores, and aeciospores of certain rusts; the mycelium, sclerotia, and rhizomorphs of certain true basidiomycetes; the seeds of parasitic phanerogams; and the eggs and cysts of nematodes. These enduring structures are usually harbored in dead suscept tissue or in the soil.

Many destructive plant pathogens are distributed in, on, or with seed, soil, or vegetative plant parts. Consequently, by shipping plants to all parts of the world, man has become the chief agent of transportation and distribution of many plant pathogens. Birds are the agents of distribution for common eastern mistletoe (Bray, 1910; see Box 3-1) and for conidia of the chestnut-blight fungus (Heald and Studhalter, 1914). Important inanimate agents of distribution are wind and water currents. There is no doubt that fungal spores capable of withstanding the drying effects of wind are carried long distances in the air (Stakman and Christensen, 1946). Urediospores of the stem-rust fungus are transported by wind currents (Stakman et al., 1923), and the rapid extension in the United States of the range of the fungus of white-pine blister rust was due to spread of aeciospores carried by the wind (Anderson and Rankin, 1914). Plant parts are sometimes transported from continent to continent by ocean currents, and pathogens might be transported in this way.

Geographical barriers, mountain ranges, and large bodies of water often limited the transport of plant pathogens. Such barriers are ineffective, however, when man, carrying infected plant materials with him, crosses oceans and mountains in his travels and migrations. As a substitute for natural barriers, man has devised the barrier of plant inspections and quarantines (McCubbin, 1946) in an effort to prevent the spread of pests and pathogens from one country to another. Despite man's efforts, however, plant pathogens continue to spread. For example, the coffee-rust pathogen (*Hemileia vastatrix*), once restricted to Asia and Africa, was reported from Brazil in 1970 (Box 4-1). Nevertheless, plant quarantines have been somewhat successful; many pathogens not yet reported from the United States (Hunt, 1946) have been intercepted and destroyed by quarantine officials at ports of entry.

Establishment and Maintenance of Plant Pathogens in Areas Previously Not Invaded

A plant pathogen may be readily transported over long distances and placed in an uninvaded area, but unless it can establish and maintain itself there, its geographical range will not be increased. It is axiomatic that plant pathogens cannot establish themselves in new areas unless susceptible plants occur there. Moreover, pathogens cannot maintain themselves in new areas unless they are able to survive the dry or cold periods between growing seasons.

Weather conditions during the growing season also influence the establishment of plant pathogens. The Texas root-rot fungus occurs only in the warm soils of the southern United States because of its inability to survive low temperature (Ezekiel, 1945); the bacterium that causes wilt (brown rot) of solanaceous plants also survives only in warm soils (Gratz, 1930). The onion-smut fungus, however, cannot survive in warm soils, and its range in the United States is restricted to the northern states (Walker and Jones, 1921). Although temperature probably is the most important environmental factor affecting distribution, lack of moisture in certain parts of the western United States limits the development of certain diseases, including bean anthracnose (Whetzel, 1908). The distribution of the organism that causes citrus canker is determined by the combined effects of temperature and moisture (Peltier and Frederich, 1926). Wherever temperature and rainfall increase as the growing season progresses—as along the coast of the Gulf of Mexico and in China and South Africa—the development of citrus canker is favored; the disease does not occur in California, where progressively less rain falls as the temperature rises during the growing season.

Maps showing the distribution of many plant pathogens have been published since 1942 by the Commonwealth Mycological Institute.

SURVIVAL OF PLANT PATHOGENS

The establishment of a plant pathogen in a geographical location presupposes its ability to survive, not only during its parasitic relations with its hosts, but also during those season in which the hosts are not growing. In temperate zones, plant pathogens must be adapted for survival over winters or over summers, like the powdery-

BOX 4-1
COFFEE RUST IN SOUTH AMERICA

Plant diseases indigenous to one part of the world often devastate suscept species in another part of the world when their pathogens are introduced. One reason for this is that strains of susceptible plant species in new regions in the geographical range of a pathogen do not usually have the tolerance or resistance that plants in the region of origin have developed through natural selection, sometimes aided by man's plant-breeding and selection programs. Also, the environment in a region into which a pathogen is introduced sometimes proves more favorable to the pathogen than that in the region whence it originally came. In any event, the threat of introduced plant pathogens is not to be taken lightly. The downy mildew of grape, introduced from America, devastated grape vines in Europe and threatened the wine industry of France in the nineteenth century. The chestnut-blight fungus from Europe literally wiped out the magnificent stands of chestnuts in the United States in the first half of the twentieth century, and elm wilt (the "Dutch elm disease") is presently eradicating the graceful American elm from forests and shade-tree plantings.

For years, agriculturists have been aware that certain plant diseases not yet introduced into particular areas of crop production would probably destroy commercially important crop species upon their introduction. Because man is the most likely agent of distribution for plant pathogens, legislation has been enacted in many countries to require inspection of plant materials at ports of entry and the destruction of any such materials found to be harboring plant pathogens. But the effectiveness of such legislation depends upon detection of pathogens at the points of inspection, and detection is not always possible.

The citrus-canker bacterium has been excluded from the southeastern United States for more than fifty years, and the South American leaf-blight fungus of rubber trees has been excluded thus far from the rubber plantations of Africa and Southeast Asia. Until very recently, the dreaded coffee-rust fungus of Asia and Africa has been kept out of South America, but the discovery of the fungus in Brazil in 1970 confirmed the worst fears of the coffee planters. Modern methods of plant inspection and quarantine have failed, and existing plantings of coffee in Brazil may be doomed.

The coffee-rust fungus, *Hemileia vastatrix,* is an obligate parasite that grows in leaves, petioles, and young twigs of coffee, in which it causes necrotic lesions. The vegetative spores of the pathogen are borne in the lesions just under the epidermis. When this covering ruptures, the spores are freed and disseminated locally by splashing rain and by wind. Long-distance dissemination occurs when diseased vegetative material for propagation is transported by man.

Obviously, coffee rust was introduced into Brazil several years before it was detected by plant pathologists; by 1970, it had broken out in many locations scattered throughout a region about 1500 miles long and several hundred miles wide. The occurrence of the disease in widely separated "pockets" suggests that young trees in one or more nurseries had been infected before they were set out in plantations.

Infection by the coffee-rust fungus leads to early defoliation followed by flushes of new growth; later, the new growth becomes defoliated also. Food reserves are thus depleted in diseased trees, which die within two or three years. The only known way to prevent epidemics of coffee rust is to plant resistant trees; however, resistant planting material is still in short supply. In addition, much time and money are required, for the replacement of millions of trees is no simple task.

mildew pathogen that attacks fall-seeded wheat. In the Torrid Zone, plant pathogens must be able to survive the dry seasons, during which susceptible plants are not growing. Pathogens may survive between crop seasons by means of specialized resting structures, by functioning as saprophytes in the soil or in plant residue, or by living in some intimate association with living plants or insects.

Survival by Means of Specialized Resting Structures

Enduring structures of plant pathogens may be as simple as conidia or as complex as perithecia. Apparently, ascospores, or conidia derived from them, serve to carry the pathogen causing peach-leaf curl (*Taphrina deformans*) over the winter (Mix, 1935). Conidia of *Alternaria solani,* the pathogen of early blight of potato and tomato, survive as long as eighteen months in dried diseased leaves (Rands, 1917). Specialized thick-walled chlamydospores of *Fusarium* and other imperfect fungi, spores of many smut fungi, and the amphiospores, urediospores, and teliospores of certain rust fungi also are important enduring structures. The hypnospores of *Plasmodiophora brassicae* may survive for as long as ten years in soils infested upon the disintegration of clubbed roots. The oospores of downy-mildew fungi survive in the soil between growing seasons. In fact, oospores of the fungus that causes onion mildew do not germinate until several years after their formation (McKay, 1935).

Some fungi survive unfavorable seasons in the form of sclerotia. Those produced by the omnivorous cottony-rot fungus, *Sclerotinia sclerotiorum,* can survive for years in a dry atmosphere (Dickson, 1930); they decay rapidly, however, in warm moist soil (Stevens and Hall, 1911). Cold-induced dormancy probably accounts for their ability to survive winters in temperate zones. Some powdery-mildew fungi and other ascomycetes survive in the form of perithecia that are not saprophytically associated with plant refuse. Parasitic phanerogams survive in the form of seeds, and the eggs, cysts, and larvae of parasitic nematodes serve as overseasoning structures.

Survival as Saprophytes

The ability to live saprophytically enables many plant pathogens to survive in the absence of growing susceptible plants. Saprophytic survival usually occurs in or on the soil. Waksman (1917) distinguished between soil inhabitants and soil invaders; the former comprise the basic fungal flora of the soil, whereas the latter are short-

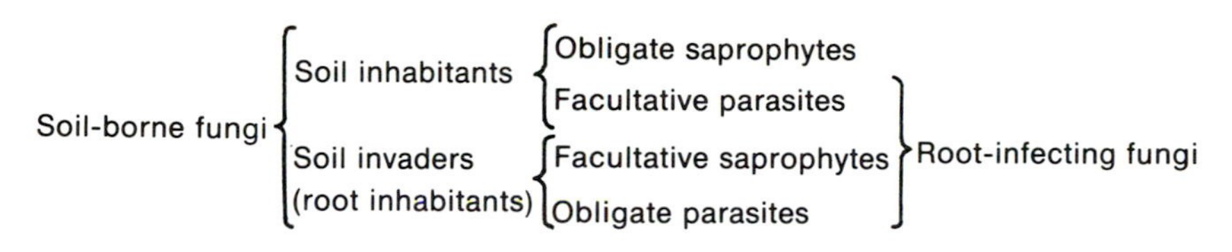

FIGURE 4-1
Interrelations of soil-borne fungi.

lived exotics. As applied to the root-infecting fungi (Garrett, 1956), **soil inhabitants** are unspecialized parasites with a wide host range that are able to survive indefinitely in the soil as saprophytes; **soil invaders (root-inhabiting fungi)** are more specialized parasites that survive in soils in close association with their hosts (Figure 4-1). Most plant-pathogenic fungi and bacteria are soil invaders, but some pathogens, notably *Rhizoctonia solani* and *Pythium debaryanum,* which cause seedling blights and root rots, live saprophytically in the soil.

The microbiological balance in the soil markedly affects the survival of saprophytic plant pathogens there. Apparently, Sanford (1926) was the first to suggest that control of potato scab by green manuring with grass might be due to the antagonistic action of saprophytic organisms flourishing on the green manure. Later, experimental evidence was obtained to support this hypothesis (Millard and Taylor, 1927), and many other instances of microbiological antagonism are now known. Not only do soil saprophytes antagonize other microorganisms by toxic action, but some, such as *Trichoderma lignorum,* actually parasitize *Rhizoctonia solani* and other soil-borne pathogens (Weindling, 1932). Despite antagonism and parasitism by other organisms, many plant pathogens survive in the soil as inhabitants or invaders. The special conditions that favor biological control of plant pathogens in sterilized soil or in culture are nonexistent in field soils, in which there is a complex microflora and a low concentration of nutrients.

Certain plant pathogens survive between growing seasons as saprophytes in the dead tissues of susceptible plants. Such organisms are only incidentally associated with the soil, and live only as long as tissues of susceptible plants are available to supply nutrients. Most plant-pathogenic bacteria and many specialized parasitic fungi survive in this way. The apple-scab pathogen (*Venturia inaequalis*) lives parasitically in leaves and fruit during the growing season, but becomes saprophytic in fallen leaves. Perithecia form in these leaves during the winter, but ascospores do not form until spring. Ascospores of certain other ascomycetes mature during the winter, but are protected from adverse conditions by perithecial walls.

Survival in Vital Association with Living Plants

Brefeld and Falk (1905) were apparently the first to recognize that the pathogen of loose smut of wheat, *Ustilago tritici,* enters the stigma and style and infects the young seed, in which it survives the winter as mycelium. The seed-infecting pathogens that cause loose smut of wheat and loose smut of barley are strikingly different from other smut fungi that attack cereal crops: most of the others survive from season to season either in nonpathogenic association with seed or as spores in the soil.

Colletotrichum lindemuthianum, the causative organism of bean anthracnose, can also infect the seed; unless the seed is killed, new infections are initiated by the fungus in newly sprouted bean seedlings (Barrus, 1921). The bacteria that cause bean blights and angular leaf-spot of cotton survive the winter in infected seed, and the mycelium of the fungus of choke disease of grasses (Sampson, 1933) survives in vital association with its perennial hosts. In Mexico, the fungus of late blight of potatoes (*Phytophthora infestans*) produces oospores (Gallegly and Galindo, 1958), but in colder regions of the world, the fungus overwinters as mycelium in diseased tubers.

Plants that harbor perennial mycelium of a parasitic fungus, or those that are infected in the seedling stage and retain the pathogen internally throughout their lives, are said to be **systemically infected** (Butler and Jones, 1949). The ability of some plant pathogens to cause systemic infections greatly improves their chances of survival between growing seasons: pathogens that cause systemic infections of perennial plants may survive in the whole plant, and those that attack annuals, biennials, and herbaceous perennials may survive in their propagative parts. This ability has been developed to the highest degree in the plant viruses.

The plant viruses that have become most important economically have survived only because they are able to invade their hosts systemically. Local invaders tend to be self-eliminating: they induce the formation of necrotic lesions, in which their insect vectors do not probe and feed. Long before viruses were known as agents of infectious diseases, growers were aware that such propagative parts as bulbs, tubers, and stem cuttings gave rise to diseased plants when those parts had been produced on diseased plants. In 1888, Smith reported that the causal agent of peach yellows is transmitted through buds from diseased trees; since that time, it has been demonstrated that many viruses attacking fruit trees are harbored in diseased propagative material. Wortley (1915) was apparently the first to publish experimental proof that the virus of potato mosaic is carried over in tubers. It is now recognized that the vegetative parts used for plant propagation are of great importance as harborers of plant viruses from season to season. Some plant pathogens, particularly those viruses that systemically invade more than one host, may survive in an alternative weed or a perennial-crop plant host. Perhaps as many as one-fourth of all plant viruses can be passed on to succeeding generations of suscepts through the seed of infected plants. Most, however, infect only a low percentage of the seeds of their host plants; two exceptions are bean-mosaic virus (Reddick and Stewart, 1919) and lettuce-mosaic virus, both of which invade a relatively high percentage of the seed.

Survival in Association with Insects

Many insects are carriers of inocula during the growing season, and several important plant pathogens survive between growing seasons within insects. Rand and Cash (1920) proved that the bacterium responsible for cucurbit wilt is carried over the winter by striped cucumber beetles, and Elliott and Poos (1934) proved that the bacterium that causes Stewart's wilt of sweet corn is harbored over the winter in the bodies of flea beetles.

The few plant viruses that multiply within their leafhopper vectors could overwinter in those insects, but perennial-crop or weed hosts and vegetative plant parts used for propagation are more likely the important harborers of viruses.

At least four plant viruses—those of wheat mosaic (McKinney, 1946) big vein of lettuce, tobacco necrosis, and tobacco ringspot—survive in the soil between crop seasons. Although tobacco-mosaic virus may survive for some time in the soil, it can hardly be thought of as a soil-borne virus; it is so resistant to high temperature, drying, and aging that it would be expected to persist in crop residue for many months. The viruses of wheat mosaic, tobacco ringspot, and tobacco necrosis, however, are not so stable, and would be expected to lose infectivity rapidly in any environment other than the living host. The way in which these viruses survive in the soil remains a mystery, although tobacco-ringspot virus is associated with the nematode *Xiphinema americana* (Fulton, 1962). The discovery that the fungus *Olpidium brassicae* is an agent of inoculation for tobacco-necrosis virus (Teakle, 1962) might explain the ease with which this virus infests the soil. Perhaps other soil-borne viruses exist in association with fungi, nematodes, or insects; certainly, nematode and fungi are agents of inoculation for several soil-borne viruses.

Literature Cited

Anderson, P. J., and W. H. Rankin, 1914. *Endothia* canker of chestnut. *Cornell Univ. Agr. Exp. Sta. Bull.* (347):529–560.

Barrus, M. F., 1921. Bean anthracnose. *Cornell Univ. Agr. Exp. Sta. Mem.* (42):101–209.

Bawden, F. C., 1964. *Plant Viruses and Virus Disease* (4th ed.). New York: Ronald Press.

Bray, W. L., 1910. The mistletoe pest in the Southwest. *U. S. Dep. Agr. Bur. Plant Ind. Bull.* (166).

Brefeld, O., and R. Falk, 1905. Die Bluteninfektion bei den Brandpilzen. *Untersuch. Gesammtgeb. Mykol.* 13:1–74.

Butler, E. J., and S. G. Jones, 1949. *Plant Pathology,* pp. 82–87. London: Macmillan.

Commonwealth Mycological Institute, 1942– . [Distribution maps of plant diseases]. Kew: Commonwealth Mycological Institute.

Dickson, F., 1930. Studies on *Sclerotinia sclerotiorum* (Lib.) deBary. Ph.D. Thesis, Cornell University.

Elliott, C., and F. W. Poos, 1934. Overwintering of *Aplanobacter stewarti. Science* 80:289–290.

Ezekiel, W. N., 1945. Effect of low temperatures on survival of *Phymatotrichum omnivorum. Phytopathology* 35:296–301.

Fulton, J. P., 1962. Transmission of tobacco ringspot virus by *Xiphinema americanum. Phytopathology* 52:375.

Gallegly, M. E., and J. Galindo, 1958. Mating types and oospores of *Phytophthora infestans* in Mexico. *Phytopathology* 48:274–277.

Garrett, S. D., 1956. *Biology of Root-infecting Fungi.* Cambridge: Cambridge University Press.

Gratz, L. O., 1930. Disease and climate as pertaining to the Florida and Maine potato sections. *Phytopathology* 20:267–288.

Heald, F. D., and R. A. Studhalter, 1914. Birds as carriers of the chestnut blight fungus. *J. Agr. Res.* 2:405–422.

Hunt, N. R., 1946. Destructive plant diseases not yet established in North America. *Bot. Rev.* 12:593–627.

McCubbin, W. A., 1946. Preventing plant disease introduction. *Bot. Rev.* 12:101–139.

McKay, R., 1935. Germination of resting spores of onion mildew (*Peronospora schleideni*). *Nature* 135:306.

McKinney, H. H., 1946. Mosaics of winter oats induced by soil-borne viruses. *Phytopathology* 36:359–369.

Millard, W. A., and C. B. Taylor, 1927. Antagonism of micro-organisms as the controlling factor in the inhibition of scab by green manuring. *Ann. Appl. Biol.* 14:202–215.

Mix, A. J., 1935. The life history of *Taphrina deformans*. *Phytopathology* 25:41–66.

Peltier, G. L., and W. J. Frederich, 1926. Effects of weather on the world distribution and prevalence of citrus canker and citrus scab. *J. Agr. Res.* 32:147–164.

Rand, L. V., and L. C. Cash, 1920. Some insect relations of *Bacillus tracheiphilus* Erw. Sm. *Phytopathology* 10:133–140.

Rands, R. D., 1917. Early blight of potato and related plants. *Wis. Agr. Exp. Sta. Res. Bull.* (42).

Reddick, D., and V. B. Stewart, 1919. Transmission of the virus of bean mosaic in seed and observations on thermal death-point of seed and virus. *Phytopathology* 9:445–450.

Sampson, Kathleen, 1933. The systemic infection of grasses by *Epichloe typhina* (Pers.) Tul. *Trans. Brit. Mycol. Soc.* 18:30–47.

Sanford, G. B., 1926. Some factors affecting the pathogenicity of *Actinomyces scabies*. *Phytopathology* 16:525–547.

Smith, E. F., 1888. Peach yellows: a preliminary report. *U. S. Dep. Agr. Div. Bot. Bull.* (9).

Stakman, E. C., and C. M. Christensen, 1946. Aerobiology in relation to plant disease. *Bot. Rev.* 12:205–253.

———, A. W. Henry, G. C. Curran, and W. N. Christopher, 1923. Spores in the upper air. *J. Agr. Res.* 24:599–606.

Stevens, F. L., and J. G. Hall, 1911. A serious lettuce disease (Sclerotiniose) and a method of control. *N. C. Agr. Exp. Sta. Tech. Bull.* (8):89–143.

Teakle, D. S., 1962. Transmission of tobacco necrosis virus by a fungus, *Olpidium brassicae*. *Virology* 18:224–231.

Waksman, S. A., 1917. Is there any fungus flora of the soil? *Soil Sci.* 3:565–589.

Walker, J. C., and L. R. Jones, 1921. Relation of soil temperature and other factors to onion smut infection. *J. Agr. Res.* 22:235–262.

Weindling, R., 1932. *Trichoderma lignorum* as a parasite of other soil fungi. *Phytopathology* 22:837–845.

Whetzel, H. H., 1908. Bean anthracnose. *Cornell Univ. Agr. Exp. Sta. Bull.* (255):428–448.

Wortley, E. J., 1915. The transmission of potato mosaic through the tuber. *Science* 42:460–461.

5

Production and Dispersal of Inocula

Inoculation is the stage of plant pathogenesis that consists of the transfer of inoculum from its source to an infection court by some agent of inoculation. **Inoculum** is any part of a plant pathogen that is capable of establishing infection. Important inocula are particles of viruses, bacterial cells, fungal spores and mycelia, seeds of parasitic phanerogams, and eggs and larvae of nematodes. The inoculum may be deposited on the surface of a plant, or it may be deposited within it, as it is when insects transmit plant viruses. Strictly speaking, the **infection court** is that part of the susceptible plant in which the pathogen establishes a disease relationship, but it is generally considered to be the part on or in which the inoculum is placed.

PRODUCTION OF INOCULA

The mere survival of plant pathogens between growing seasons does not insure an abundance of inocula. Inocula for primary cycles of a pathogen are not always the overseasoning structures, but must often be produced at their sources by the structures of survival. It has been pointed out, for example, that the ascospores of the apple-scab pathogen (*Venturia inaequalis*) do not form until the spring; the fungus survives the winter as mycelium and perithecia in fallen leaves. The pathogen of late blight of potato (*Phytophthora infestans*) usually overwinters as mycelium in diseased tubers; in the spring, this mycelium becomes active in sprouts and produces the sporangia that serve as inocula. Sclerotia that survive between seasons often produce elaborate fruiting bodies that bear spores. *Sclerotinia* spp. produce asci and ascospores in apothecia that arise from sclerotia, and the ergot fungus, *Claviceps purpurea,* produces asci and spores in perithecia that are embedded in stalked stromata that grow from overwintered sclerotia. Nevertheless, some overseasoning structures of plant pathogens also serve as inocula. These include teliospores of certain rusts, some smut spores, soil-borne spores and mycelia, viruses, bacterial cells, fungal spores, and nematodes that survive in association with overseasoning hosts.

Inocula involved in the first cycle of the pathogen after a period of seasonal inactivity of susceptible plants and pathogens are those of **primary cycles,** and may be the same as, or different from, those for **secondary cycles** of the pathogen. A recently diseased plant is the source of inocula for secondary cycles, whereas soil, seed, decayed plant tissue, and so forth, are sources of inocula for primary cycles. The only inocula of viruses and bacteria are infectious particles and cells, respectively. Many fungi produce more than one kind of spore, and different spores may initiate the different cycles. In many apple-growing regions, ascospores initiate the primary cycles of the apple-scab fungus, whereas conidia produced in lesions during the growing season are the inocula for secondary cycles. In other areas in which apples are grown, notably parts of England and the Pacific Northwest of the United States, conidia and mycelium survive in buds of dormant trees. Conidia, then, may be inocula for primary cycles. *Pseudopeziza medicaginis* does not produce conidia, and ascospores are inocula for both primary and secondary cycles of the alfalfa-leafspot fungus.

AGENTS OF DISPERSAL AND INOCULATION OTHER THAN INSECTS

The spread of plant pathogens into new geographical areas and their establishment there constitute their distribution. The spread of a plant pathogen within the general area in which it is established is termed dissemination or dispersal. Contamination, the movement of inoculum to a susceptible plant, is a kind of dissemination. Moving the inoculum only a few inches and transporting it for hundreds of miles may both constitute dispersal or dissemination. Wind, water, insects, man, and animals other than insects are the most important agents of dispersal.

Wind

Fungal spores are the most typical wind-borne inocula. Wind may spread spores only locally or it may disperse them over long distances. Inoculation of plants by wind-blown spores is influenced by the quantity of spores produced at the source, the numbers of spores that reach the air, the direction and speed of the wind, the ability of the spores to survive the drying effects of exposure, and the availability of infection courts. *Venturia inaequalis, Phytophthora infestans,* and *Puccinia graminis* are among the plant-pathogenic fungi for which wind is an important agent of dispersal.

In most apple-growing regions of the world, ascospores of *Venturia inaequalis,* the apple-scab pathogen, mature in the spring at about the time the apple blossoms are in the pink stage. The ascospores are forcibly ejected into the air and are carried by wind currents to infection courts. The mechanism by which spores are shot from asci is not completely understood, but a plausible explanation can be given on the basis of observations of several different fungi. Ascospore ejection occurs only during moist periods, and hydrostatic pressure built up within the asci undoubtedly provides the propulsive force. It is possible that water is simply imbibed by perithecia, but there is some evidence that the great increase in hydrostatic pressure in asci, paraphyses, and periphyses within asci is due to the passage of water through the membranes of those structures by the process of osmosis. Immature asci are rich in glycogen, which is not osmotically active. In the presence of water, glycogen may be enzymatically hydrolyzed to sugars, which greatly increase the osmotic concentration of the cells. The increased pressure is certainly responsible for the expulsion of the glebal mass of *Sphaerobolus* (Walker and Anderson, 1925), and it is likely that such pressure results in the sequential protrusion of the asci of *Venturia inaequalis* through the neck of the ostioles and the shooting of ascospores through the ruptured tips of the asci.

Simultaneous expulsion of ascospores (Figure 5-1) occurs in many discomycetes, which are ascomycetes that produce apothecia (Buller, 1934; Ingold, 1953, 1965). If an apothecium containing mature ascospores is kept for a time under a bell jar, a visible cloud of spores

FIGURE 5-1
Clouds of ascospores forcibly ejected by the fungus *Monilinia fructicola.*

will be shot out upon removal of the bell jar. Asci of some discomycetes dehisce by means of a lid at the top, whereas those of others (*Monilinia fructicola,* for example) are equipped with a small plug of glycogen, which is digested enzymatically, leaving a pore through which the ascospores are forced (Figure 5-1). Some pyrenomycetes (ascomycetes that produce perithecia) bear asci that have double walls. The outer wall, the ectoascus, is relatively thick and rigid, whereas the inner wall, the endoascus, consists of a thin elastic membrane. As water pressure builds up, the ectoascus ruptures and the endoascus elongates rapidly. It is from the stretched endoascus, or "jack-in-the-box," that the ascospores are shot (Wolf and Wolf, 1947). Thus, among the ascomycetes, there are several different mechanisms for forcibly ejecting ascospores; these mechanisms are important to those plant pathogens—including *Venturia inaequalis*—that actually deliver the inocula for the primary cycle to the wind for dispersal.

Wind is also the agent of dispersal for some conidia. Conidia of many fungi simply fall off the conidiophores at maturity or they are blown away. The mature conidia of the apple-scab fungus, however, cling to the conidiophores, and are not easily dislodged by wind unless the humidity is high.

The best evidence indicates that *Phytophthora infestans,* the pathogen of late blight of potato, is locally dispersed largely by wind. The sporangia are produced during humid weather on long sporangiophores that protrude through stomata of infected leaves. A slight change in the relative humidity of the air around sporangiophores bearing mature sporangia causes a twisting motion of the sporangiophore tips, thereby dislodging the sporangia (deBary, 1887). It is not known just how far the sporangia of *P. infestans* may be blown and still remain viable. Possibly, the pathogen is wind-blown in stepwise fashion from the southeastern United States northward along the Atlantic Seaboard during the spring. This seems to be true for *Peronospora tabacina,* a related fungus that causes downy mildew (blue mold) of tobacco.

In contrast to *Phytophthora*, the downy-mildew fungi *Sclerospora philippinensis* and *S. graminis* expel their sporangia forcibly (Weston, 1923). In these fungi, a double wall occurs at the juncture between the sporangium and sporangiophore. As the developing sporangium becomes turgid, the adhesive force that holds the two membranous walls together is suddenly overcome; both walls bulge outward with a snap, and the sporangium is catapulted into the air.

Like other rust fungi, *Puccinia graminis,* the pathogen of stem rust of wheat, is pleomorphic; it provides an opportunity, therefore, for the study of different kinds of spore liberation in a single species. The urediospores of *P. graminis* may be carried by the wind for many miles from their source before being deposited in infection courts; this is an outstanding example of wind as the agent of long-distance dispersal of inoculum.

Basidiospores (sporidia) of the smut and rust fungi and basidiospores of hymenomycetes are wind-blown over relatively short distances. All except those of the smut fungi are forcibly discharged, but there is no satisfactory explanation of the mechanism of their discharge. The spore is attached asymmetrically to a fine sterigma of the basidium (promycelium) and a drop of liquid or bubble of gas forms in the region of a small projection, the **hilum,** just before spore discharge. Such spores are called **ballistospores** (Ingold, 1953, 1965), and the method of expulsion is termed the drop-excretion mechanism.

Several theories have been proposed to account for the forcible discharge of ballistospores. Buller (1922) suggested that the bursting of the turgid sterigma might provide the force for propulsion, and

Ingold (1953) postulated that the surface energy of the drop of liquid excreted at the hilum somehow brought about spore discharge. Ingold's theory does not explain the phenomenon, but may prove to be true. One serious objection to Buller's theory is that it would require a rapid sealing of the burst tip of a sterigma after spore discharge to maintain turgor within the basidium to provide for the discharge of the next spore in succession. Prince (1943) concluded that the sporidium of the rust fungus *Gymnosporangium nidus-avis* is propelled from its sterigma in much the same way that sporangia of *Sclerospora* are discharged from their sporangiophores. A flat crosswall separates the sporidium from its sterigma. In this fungus, the beak is at the point of spore attachment, rather than at the hilum. Immediately after spore discharge, the surfaces of the spore beak and sterigma tip are rounded rather than flat, and Prince suggests that the sudden rounding off of these previously flat surfaces accounts for the discharge of the spore. Prince also observed that spores were forcibly ejected in water mounts and that drop excretion occurs in water, indicating that the excreted drop of liquid may be of no value in spore discharge and that it is somehow prevented from mixing with water. The mechanism of discharge in *Gymnosporangium* may not be applicable to all ballistospores; it does not explain the asymmetrical attachment of spores and the significance of the drop excretion.

Olive (1964) stated that the "drop" at the base of spores of a basidiomycetous yeast was actually a bubble of gas, presumably carbon dioxide. The explosion of this bubble and the pressure of the residual gas between the inner wall and outer membrane of the spore apparatus were postulated to effect forcible discharge of the spore. Saville (1965) argued that this would explain severance of the spore from the sterigma, but not propulsion; he suggested that forcible discharge could occur as a result of the repulsion of like electrical charges in the base of the spore and the tip of the sterigma. After a thorough study of the basidiomycete *Schizophyllus commune,* Wells (1965) postulated a convincing theory to account for discharge of its ballistospores: A drop of liquid forms within an extension of the wall of the sterigma that surrounds the pointed base of the spore. The pressure of the enlarging drop and the action of enzymes rupture the connection between spore and sterigma. Finally, the flattened tip of the sterigma suddenly rounds off under the hydrostatic pressure within the basidium. Pressures developing in the excreted drop and in the basidium combine to propel the spore outward.

The different theories of forcible release of ballistospores are based on studies of different fungi, suggesting that the actual mechanism of spore discharge varies with groups of fungi. Whatever the mechanism of ballistospore discharge, the fact is that these spores are actually delivered to the agent of dispersal. Apparently, most ballistospores are not well adapted to long-distance dissemination. They are small and thin-walled, and do not withstand desiccation. Nevertheless, local dispersal by wind is often effective.

In the Pucciniaceae, aeciospores alternating with separating cells are cut off in closely packed chains within the aecium or cluster cup. Upon degeneration of the separating cells, aeciospores are held together by the sticky material that remains. The spores are packed tightly, and each assumes a polyhedral form under the pressure. In moist weather, they absorb water and become spherical, particularly in the outer region of the aecium. These spores round off suddenly, and spring out of the aecium singly or in groups (Ingold, 1953). These liberated spores are then dispersed by wind, which is the local dispersal agent for aeciospores of rust fungi that are produced on alternate dicotyledonous hosts and that later infect small grains.

Mature urediospores of *Puccinia graminis* are dislodged by wind; they may be wind-dispersed locally, as are the spores of many other plant-pathogenic fungi. As the following two paragraphs relate, however, circumstances in North America are such that viable urediospores of *P. graminis* are frequently carried for as much as 600 miles by wind.

The distant dispersal of the urediospores of *P. graminis* has been discussed by Stakman and Harrar (1957). The inoculum for the primary cycle of the stem-rust fungus may be overwintered urediospores or aeciospores from nearby infected barberry bushes. The fungus produces few teliospores in some parts of western Canada; subsequently, the barberries there are only mildly infected, and the fungus produces relatively few aeciospore inocula. In many parts of the wheat belt of North America, moreover, the barberry has been largely eradicated. Relatively few urediospores survive northern winters. Thus, urediospores blown from the southern part of the wheat belt would seem to make up most of the inocula. This idea is strongly supported and documented. Events in 1934 and 1935 strikingly illustrate the importance of wind as the agent of dispersal for stem rust. The epiphytotic of 1935 was due to the north-to-south movement of urediospores in the fall of 1934. Urediospores are vulnerable to the hot, dry summer weather in Mexico and Texas, and may cause only light infections of winter wheat in the autumn; but often, as in 1934, winds from the north carry urediospore inocula into Texas and Mexico, where young winter-wheat plants are infected.

An abundance of rainfall in Texas during May of 1935 led to a delay in the maturity of the crop and to an unusually heavy epiphytotic of rust. On May 11 and 12, 1935, a large mass of air moved northward from southern Texas, and on May 13, there was air movement northeastward from western Texas to Missouri. The next day in Nebraska, spores were caught on spore traps (vaseline-covered glass slides); subsequent development of disease in wheat indicated that spores had been dropped, at about that time, also in Oklahoma and Kansas. From May 24 to May 26 there was another massive movement of air from Texas northward, and spores were caught during this time in North Dakota, about a thousand miles from their source in Texas. The summer of 1935 remained wet and cool in many parts of the north-central United States, and stem rust developed rapidly as a consequence of both local and longdistance dissemination of inocula by wind.

The north-to-south movement of wind-blown urediospores in the autumn and their south-to-north movement in the following spring and summer probably account for most of the primary cycles of the stem-rust pathogen. Unusual air movements, coupled with cool wet weather—which makes for delayed ripening of the crop—favor such disease outbreaks as the epiphytotic of 1935 in the United States.

Water

Studies of the liberation of fungal spores and their dispersal are not only fascinating but are basic to an understanding of plant pathology. Much attention has been given to wind as an agent of dispersal, but dispersal and inoculation by water, though less dramatic, is often more important than wind.

Surface water—such as that in streams, drainage ditches, and irrigation systems—may disperse and distribute plant pathogens. Although it is not usually an agent of inoculation, surface water may occasionally be responsible for depositing nematodes and soil-borne fungal inocula in various infection courts.

Water—in the form of rain, overhead irrigation, or insecticidal or fungicidal sprays—serves as the agent of inoculation and short-distance dispersal for many plant pathogens: typically, the inocula are spread about over the surface of the plant or splashed a few inches or feet to adjacent plants. Bacterial cells from the ooze that is often deposited on the surfaces of diseased tissues may be carried by splashing or flowing water. Conidia are often produced in slimy masses, as in the anthracnose fungus, *Colletotrichum lindemuthianum,* which produces conidia in pink masses on diseased bean pods; when these are rained upon, the spores are spattered onto healthy pods, where they may cause infection. Some pyrenomycetes do not forcibly eject ascospores into the air, but free their mature spores to mix with gelatinous material within the perithecia. Water is absorbed by the perithecia, and the spores, mixed with the gelatinous material, are squeezed out through the ostiole, whence they can then be splashed about by water drops falling on the slime masses (Ingold, 1953). Spores are extruded similarly from the pycnia of many rust fungi and from the pycnidia of certain imperfect fungi. Foliar nematodes and the spores of soil-borne pathogenic fungi may be splashed onto above-ground infection courts by rain. The conidia of *Venturia inaequalis* are dislodged in wet weather and are dispersed by wind-blown rain.

Dimock (1951a, 1951b) showed that water used as the diluent for insecticidal and fungicidal sprays can serve as an agent of inoculation. Chrysanthemum and snapdragon leaves were successfully inoculated with *Septoria obesa* and *Puccinia antirrhini,* respectively, by spraying with spores suspended in preparations of insecticides at recommended dosage levels. *P. antirrhini* was able to initiate infections even after having been applied to snapdragon leaves in the form of spore suspensions in fungicidal sprays. Thus, the use of an insecticide, or of a fungicide that is inadequate or is used under environmental conditions unfavorable to its action, may actually result in an intensification of certain diseases.

Man and Other Animals

Man is more important as an agent of distribution and simple dispersal than as an agent of inoculation. Those inocula that are suited for dissemination by water may be spread from diseased to healthy plants by man as he manages his crops: growers are advised, for example, not to cultivate or harvest beans when the plants are wet, so as to avoid effective dispersal of blight bacteria and the spores of the anthracnose fungus.

Man is probably the chief inoculating agent for tobacco-mosaic virus; the virus is not carried by insects under field conditions. When juice from diseased plants is rubbed over the surface of healthy plants, infection of the latter is likely. The virus is unusually resistant to heat, drying, and aging, and persists even in processed tobacco products. Tobacco users whose fingers may have become contaminated with the virus are likely to inoculate small plants handled at transplanting time.

The possibility that nematodes might be agents of inoculation has long been considered, but little experimental evidence has been presented on this subject. Atanasoff (1925) demonstrated that the wheat nematode, *Tylenchus tritici,* is an agent of inoculation for the fungus *Dilophospora alopecuri.* The fungus must reach the growing point of susceptible plants in order to exert its parasitism, and it is carried there on the body surfaces of the nematodes. Conidia attach themselves firmly to the nematodes by means of branched appendages that occur at both ends of the cylindrical spores. *T. tritici* not only acts as the agent of inoculation, but its presence appears to enhance the fungal infection. In nature, the fungus apparently does not initiate disease in nematode-free plants, nor does it continue its pathogenic activities in those infected plants in which the nematodes die or form galls. Under controlled conditions, however, Schaffnit and Wieben (1928) successfully inoculated young tissues of germinating seeds with spores of the fungus alone.

Xiphinema index was the first nematode to be implicated as the agent of inoculation for a plant virus. Hewitt et al. (1958) presented evidence that the soil-borne virus of fanleaf of grapevines can be transmitted to previously healthy plants by *X. index.* Since that time, several other ectoparasitic nematodes have been found to transmit a total of nine other viruses, including those of tobacco rattle and tobacco ringspot. Cadman (1963) has discussed the biology of soil-borne plant viruses.

Fungi

The fungus *Olpidium brassicae* is an agent of transmission for the viruses of tobacco necrosis, big vein of lettuce, and tobacco stunt; it is probable that other fungi transmit potato virus X and the mosaic viruses of wheat, barley, and oats. The subject of fungal transmission of plant viruses has been reviewed by Grogan and Campbell (1966).

INSECTS AS AGENTS OF DISPERSAL AND INOCULATION

Insects serve as agents of dispersal and inoculation for hundreds of plant pathogens, but their importance as such varies from pathogen to pathogen. Insects are the chief means by which field-grown plants are inoculated with viruses; although insects are also agents of inoculation for many plant-pathogenic bacteria and fungi, other agents are often more effective. Relations between insects and plant diseases are summarized in Box 5-1.

Vectors of Plant-Pathogenic Viruses

Most plant viruses are insect-transmissible; of these, the majority are transmitted by insects with piercing-sucking mouthparts. A few plant viruses are transmitted, apparently mechanically, by various members of the orders Orthoptera (grasshoppers) and Coleoptera (beetles), insects with chewing mouth parts. Thrips (order Thysanoptera), which feed by a rasping-sucking action, transmit the tomato spotted-wilt virus (Samuel and Bald, 1931). Although the true bugs (order Hemiptera) have piercing-sucking mouth parts, few of them are agents of inoculation for plant viruses. Some capsids (the tarnished plant bug, for example) transmit plant viruses that are readily transmitted by mechanical means, and lace bugs have been incriminated in the transmission of crinkle-leaf disease and savoying disease of sugar beet (Coons et al., 1937; Wille, 1928). The saliva of many capsids is toxic to plant cells, and the punctures made by these insects while feeding become surrounded by necrotic areas. Viruses are obligately parasitic pathogens, and cannot become established in necrotic tissues: a zone of dead cells around possible points of viral entry would prevent virus particles from moving into the protoplasts of living cells outside of the necrotic zone. Smith (1951) has suggested, therefore, that the phytotoxicity of the saliva of certain hemipterous insects, notably the capsids, might account for their inability to transmit viruses effectively. Another hypothesis to account for the usual ineffectiveness of these insects as vectors of plant viruses is that their saliva inactivates viruses as they are picked up from diseased plants (Leach, 1940).

Homopterous insects are the vectors of more than 90% of the insect-transmitted plant viruses. White flies have been shown to transmit several plant viruses (Golding, 1930), and mealy bugs are known to transmit viruses of cacao trees (Posnette and Robertson, 1950), but aphids and leafhoppers are by far the most important homopterous vectors.

Plant viruses can be separated into two groups on the basis of their association with insect vectors. The larger group comprises those plant viruses that are carried externally, the **stylet-borne** viruses (Bradley, 1964), and the smaller group comprises those that are carried internally, the **circulative** viruses (Maramorosch, 1964). Viruses in the latter group are classed as **propagative** or **nonpropaga-**

tive, depending upon whether they multiply within the insect vector. Circulative and stylet-borne viruses were formerly classified as **persistent** and **nonpersistent** viruses, respectively. A circulative (persistent) virus is retained by the insect for a long period of time—often throughout the entire life of the insect—whereas a stylet-borne (nonpersistent) virus is lost rapidly after its acquisition by the insect. All known leafhopper vectors and a few aphids transmit circulative viruses; most aphids transmit stylet-borne viruses.

Aphids. Stylet-borne viruses that are transmitted by aphids usually reach high concentrations in their plant hosts, and many such viruses

BOX 5-1
INSECTS IN RELATION TO PLANT DISEASES

Insects may be related directly or indirectly to plant diseases. Although most insects that attack plants cause injury, which is a discontinuous or transient dysfunction, some cause diseases directly. Notable examples are the insect gall diseases, which are caused by members of the order Hymenoptera, and hopperburn of potato, psyllid yellows of potato, alfalfa yellows, and mealy-bug wilt of pineapple, all of which are caused by members of the order Homoptera.

Insects affect plant diseases indirectly in several ways: some insects are essential to the survival of plant pathogens between crop seasons, some act as agents of dispersal, some are agents of inoculation or penetration, and others act in all three capacities. For example, flea beetles not only preserve the bacteria of sweet-corn wilt through the winter but, in the spring, disperse them to corn fields, where they inoculate them into the leaves of young plants. All insects that act as vectors of plant viruses or mycoplasmalike bodies serve as agents of dispersal and inoculation.

As with flea beetles and the bacteria of sweet-corn wilt, some insects disseminate pathogens biologically. But insect dispersal of plant pathogens is often only mechanical: thus, fire-blight bacteria are spread from flower to flower on contaminated pollinating insects, and the fungus of Dutch elm disease is dispersed by young can be transmitted mechanically. Some stylet-borne viruses occur in high concentrations in epidermal cells (Bawden et al., 1954), and aphids acquire them as a result of probing with their mouthparts in or between those cells. It is well known that starved aphids acquire stylet-borne viruses more efficiently and lose them more rapidly than well-fed aphids (Watson, 1938). Bradley (1953) observed that starved aphids made many feeding punctures of a probing nature before settling down to continuous feeding. Well-fed aphids, however, did not begin to feed until 5–10 minutes after having been placed on a plant. Virus is rapidly acquired from epidermal cells of infected plants by starved aphids early in the period of probing, and

bark beetles whose bodies have been contaminated by the sticky masses of conidia.

When acting as agents of inoculation and penetration, insects may simply make wounds through which pathogens may enter after dispersal by another agent. For example, entrance into plant foliage by such facultative parasites as certain species of *Colletotrichum*, *Fusarium*, and *Alternaria* often occurs through wounds, many of which are made by insects of the orders Coleoptera, Hymenoptera, and Hemiptera. The brown-rot fungus, after dispersal by wind or splashing rain, enters the fruit of plums through wounds made by the plum curculio.

In many important insect-disease relationships, however, the plant is wounded by the insect that is also the agent of pathogen dispersal. This is true for the more than two hundred viral diseases that are transmitted by aphids in a stylet-borne manner, as well as for the circulative viruses and the mycoplasmalike organisms that cause the "yellows" diseases. Other pathogens that enter through wounds made by their insect agent of dispersal include the bacteria that cause sweet-corn wilt, crown gall of woody plants, and blackleg of potato, and the fungi that cause Dutch elm disease and oak wilt.

viruliferous aphids apparently inoculate healthy plants while probing them. Bradley noted that, during the short effective inoculation feedings, the aphids did not secrete saliva from their small ventral maxillary ducts; this led him to suggest that the virus is taken into (and released from) the dorsal duct through which plant juices are drawn during feeding. These observations provide the best explanations both for the greater efficiency of starved aphids as vectors and for their more rapid loss of infectivity.

There is no satisfactory explanation, however, for the specificity of some stylet-borne viruses for their aphid vectors. Generally speaking, this specificity applies more to families of insects than to species, and more to specific viruses than to groups of related viruses. Some closely related species of aphids, however, are unable to transmit the same stylet-borne viruses. For example, *Myzus persicae* will transmit the viruses of cauliflower mosaic and cabbage black ringspot from cauliflower seedlings infected with both, but *Myzus ornatus* will transmit only the cauliflower-mosaic virus from similarly infected seedlings (Kvicala, 1945). Other anomalies, such as the fact that tobacco-mosaic virus and potato virus X are not regularly transmitted by insects, have not been explained. Both viruses reach high concentrations in their hosts, and are easily transmitted by mechanical means.

Aphids transmit a few circulative viruses. These have more specific vector relationships than do the stylet-borne viruses, are not usually transmissible by mechanical inoculations, and undergo a latent period in the vector. Relatively long feeding periods are required for the vectors to acquire the viruses and starving the vectors prior to acquisition feeding does not influence their ability to acquire them (Black, 1954).

The mechanism of aphid transmission of circulative plant viruses may differ from the mechanism of transmission of stylet-borne viruses in that the host tissues fed upon by aphids and invaded by the virus may be different for the two kinds of viruses (Kirkpatrick and Ross, 1952). Some viruses persist for many days in aphids, even through molts (Black, 1954); such persistence could result from passage of virus through the body of the insect to the salivary glands, and is probably similar to the persistence of the virus of curly top of sugar beet in its leafhopper vector.

Leafhoppers. Most circulative plant viruses are transmitted by leafhoppers; no leafhopper is known to transmit a stylet-borne virus.

Some leafhopper-borne viruses multiply inside their vectors, while others do not.

At this point, we should remind the reader that mycoplasmalike organisms have been detected in tissues of plants showing symptoms of the so-called yellows diseases (Chapter 3). This is presumptive evidence—not proof—of the plant-pathogenicity of such organisms. The pathogens of aster yellows, rice stunt, and club-leaf of clover, discussed in the next two paragraphs, may prove to be mycoplasmalike organisms rather than viruses, but this would not alter the facts presented below.

One explanation for the persistence, and for the latent period, of a pathogen in its leafhopper vector is that the pathogen multiplies in the leafhopper. Kunkel (1926) found that aster-yellows pathogen may persist throughout the life (as long as 100 days) of an infective leafhopper fed on insusceptible plants; it seemed logical to assume that the small amount of the pathogen acquired during a relatively short period of feeding on a diseased plant would soon be lost in subsequent feedings on insusceptible plants unless the pathogen multiplied within the body of the insect. Kunkel (1937) obtained further evidence to support this theory. The pathogen was completely inactivated within infective leafhoppers kept at 31–32°C for twelve days. When the insects were exposed to this temperature for periods of from one to eleven days, however, they lost the ability to transmit for only a few hours or days. These results were interpreted as strong circumstantial evidence that the aster-yellows pathogen multiplies in its insect vector. Maramorosch (1952) obtained direct evidence for multiplication of the pathogen in its vector. Leafhoppers were continuously kept on rye, a plant immune from disease caused by the aster-yellows pathogen. Suspensions of the pathogen prepared from triturated infective insects were injected into noninfective ones, which later became infective. The procedure was continued through ten serial passages of the pathogen through the insects. If it is assumed that there is no multiplication of the pathogen in the leafhoppers, the dilution of the pathogen at the end of the experiment would have been far too great to have allowed for the infectivity that was demonstrated. Black and Brakke (1952), who also used an injection technique, demonstrated that the wound-tumor pathogen of clover multiplies in its leafhopper vector.

A different line of evidence supporting the hypothesis that some persistent pathogens multiply in their vectors was obtained by Fukushi (1940), working with the agent of rice stunt, and by Black (1950) who studied the agent of clover club-leaf. The two pathogens in question are unusual in that they are transmitted by infective female vectors to their progeny through the eggs. The rice-stunt pathogen was retained in seven generations of leafhoppers that were the progenies of a single infective female. Because not all infective females transmitted the pathogen through eggs to their progeny, and because there were no data on the possible correlation between the length of acquisition feeding and the inheritance of the pathogen, Smith (1951) suggested that transovarian passage might depend upon the amount of pathogen acquired by the original female, not upon its multiplication in her progeny. There is no doubt, however, that transovarian passage of the clover club-leaf pathogen in its vector is dependent upon multiplication of the pathogen in the insect (Black, 1950). The pathogen was passed through twenty-one generations of insects that were caged on insusceptible plants. During the five-year course of the experiment, the pathogen present in the original female insect would have been diluted to approximately $1:2.8 \times 10^{26}$, if we assume that it did not multiply in the succeeding generations of insects. The data cannot be explained on the assumption that the original insect acquired all the pathogen necessary for its detection in the twenty-first generation. Apparently, the pathogen can be maintained indefinitely in succeeding generations of insects because of its ability to multiply within the bodies of the insects.

Not all leafhopper-transmitted pathogens are capable of multiplying in their vectors. Two pathogens, both presumed to be viruses rather than mycoplasmalike organisms, illustrate this point. Beet leafhoppers that have acquired the curly-top virus sometimes lose their infectivity, but they can become viruliferous again if they feed on diseased plants. Moreover, the efficiency and duration of transmission is increased with a lengthening of the acquisition feeding. Apparently, the curly-top virus does not multiply in its vector; certainly, the virus cannot be maintained indefinitely in the insects (Bennett and Wallace, 1938; Freitag, 1936). Some circulative viruses have short latent periods in the insects, and some may have no latent period at all; Severin (1949) found it impossible to detect a latent period for the alfalfa-dwarf virus in its vectors.

It is not always necessary to assume multiplication of viruses in their vectors to account for incubation (latent) periods. Working with corn-streak virus, Storey (1938) demonstrated that a strain of the vector, *Cicadulina mbila,* was unable to transmit the virus because its gut wall was impermeable to it. Some individuals of this strain became infective, however, after the intestinal wall had been punctured with a fine needle. Apparently, acquired virus normally diffuses from the digestive tract into the blood, whence it moves into the salivary glands, thereby making the insect infective. The time required for this course of events to be completed is sufficient to explain the latent period of some viruses in insects.

Mites. Although they are not insects, mites are mentioned here because they are vectors of several plant viruses. Black-currant reversion virus is transmitted by the big-bud mite, *Eriophyses ribis* (Amos et al., 1927). According to Slykhuis (1953), wheat streak-mosaic virus, which is transmitted by *Aceria tulipae,* persists for several days in viruliferous mites and is retained through periods of molting, but is not transmitted through eggs.

Vectors of Plant-Pathogenic Bacteria

Insects are the most important agents of inoculation of several plant-pathogenic bacteria. The small bacterial cells occurring in nectaries or in ooze from the surfaces of lesions and cankers may be readily picked up by insects. The bacteria may adhere to the legs and bodies of insects, or, as with the cucurbit-wilt and corn-wilt bacteria, they may be ingested by their vectors.

Apparently, the first experimental proof that insects act as agents of inoculation for plant-pathogenic bacteria was published in 1891 by Waite, who showed that bees and wasps carry the fire-blight bacteria from flower to flower. Oddly enough, the exact relationship between the epidemiology of fire blight and various kinds of insect vectors remains obscure. The best evidence indicates that the inoculum for primary cycles originates in hold-over cankers, whence it is carried by ants, flies, bees, and other insects to the blossoms (Thomas and Ark, 1934; Hildebrand and Phillips, 1936; Parker, 1936).

The soil-borne bacteria that cause soft rots of plants are often deposited in infection courts by insects. The potato-blackleg pathogen, for example, is carried by the seed-corn maggot to potato seed pieces (Leach, 1940), and cruciferous plants are inoculated with the soft-rot bacterium by the cabbage maggot (Johnson, 1930). The role of insects in the survival of the bacteria of corn wilt and cucurbit wilt has already been mentioned. These bacteria are also effectively dispersed by the insects in which they survive between crop seasons. Important bacterial plant pathogens for which insects are the agents of inoculation include those of such diseases as bacterial rot of apples, olive knot, sugar-cane gummosis, and bacterial wilt of potatoes and other solanaceous plants (Leach, 1940).

Vectors of Plant-Pathogenic Fungi

Although wind and water are the usual agents of dispersal for most fungi, insects often play an important role in this regard. In fact, some plant-pathogenic fungi are entirely dependent upon insects for their dissemination. Spores are the most common fungal inocula, and many spores seem well adapted to insect dispersal. Although spores of many fungi are produced in sticky masses that facilitate their adherence to insects, dry spores may adhere to parts of insects from which they may be easily dislodged. The mechanism by which dry spores adhere to apparently dry surfaces of insect bodies is not known. It has been suggested that the spores may be attracted to a film of water on the surface of the insect (Leach, 1940). Because dry spores will adhere to the dry surface of a needle, regardless of the electrical charge of the needle (Hanna, 1924), it seems unlikely that their adherence to insects is correlated with positive or negative electrical charge.

The *Ceratocystis* species that cause Dutch elm disease (Figure 5-2) and the blue-stain diseases of conifers depend exclusively upon bark beetles for their effective dispersal (Figure 5-3). These fungi sporulate in the pupal chambers of the bark beetles in dead trees. When the adult beetles emerge in the spring, they are infested with spores, which are deposited in the wounds that the beetles make while feeding on healthy trees (Leach et al., 1934). *Ustilago violacea,* the pathogen of the anther smut of pinks (Brefeld and Falk, 1905), and *Botrytis anthophila,* the cause of blossom blight of red clover (Silow, 1933) are also dependent upon insects for their effective dispersal.

Secondary cycles of the ergot fungus of rye and other members of the grass family are initiated by the conidia of *Claviceps purpurea.* These conidia are produced abundantly on the convoluted surfaces of young sclerotia that have replaced the ovaries of diseased flowers. The spores occur in droplets of a foul smelling "honey-dew," a sugary solution that may be of fungal or insect origin. Although rye is not insect-pollinated, insects, attracted to the "honey-dew," probably disperse the conidia of *C. purpurea.* A fungus gnat carries the spores externally and internally (Mercier, 1911), but the conidia are also well adapted to dispersal by splashing rain or by such mechanical means as the brushing of a diseased head of rye against a healthy one (Ingold, 1953).

Ascospores of *C. purpurea* may initiate primary cycles of the ergot

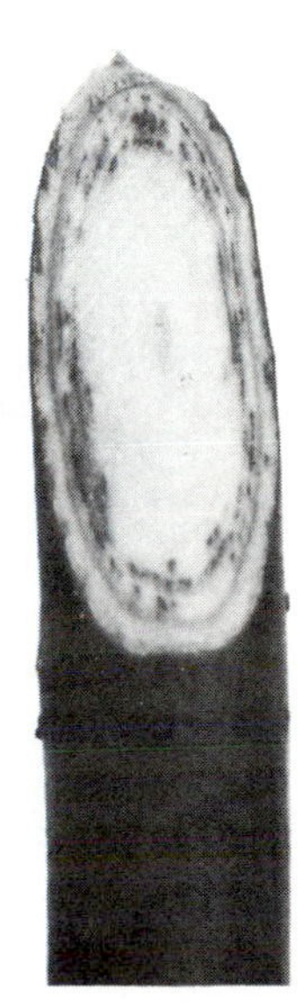

FIGURE 5-2
Dutch elm disease: *left*, healthy and diseased elm trees; *right*, vascular discoloration in a stem of a diseased tree.

fungus after having been dispersed by insects; apparently, however, these spores are not entirely dependent upon insects for their effective dissemination. Some ascospores are forcibly discharged, but others exude from perithecia and are held on the stroma by a sticky liquid (Wilson, 1875). Thus, the ascospores of *C. purpurea* seem well adapted to dispersal both by wind and by insects.

Insects may be indirectly involved in the dispersal of spores that are forcibly discharged into the air. It is well known that spore puffing in discomycetes may be stimulated by radiant heat (Falk, 1916) or by touch (Falk, 1923). It is possible that the weight of an insect, as it lights on a fruiting body, is sufficient to trigger the puffing mechanism (Leach, 1940). The discharged spores might then be dispersed by the contaminated insect as well as by wind currents.

For certain fungi, insects are important as agents of hybridization. The pycniospores of most rusts ooze out of the pycnia in a sweet-scented sugary "nectar" that is attractive to insects. Because rust fungi that develop pycnia are probably all heterothallic, the pycniospores from pycnia of one mating type must fuse with receptive hyphae from pycnia of the opposite mating type before mature di-

FIGURE 5-3
Above, feeding galleries of elm-bark beetles in dead elm wood (*left*) and conidial fructifications of *Ceratocystis ulmi* in dead elm wood (*right*); *below,* adults (*left*) and a pupa (*right*) of elm-bark beetles.

caryotic aeciospores can be produced. Craigie (1927) first proved heterothallism in *Puccinia graminis* and *P. helianthi,* and showed that dicaryotization (plasmogamy) did not occur when insects were excluded from the leaves bearing pycnia, unless the pycniospores of one mating type were manually transferred to pycnia of the other type. Flies, beetles, ants, and many other insects visit pycnia and may mix nectar from different monosporidial infections (Rathay, 1883).

Literature Cited

Amos, J., R. G. Hatton, R. C. Knight, and A. M. Massee, 1927. Experiments in the transmission of "reversion" in black currants. *East Malling Res. Sta. Annu. Rep.* 1925(Suppl. 2):126–150.

Atanasoff, D., 1925. The *Dilophospora* disease of cereals. *Phytopathology* 15:11–40.

Bawden, F. C., B. M. G. Hamlyn, and M. A. Watson, 1954. The distribution of viruses in different leaf tissues and its influence on virus-transmission by aphids. *Ann. Appl. Biol.* 41:229–239.

Bennett, C. W., and H. E. Wallace, 1938. Relation of the curly top virus to the vector, *Eutettix tenellus. J. Agr. Res.* 56:31–51.

Black, L. M., 1950. A plant virus that multiplies in its insect vector. *Nature* 166:852–853.

———, 1954. Arthropod transmission of plant viruses. *Exp. Parasitol.* 3:72–104.

———, and M. K. Brakke, 1952. Multiplication of wound-tumor virus in an insect vector. *Phytopathology* 42:269–273.

Bradley, R. H. E., 1952. Studies on the aphid transmission of a strain of henbane mosaic virus. *Ann. Appl. Biol.* 39:78–97.

———, 1953. Infectivity of aphids after several hours on tobacco infected with potato virus Y. *Nature* 171:755–756.

———, 1964. Aphid transmission of stylet-borne viruses. *In* M. K. Corbett, and H. D. Sisler, eds., *Plant Virology,* pp. 148–174. Gainesville, Fla.: University of Florida Press.

Brefeld, O., and R. Falk, 1905. Die Bluteninfektionen bei den Brandpilzen und die naturliche Verbreitung der Brandkrankheiten. *Untersuch. Gesammtgeb. Mykol.* 13:1–74.

Buller, A. H. R., 1909–1934. *Researches on Fungi,* vol. I (1909), vol. II (1922), vol. III (1924), vol. V (1933), vol. VI (1934). New York: Longmans, Green.

Cadman, C. H., 1963. Biology of soil-borne viruses. *Annu. Rev. Phytopathol.* 1:143–172.

Coons, G. H., J. E. Kotila, and D. Stewart, 1937. Savoy, a virus disease of beet transmitted by *Piesma cinerea. Phytopathology* 27:125. (Abstr.)

Craigie, J. H., 1927. Discovery of the function of the pycnia of the rust fungi. *Nature* 120:765–767.

deBary, A., 1887. *Comparative Morphology and Biology of Fungi, Mycetozoa, and Bacteria.* Oxford: Clarendon Press.

Dimock, A. W., 1951a. The dispersal of viable fungus spore by insecticides. *Phytopathology* 41:152–156.

———, 1951b. The dispersal of viable spores of phytopathogenic fungi by fungicidal sprays. *Phytopathology* 41:157–163.

Falk, R., 1916. Über der Sporenverbreitung bei den Ascomyceten, I. Die Radiosensiblen Discomyceten. *Mykol. Untersuch. Ber.* 1:77–145.

———, 1923. Über der Sporenverbreitung bei den Ascomyceten, II. Die taktiosensiblen Discomyceten. *Mykol. Untersuch. Ber.* 1:370–403.

Freitag, J. H., 1936. Negative evidence on multiplication of curly-top virus in the beet leafhopper, *Eutellix tenellus. Hilgardia* 10:305–342.

Fukushi, T., 1940. Further studies on the dwarf disease of rice plant. *J. Fac. Agr. Hokkaido Univ.* 45:83–154.

Golding, F. D., 1930. A vector of leaf curl of cotton in Southern Nigeria. *Empire Cotton Grow. Rev.* 7:120–126.

Grogan, R. G., and R. N. Campbell, 1966. Fungi as vectors and hosts of viruses. *Annu. Rev. Phytopathol.* 4:29–52.

Hanna, W. F., 1924. The dry-needle method of making monosporous cultures of Hymenomycetes and other fungi. *Ann. Bot. (London)* 38:791–795.

Hewitt, W. B., D. J. Raski, and A. C. Goheen, 1958. Nematode vector of soil-borne fanleaf virus of grapevines. *Phytopathology* 48:586–595.

Hildebrand, E. M., and E. F. Phillips, 1936. The honeybee and the beehive in relation to fire blight. *J. Agr. Res.* 52:789–810.

Ingold, C. T., 1953. *Dispersal in Fungi.* Oxford: Clarendon Press.

———, 1965. *Spore Liberation.* Oxford: Clarendon Press.

Johnson, D. E., 1930. The relation of the cabbage maggot and other insects to the spread and development of soft rot of Cruciferae. *Phytopathology* 20:857–872.

Kirkpatrick, H. C., and A. F. Ross, 1952. Aphid transmission of potato leafroll virus to solanaceous species. *Phytopathology* 42:540–546.

Kunkel, L. O., 1926. Studies on aster yellows. *Amer. J. Bot.* 13:646–705.

———, 1937. Effect of heat on ability of *Cicadula sexnotata* (Fall.) to transmit aster yellows. *Amer. J. Bot.* 24:316–327.

Kvicala, B., 1945. Selective power in virus transmission exhibited by an aphis. *Nature* 155:174.

Leach, J. G., 1940. *Insect Transmission of Plant Diseases.* New York: McGraw-Hill.

———, L. W. Orr, and C. Christensen, 1934. The interrelationships of bark beetles and blue-staining fungi in felled Norway pine timber. *J. Agr. Res.* 49:315–342.

Maramorosch, K., 1952. Direct evidence for the multiplication of aster-yellows virus in its insect vector. *Phytopathology* 42:59–64.

———, 1964. Virus-vector relationships: vectors of circulative and propagative viruses. *In* M. K. Corbett and H. D. Sisler, eds., *Plant Virology,* pp. 175–193. Gainesville, Fla.: University of Florida Press.

Mercier, L., 1911. Sur le role des insectes comme agents de propagation de l' "ergot" des graminees. *C. R. Hebd. Soc. Biol. (Paris)* 70:300–302.

Olive, L. S., 1964. Spore discharge mechanism in Basidiomycetes. *Science* 146:542–543.

Parker, K. G., 1936. Fire blight: overwintering, dissemination, and control of the pathogene. *Cornell Univ. Agr. Exp. Sta. Mem.* (193).

Posnette, A. F., and N. F. Robertson, 1950. Virus diseases of cacao in West Africa, VI. Vector investigations. *Ann. Appl. Biol.* 37:363–377.

Prince, A. E., 1943. Basidium formation and spore discharge in *Gymnosporangium nidus-avis. Farlowia* 1:79–93.

Rathay, E., 1883. Untersuchungen über die Spermagonien der Rostpilze. *Denkschr. Kaiserl. Akad. Wiss. Wien* 46:1–51.

Samuel, G., and J. C. Bald, 1931. *Thrips tabaci* as a vector of plant virus disease. *Nature* 128:494.

Saville, D. B. O., 1965. Spore discharge in Basidiomycetes: a unified theory. *Science* 147:165–166.

Schaffnit, E., and M. Wieben, 1928. Untersuchungen über den Erreger der Federbuschsporenkrankheit, *Dilosophora alopecuri* (Fr.) Fr. *Forsch. Gebiet Pflanzenkr.* 5:5–38.

Severin, H. H. P., 1949. Transmission of the virus of Pierce's disease of grapevines by leafhoppers. *Hilgardia* 19:190–206.

Silow, R. A., 1933. A systemic disease of red clover caused by *Botrytis anthophila* Bond. *Trans. Brit. Mycol. Soc.* 18:239–248.

Slykhuis, J. T., 1953. The relation of *Aceria tulipae* Kiefer to streak mosaic and other chlorotic symptoms on wheat. *Phytopathology* 43:484–485. (Abstr.)

Smith, K. M., 1951. *Recent Advances in the Study of Plant Viruses* (2nd ed.). Philadelphia: Blakiston.

Stakman, E. C., and J. G. Harrar, 1957. *Principles of Plant Pathology,* pp. 207–257. New York: The Ronald Press.

Storey, H. H., 1938. Investigations of the mechanism of the transmission of plant viruses by insect vectors, II. The part played by puncture in transmission. *Proc. Roy. Soc. London* (B) 125:455–477.

Thomas, H. E., and P. A. Ark, 1934. Fire blight of pears and related plants. *Calif. Agr. Exp. Sta. Bull.* (586).

Waite, M. B., 1891. Results from recent investigations in pear blight. *Bot. Gaz.* 16:259.

Walker, L. B., and E. N. Anderson. 1925. Relation of glycogen to spore-ejection. *Mycologia* 17:154–159.

Watson, M. A., 1938. Further studies on the relationship between *Hyoscyamus* virus 3 and the Aphis *Myzus persicae* (Sulz.) with special reference to the effects of fasting. *Proc. Roy. Soc. London* (B) 125:144–170.

Wells, K., 1965. Ultrastructural features of developing and mature basidia and basidiospores of *Schizophyllum commune*. *Mycologia* 57:236–261.

Weston, W. H., 1923. Production and dispersal of conidia in the Philippine Sclerosporas of maize. *J. Agr. Res.* 23:239–278.

Wille, J., 1928. Die durch die Rübenblattwanze erzeugte Krauselkrankheit der Rüben. *Arb. Biol. Reichsanst. Land- Forstwirt.* 16:115–167.

Wilson, A. S., 1875. Observations and experiments on ergot. *Gard. Chron.* 4:774–775.

Wolf, F. A., and F. T. Wolf, 1947. *The Fungi,* vol. II, pp. 166–209. New York: Wiley.

6

Prepenetration Phenomena and Entrance of Plants by Pathogens

Pathogens deposited in tissues of susceptible plants are likely to induce immediate pathologic reactions. Most fungal inocula, however, are deposited on the surfaces of plants; these inocula, usually spores, may go through a period of inactivity (**quiescence**) in the infection court, or may begin their germination and growth (**development**) immediately after their deposition (Whetzel, 1929). When the developing pathogen gains **ingress** (entrance) into the plant, the first physiological response of the attacked plant occurs, marking the onset of disease.

PREPENETRATION PHENOMENA

Germination of Fungal Spores

The process of germination in fungal spores is analogous to that of seed germination. First, water is absorbed. Then, in the presence of the absorbed water, hydrolytic enzymes are activated and reserve food materials are broken down, thereby providing the energy and materials for growth. Soon, a protuberance emerges from the spore and develops into a short hypha, or germ tube. The germ tube may emerge in any of several ways, depending on the species of fungus. In spores with both endospores and exospores, the tube arises from the endospore and must somehow make its way through the exospore. The urediospores of rust fungi contain germ pores, which are small circular holes in the exospore. Promycelia of the smut fungi, however, literally break through the exospore wall. In spores with membranous walls, germ tubes apparently emerge at weak points in the thin walls. As the germ tube develops, it lays down its own thin wall, and continues to grow as long as a supply of food is available and environmental conditions remain favorable. Most fungal spores contain food materials that support development of germ tubes into hyphae that may become many times as long as the spore. The time required for germination varies, but most mature spores germinate within approximately one to six hours when environmental conditions are favorable.

Factors Influencing Spore Germination

Water. Spores of the apple-scab pathogen (Wiltshire, 1915) and the downy-mildew fungi germinate only when immersed in a film of water, such as might be provided by dew or rain. A film of water is not required by all fungi for germination (Lauritzen, 1919); the powdery-mildew fungi, for example, germinate best on dry surfaces in moist air (Yarwood, 1936). The ability of conidia to germinate on dry surfaces seems to be directly proportional to the water content of the spores. Yarwood (1950) reported the water content of conidia as expressed by a percentage of spore weight: spores of *Erysiphe polygoni* contained 72% water, those of *E. graminis* 75%, and those of *E. cichoracearum* 52%; by contrast, the spores of *Uromyces phaseoli* contained 12% water, those of *Peronospora destructor* and *Botrytis cinerea* each contained 17%, and those of *Monilinia fructicola* were 25% water by weight. Obviously, moisture is essential to spore germination, but the amount required in the external environment varies, depending on the fungus, from a film of liquid water to a high relative humidity of the surrounding air.

Temperature. For every fungal spore, there is a temperature at which germination is most rapid; there are also minimum and maximum temperatures below and above which germination will not occur. Spores may germinate over a wide temperature range, but those of most plant pathogens germinate best at 15–30°C. *Cystopus candidus,* the pathogen of white "rust" of crucifers, is unusual in that the optimum temperature for spore germination is near 10°C; the lower and upper limits for germination are about 0°C and 25°C, respectively. At the other extreme, some species of *Aspergillus* produce spores that germinate best at temperatures near 35°C.

Temperature determines the kind of spore germination of many downy-mildew fungi (Brodie, 1945). For example, at 12–13°C, sporangia of *Phytophthora infestans* germinate by the production of eight to sixteen motile zoospores, each of which can germinate later and initiate infection. Above 18°C, however, relatively few zoospores are produced. Instead, most sporangia function as conidia, each of which produces a single germ tube.

The optimum temperature for germination of spores of a given species may not always be the same, but may vary under the influence of other environmental factors. Infection of oats by *Ustilago kolleri,* an organism causing covered smut, was greatest between 5°C and 20°C when plants were grown in soil in which the moisture content was only 15% of the water-holding capacity; at 20% moisture, however, most infections occurred in plants kept between 20°C and 30°C (Reed and Faris, 1924). Similarly, when barley plants were grown in soil moistened to 50% of its water-holding capacity, the optimum temperature for the development of covered smut due to *U. hordei* varied with the *p*H of the soil (Faris, 1924). It is likely that these effects were largely on spore germination, because the spores of both *Ustilago kolleri* and *U. hordei* contaminate the seed and attack seedling stems, and the infections begin in the soil or at its surface.

Light. The germination of spores is affected much less by light than by moisture and temperature, although there is evidence that light is important to spore germination in certain fungi. Some light seems necessary for the production of spores by *Alternaria* (Klaus, 1941) and certain other fungi, and sporangia of *Physoderma maydis,* which causes brown spot of corn, did not germinate either in the dark or in direct sunlight (Voorhees, 1933), low intensities of blue light being most favorable to spore germination (Hebert and Kellman, 1958). Conidia of *Erysiphe polygoni* collected between noon and 7:00 PM germinated equally well in light and in darkness, whereas those collected during the night and early morning germinated better in light than in darkness (Yarwood, 1936). Fungal spores are much more resistant to direct sunlight than are bacterial cells.

Hydrogen-Ion Concentration. Club root of crucifers, caused by *Plasmodiophora brassicae,* is more severe in acid soils than in neutral or slightly alkaline soils (Chupp, 1928); it is not known, however, whether the *p*H of the soil affects spore germination alone, or both germination and growth of the pathogen, or whether it increases the susceptibility of cruciferous plants. Most fungal spores will germinate over a relatively wide range of *p*H values, but for each species in a given environment, there is a definite minimum, optimum, and maximum.

Oxygen and Carbon Dioxide. Although oxygen apparently is required for spore germination, many spores can germinate over a range of O_2 concentration from 1% to 60% (Brown, 1936). Similarly, spores will germinate over a wide range of concentration of CO_2. Some smut spores of *Ustilago zeae* germinate in atmospheres of 50% CO_2, and the highest percentage of germination occurs at 15% CO_2 (Platz et al., 1927). It is doubtful that concentrations of O_2 and CO_2 usually occurring in nature exert any noticeable influence on the germination of the spores of plant-pathogenic fungi, except possibly those spores that germinate in the soil, where concentrations of these gases may be unusually low under certain conditions.

Nutrients and Special Stimuli. The germination of fungal spores may be stimulated by solutes that diffuse from within the plants into the water film over the infection court. Distilled water that has lain on the surfaces of leaves and petals of a number of plants contains more electrolytes than distilled water that has lain for the same length of time on glass slides, and the spores of *Botrytis cinerea* usually germinate better in water containing electrolytes (Brown, 1922a); the spores of *Monilinia fructicola* also require electrolytes for spore germination (Lin, 1945). As early as 1895, Ravaz showed that spores of *Botrytis* from grapes did not germinate as well in rain water as in distilled water, but that a high percentage germinated when placed in juice from grapes. Exudates from tissues of susceptible plants are known to stimulate spore germination in such plant pathogens as *Fusarium oxysporum* (Schroth and Snyder, 1961). Soil or manure infusions and various chemicals stimulate germination in many species. On the other hand, some plant tissues contain inhibitors of spore germination, among them orange peel, potato tubers, and onion tissues, which inhibit germination of the conidia of *Botrytis cinerea* (Brown, 1922b).

ENTRANCE OF PLANTS BY PATHOGENS

Ingress is the act of entering; as a part of the penetration stage of pathogenesis, it refers to the entrance of a plant pathogen into a plant. The subject has been reviewed by Flentje (1959). Pathogens may enter plants through wounds, through natural openings, or by direct penetration of uninjured plant surfaces. All plant viruses and many bacteria that attack plants gain ingress through wounds, bacteria and certain fungi enter plants through natural openings, and many fungi penetrate the intact surfaces of plants. Certain nematodes literally force their way into plants through openings resulting from the piercing action of their stylets.

Entrance Through Wounds

Mechanical Injuries. Mechanically transmitted plant viruses often find their way into wounds made by persons who handle plants or by the rubbing together of plants in the wind. The cracking of root and crown tissues of alfalfa as a consequence of winter injury affords an avenue of ingress for the wilt-inducing bacterium *Corynebacterium insidiosum* (Jones, 1928). This organism also enters plants through the clipped stems at haying time, and probably is spread on the cutter bar of the harvester. The pathogen of silver-leaf of fruit trees, *Stereum purpureum*, enters through pruning wounds, being literally sucked into exposed xylem vessels (Brooks and Moore, 1923). Organisms responsible for the seed decay of cotton enter small wounds near the hilum (Christensen et al., 1949). An unusual example of ingress through wounds was reported by Baker et al. (1954), who observed that *Botrytis* entered the leaves of stocks through tissues that had been killed by the high concentration of soluble salts in guttation water.

Bruising of fruit, roots, tubers, and other marketable plant products occurs during harvesting and shipping and facilitates the ingress of many destructive pathogens. For example, the soft-rot bacteria and fungi gain ingress into potato tubers through bruises and wounds incurred at and after harvest.

Naturally Occurring Ruptures in Plant Surfaces. During the growth of healthy plants, small ruptures occur in the surfaces of roots at the points of emergence of secondary roots. Before they heal, such wounds afford avenues of ingress for soil-borne fungi. *Thielaviopsia basicola* initiates root rot of tobacco at the junctures of secondary and primary roots (Conant, 1927), apparently after entering the ruptured primary roots at those points; it is conceivable that other fungi gain ingress in much the same way. Primary cycles of the canker pathogen of camellia are initiated after the causal organism, *Glomerella cingulata,* has entered through the ruptures at leaf scars (Baxter and Plakidas, 1954).

Wounds Made by Insects. Many insects that are agents of inoculation also make the wounds necessary for ingress. Insects, then, may be entirely responsible for the survival of a pathogen between growing seasons, for its dissemination and inoculation, and for providing it with an avenue of ingress. Two examples are bacterial wilt of sweet corn and bacterial wilt of cucurbits. The flea beetles that carry the corn-wilt pathogen deposit the bacteria in their feeding punctures in corn leaves, and the striped cucumber beetles feed on cucurbits and expose xylem tissues to infection by the cucurbit-wilt bacteria that they carry.

Even insects that are not agents of inoculation may foster the spread of pathogens by making wounds in plants that allow ingress of pathogens. In an early investigation of brown rot of stone fruit, it was found that the pathogen had entered more than 90% of the infected fruit through punctures made by the plum curculio (Scott and Ayres, 1910). Crown-gall bacteria enter plants through wounds made by arthropods: in one experiment (Banfield, 1931), little disease developed in 118 plants grown in soil infested with the crown-gall pathogen but from which root-feeding arthropods had been excluded, but in soil infested both by the pathogen and by arthropods—mostly white grubs—nearly 90% of the plants became diseased. The crown gall that occurred in the absence of insects was attributed to infections that developed after the bacteria had entered through mechanically injured tissues.

Wounds Made by Other Pathogens. Some plant pathogens that cannot invade sound plants, or do so only with difficulty, gain ingress through breaks in plant surfaces created by other plant pathogens. Although the wilt fungus can penetrate roots of cotton in the absence of other organisms (Young, 1928), ingress is favored by root-knot nematodes, presumably because they make wounds that provide avenues of ingress for the fungus. Root-knot nematodes in combination with the black-shank fungus invariably increased the incidence of black shank of tobacco, although the fungus can penetrate in the absence of nematodes. It has also been suggested (Sasser et al., 1955) that nematode infections may actually increase the susceptibility of the tobacco plant to the fungal disease. Attack by nematodes may also predispose plants to attack by other pathogens, notably the vascular wilt-inducing organisms.

Some fungi enter plants through wounds made by other fungal species. *Colletotrichum pisi,* for example, enters stems of the garden pea through mechanical injuries or through the lesions induced by *Mycosphaerella pinodes* (Ou and Walker, 1945), and blight-inducing *Fusarium* spp. enter snapdragon leaves through the pustules produced by the rust fungus *Puccinia antirrhini* (Dimock and Baker, 1951).

Entrance Through Natural Openings

Structures that provide openings in the surface of a plant may allow the ingress of plant pathogens (see Figure 19-3). Most leaf-spotting bacteria enter through stomata (Burkholder, 1948), as do certain downy-mildew fungi (deBary, 1886; Gregory, 1912) and the penetration tubes from the uredial appressoria of the rust fungi (Pole-Evans, 1907). Rot-inducing *Penicillium* spp. can enter the stomata of orange fruit, where they can initiate infection, provided that the pectic substances in the yellow rind are hydrolyzed by fungal enzymes (Green, 1932). When plants are grown in warm, moist soil, water droplets are produced at stomata when the ambient air is cool and humid. Under such conditions, *Pleospora* spp. and *Botrytis* spp. may enter leaves through their stomata. Zoospores of the grape downy-mildew fungus, *Plasmopara viticola,* congregate at stomata, and their germ tubes invariably grow through stomatal openings (Gregory, 1912). The nature of the stimulus governing the direction of germ-tube growth is not known. Entrance through stomata may be a positive hydrotropic response (Dickinson, 1949), but not all plant-pathogenic fungi make such a response; many organisms that gain ingress by direct penetration of the cuticle of a leaf produce germ tubes that grow directly over stomata but never enter the leaf through stomatal openings.

The bacterium that causes black rot of crucifers enters leaves through hydathodes (Smith, 1897), and the potato-scab bacterium enters tubers through lenticels; the fire-blight bacterium gains ingress through the nectaries of pears and other susceptible plants.

Direct Penetration of Unbroken Plant Surfaces

Most plant-pathogenic fungi and all plant-pathogenic nematodes are capable of penetrating directly through the unbroken surfaces of plants. Kühn (1858) was apparently the first to illustrate such direct penetration; he showed that the wheat-bunt fungus penetrates seedling stems. Since that time, many other examples of the direct penetration of pathogenic fungi through unbroken plant surfaces have been given (see Figure 19-3).

Penetration of Noncuticularized and Nonsuberized Surfaces. The relatively unspecialized surfaces of such plant parts as seedling stems, flowers, and root tips are readily penetrated by certain fungi. *Plasmodiophora brassicae* penetrates the root hairs of crucifers (Kunkel, 1918), and the wilt-inducing *Fusarium vasinfectum* penetrates cotton plants in the root-tip region (Fahmy, 1930). *Rhizoctonia solani* penetrates the surfaces of potato sprouts, but the mechanism of penetration seems to vary with the part of the sprout that is first invaded. Inoculation occurs when hyphae from the soil or from tuber-borne sclerotia contact the potato sprout and grow along its surface. Branches of these hyphae penetrate the sprout. When the attack is made at the tip of the sprout, a penetration hypha grows out from the appressed tips of the mycelium and enters the plant tissues, evoking no apparent response until the intercellular hyphae have reached cortical tissues; when the attack is made on the differentiated shoot, however, the fungus does not enter the plant until the underlying epidermal cells have been killed, apparently as a result of the action of a toxin secreted by the mycelium on the surface of the shoot (Müller, 1924). This exemplifies as unusual situation in which disease is initiated prior to ingress.

Penetration of Cuticularized and Suberized Surfaces. Numerous plant-pathogenic fungi can penetrate the protective coverings of green plants. Under favorable environmental conditions, the spore germinates in the infection court and produces a tube that develops into a hypha. The tip of the young hypha adheres to the plant surfaces, often as a result of the adhesive action of gelatinous substances secreted by the hypha. In some fungi, the hyphal tip enlarges and forms an entire or branched discoidal structure, the **appressorium,** which secretes gelatinous substances that hold it fast to the plant surface. A tiny branch, the **penetration peg,** develops from this region of the hypha and penetrates through the cuticularized surface of the plant. After having penetrated the covering of the plant, the penetration peg increases in diameter and develops as a hypha either inside or between the cells of the attacked plant. Two hypotheses have been proposed to explain the mechanism of direct penetration. According to one hypothesis, the penetration peg gains ingress through mechanical pressure; according to the other, the fungus penetrates after having exerted chemical action that partially destroys the protective covering of the plant.

The first direct evidence in support of the mechanical-pressure hypothesis was published by Miyoshi (1895), who found that fungi can penetrate gold foil and paraffin-wax membranes of varying thicknesses. Results of some of Miyoshi's experiments were interpreted as indications that a chemical stimulus was responsible for the growth of the penetration tubes in the direction of the gold foil or the paraffin-wax membranes, but the mode of penetration was considered to be purely mechanical. Brown and Harvey (1927) studied the penetration of paraffin wax and gelatin membranes by *Botrytis cinerea.* The fungus penetrated the highly permeable gelatin membranes regardless of the original distribution of nutrients in the solution on which the membranes were floated. The paraffin-wax membranes were also readily penetrated, even though they were impermeable to the simple electrolytes in the medium on which they had been placed. Further, thoroughly washed scales of onion or membranes of *Eucharis* leaf epidermis were penetrated by the fungus. These results, and those obtained by Dickinson (1949a; 1949b) led to the conclusion that a contact stimulus, and not a chemical one, governs the direction of growth of fungal penetration pegs. Ingress, at least by *Botrytis cinerea,* can be accomplished through mechanical pressure alone (Brown and Harvey, 1927).

Two lines of indirect evidence may be cited in support of the mechanical-pressure hypothesis of direct penetration. The first is that fungal penetration may not occur through especially thick layers of cuticle. The ability of certain fungi to attack tomato fruit (Rosenbaum and Sando, 1920; Rowell, 1953) and that of *Puccinia graminis* to penetrate leaves of barberry (Melander and Craigie, 1927) are inversely proportional to the ability of the plant surfaces to resist pressure. Resistance of plants to ingress by *Botrytis* is positively correlated with the thickness of the cuticle (Ainsworth et al., 1938).

The second line of indirect evidence in support of the mechanical-pressure hypothesis is based on the physical nature of the fungal structures that may form in the infection court. Regardless of whether they form appressoria at their tips, germ tubes are often held fast to the surfaces of substrates by gelatinous secretions. This is true for *Botrytis cinerea* (Blackman and Welsford, 1916) and *Colletotrichum lindemuthianum*, in which a sticky gelatinous sheath develops along the walls of the germ tube and the appressorium (Dey, 1919). The surface area of fungal structures in contact with the plant is greatly increased when appressoria form; *Sclerotinia libertiana* produces branched appressoria, and a penetration peg may develop from each of the several branches (Boyle, 1921). The germ tube or appressorium, being stuck fast to the surface of the infection court, provides a firm base for the penetration peg, making credible, but not proving, the mechanical-pressure hypothesis of fungal penetration. It can be concluded that at least certain fungi are capable of penetrating unbroken plant coverings as a consequence of purely physical pressure.

Mycelial penetration by *Rhizoctonia solani* is not fundamentally different from penetration by other fungi by means of penetration pegs from germ tubes or appressoria. Mycelial penetration by *Armillaria mellea*, however, is unusual in that the entire tip of the attacking rhizomorph may penetrate susceptible roots through unbroken suberized surfaces (Thomas, 1934). Strands of mycelium that make up the inner portion of the rhizomorph (see Figure 12-4) bore as a unit into the root. The rhizomorph tip secretes a gelatinous substance that holds it fast to the surface of the root, thus providing an anchor for the mechanical penetration of the hyphal strands. Thomas (1934) obtained evidence, however, that ingress of *A. mellea* is aided by the digesting action of secreted enzymes.

Results of several investigations indicate that direct penetration by plant-pathogenic fungi may not always be by mechanical pressure alone. The chemical actions that occur during the ingress of *Rhizoctonia solani* into differentiated potato-sprout tissue and during the penetration of roots by rhizomorphs of *Armillaria mellea* have been mentioned. Although no cuticle-digesting enzyme of fungal origin has been demonstrated, there is circumstantial evidence that *Venturia inaequalis* may gain ingress partly as a result of chemical action. The fungus remains in close association with the cuticle of living leaves, and may derive some nourishment from breakdown products of the cuticle. Wiltshire (1915) demonstrated the relation of the fungus to the cuticle, and O'Leary (1939) found that the fungus was able to grow in culture media prepared from partial-breakdown products of waxes that occur in cutin.

It is unnecessary, however, to postulate production of cutin-digesting enzymes by *V. inaequalis* to account for its penetration by chemical action. Leaves of the apple cultivar 'McIntosh' are not covered with a continuous layer of cutin, which would prevent the absorption of water (Roberts et al., 1948); the cutin exists in lamellae parallel to the outer walls of epidermal cells, and these lamellae are interspersed with pectic compounds. The pectic substances form a continuous path from the outside through the cuticle, and probably make possible the absorption of solutes and the loss of some water through the cuticle. It has long been known that numerous plant-pathogenic fungi produce pectolytic enzymes, and these could play an important part in the direct penetration of plants by some fungi. By enzymatic action, fungi might digest a pathway through the pectic portion of the cuticle. Moreover, the hydrolytic products of the pectic substances might provide nourishment for the attacking fungus.

Literature Cited

Ainsworth, G. C., E. Oyler, and W. H. Read, 1938. Observations on the spotting of tomato fruits by *Botrytis cinerea*. *Ann. Appl. Biol.* 25:308–321.

Baker, K. F., O. A. Matkin, and L. H. Davis, 1954. Interaction of salinity injury, leaf age, fungicide application, climate, and *Botrytis cinerea* in a disease complex of common stock. *Phytopathology* 44:39–42.

Banfield, W. M., 1931. The relation of root-feeding arthropods to crown-gall infection on raspberry. *Phytopathology* 21:112–113. (Abstr.)

Baxter, L. W., and A. G. Plakidas, 1954. Dieback and canker of camellias caused by *Glomerella cingulata*. *Phytopathology* 44:129–133.

Blackman, V. H., and E. J. Welsford, 1916. Studies in the physiology of parasitism, II. Infection by *Botrytis cinerea*. *Ann. Bot. (London)* 30:389–398.

Boyle, C., 1921. Studies on the physiology of parasitism, VI. Infection by *Sclerotinia libertiana*. *Ann. Bot. (London)* 35:337–347.

Brodie, H. J., 1945. Further observations on the mechanism of germination of the conidia of various species of powdery mildew at lower temperatures. *Can. J. Res.* (C)23:198–211.

Brooks, F. T., and W. C. Moore, 1923. On the invasion of woody tissues by wound parasites. *Proc. Cambridge Phil. Soc. Biol. Sci.* 1:56–58.

Brown, W., 1922a. Studies on the physiology of parasitism, VIII. On the exosmosis of nutrient substances from the host tissue into the infection sap. *Ann. Bot. (London)* 36:101–119.

Brown, W., 1922b. Studies on the physiology of parasitism, IX. The effect on the germination of fungal spores of volatile substances arising from plant tissues. *Ann. Bot.* (*London*) 36:285–300.

———, 1936. The physiology of the host-parasite relation. *Bot. Rev.* 2:236–381.

———, and C. C. Harvey, 1927. Studies on the physiology of parasitism, X. On the entrance of parasitic fungi into the host plant. *Ann. Bot.* (*London*) 41:643–662.

Burkholder, W. H., 1948. Bacteria as plant pathogens. *Annu. Rev. Microbiol.* 2:389–412.

Christensen, C. M., J. H. Olafson, and W. F. Geddes, 1949. Grain storage studies, VIII. Relation of molds in moist stored cotton-seed to increased production of carbon dioxide, fatty acids, and heat. *Cereal Chem.* 26:109–128.

Chupp, C., 1928. Club root in relation to soil alkalinity. *Phytopathology* 18:301–306.

Conant, G. H., 1927. Histological studies of resistance in tobacco to *Thielavia basicola*. *Amer. J. Bot.* 14:457–480.

deBary, A., 1886. Über einige Sclerotineen und Sclerotineenkrankheiten. *Bot. Ztg.* 44:377–387, 393–404, 409–426, 433–441, 449–461, 465–474.

Dey, P. K., 1919. Studies in the physiology of parasitism, V. Infection by *Colletotrichum lindemuthianum*. *Ann. Bot.* (*London*) 33:305–312.

Dickinson, S., 1949a. Studies in the physiology of obligate parasitism, I. The stimuli determining the direction of growth of the germ-tubes of rust and mildew spores. *Ann. Bot.* (*London*) (N.S.) 13:89–104.

———, 1949b. Studies in the physiology of obligate parasitism, II. The behaviour of the germ-tubes of certain rusts in contact with various membranes. *Ann. Bot.* (*London*) (N.S.) 13:219–236.

Dimock, A. W., and K. F. Baker, 1951. Effect of climate on disease development, injuriousness, and fungicidal control, as exemplified by snapdragon rust. *Phytopathology* 41:536–552.

Fahmy, T., 1930. Étude de la penetration du champignon *Fusarium vasinfectum* var. *aegyptiacum* (Atk.) T. Fahmy dans les racines du cotonnier. Doctoral Thesis (Natural Sciences), University of Geneva (thesis no. 881), pp. 1–70.

Faris, J. A., 1924. Factors influencing infection of *Hordeum sativum* by *Ustilago hordei*. *Amer. J. Bot.* 11:189–214.

Flentje, N. T., 1959. The physiology of penetration and infection. *In* C. S. Holton, et al., eds., *Plant Pathology, Problems and Progress, 1908–1958,* pp. 76–87. Madison, Wis.: University of Wisconsin Press.

Green, F. M., 1932. The infection of oranges by *Penicillium*. *J. Pomol. Hort. Sci.* 10:184–215.

Gregory, C. T., 1912. Spore germination and infection with *Plasmopara viticola*. *Phytopathology* 2:235–249.

Hebert, T. T., and A. Kelman, 1958. Factors influencing the germination of resting sporangia of *Physoderma maydis*. *Phytopathology* 48:102–106.

Jones, F. R., 1928. Development of the bacteria causing wilt in the alfalfa plant as influenced by growth and winter injury. *J. Agr. Res.* 37:545–569.

Klaus, H., 1941. Untersuchungen über *Alternaria solani* Jones et Grout, Insbesondere über seine Pathogenität an Kartoffelknollen in Abhängigkeit den Aussenfaktoren. *Phytopathol. Z.* 13:126–195.

Kühn, J., 1858. *Die Krankheiten der Kulturgewächse; ihre Ursachen und ihre Verhütung*. Berlin: Bosselmann.

Kunkel, L. O., 1918. Tissue invasion by *Plasmodiophora brassicae*. *J. Agr. Res.* 14:543–572.

Lauritzen, J. A., 1919. The relation of temperature and humidity to infection by certain fungi. *Phytopathology* 9:7–35.

Lin, C. K., 1945. Nutrient requirements in the germination of the conidia of *Glomerella cingulata*. *Amer. J. Bot.* 32:296–298.

Melander, L. W., and J. H. Craigie, 1927. Nature of resistance of *Berberis* spp. to *Puccinia graminis*. *Phytopathology* 17:95–114.

Miyoshi, M., 1895. Die Durchbohrung von Membränen durch Pilzfaden. *Jahrb. Wiss. Bot.* 28:269–289.

Müller, K. O., 1924. Untersuchungen zur Entwickelungsgeschichte und Biologie von *Hypochnus solani* P. u. D. (*Rhizoctonia solani* K.). *Arb. Biol. Reichsanst. Land- Forstwirt.* 13:197–262.

O'Leary, D. K., 1939. The utilization of cuticular compounds with reference to penetration by fungi. Ph.D. Thesis, Cornell University.

Ou, S. H., and J. C. Walker, 1945. Anthracnose of garden pea. *Phytopathology* 35:565–570.

Platz, G. A., L. W. Durrell, and M. F. Howe, 1927. Effect of carbon dioxide upon the germination of chlamydospores of *Ustilago zeae* (Beckm.) Ung. *J. Agr. Res.* 34:137–147.

Pole-Evans, I. B., 1907. The cereal rusts, I. The development of their uredomycelia. *Ann. Bot.* (*London*) 21:441–466.

Ravaz, L., 1895. La Pourriture des raisins. Reprint from *Revue de Viticulture*.

Reed, G. M., and G. A. Faris, 1924. Influence of environal factors on the infection of sorghums and oats by smuts, I. Experiments with covered and loose kernel smuts of sorghum. *Amer. J. Bot.* 11:502–512.

Roberts, E. A., M. D. Southwick, and D. H. Palmiter, 1948. A microchemical examination of McIntosh apple leaves showing relationship of cell wall constituents to penetration of spray solutions. *Plant Physiol.* 23:557–559.

Rosenbaum, J., and C. E. Sando, 1920. Correlation between size of the fruit and the resistance of the tomato skin to puncture and its relation to infection with *Macrosporium tomato* Cooke. *Amer. J. Bot.* 7:78–82.

Rowell, J. B., 1953. Leaf blight of tomato and potato plants—factors affecting the degree of injury incited by *Alternaria dauci* f. *solani. Rhode Island Univ. Agr. Exp. Sta. Bull.* (320).

Sasser, J. N., G. B. Lucas, and H. R. Powers, Jr., 1955. The relationship of root-knot nematodes to black-shank resistance in tobacco. *Phytopathology* 45:459–461.

Scott, W. M., and T. W. Ayres, 1910. The control of peach brown rot and scab. *U. S. Dep. Agr. Bur. Plant Ind. Bull.* (174).

Schroth, M. N., and W. C. Snyder, 1961. Effect of host exudates on chlamydospore germination of the bean root rot fungus, *Fusarium solani* f. *phaseoli. Phytopathology* 51:389–393.

Smith, E. F., 1897. *Pseudomonas campestris* (Pammel), the cause of a brown rot in cruciferous plants. *Zentralb. Bakteriol. Parasitenk. Infektskr.* (2)3:284–291.

Thomas, H. E., 1934. Studies on *Armillaria mellea* (Vahl.) Quel., infection, parasitism, and host resistance. *J. Agr. Res.* 48:187–218.

Voorhees, R. K., 1933. Effect of certain environmental factors on the germination of the sporangia of *Physoderma zeae-maydis. J. Agr. Res.* 47: 609–615.

Whetzel, H. H., 1929. The terminology of phytopathology. *Proc. Int. Congr. Plant Sci.* 2:1204–1215.

Wiltshire, S. P., 1915. Infection and immunity studies on the apple and pear scab fungi (*Venturia inaequalis* and *V. pirini*). *Ann. Appl. Biol.* 1:335–349.

Yarwood, C. E., 1936. The tolerance of *Erysiphe polygoni* and certain other powdery mildews to low humidity. *Phytopathology* 26:845–859.

———, 1950. Water content of fungus spores. *Amer. J. Bot.* 37:636–639.

Young, V. H., 1928. Cotton wilt studies, I. The relation of soil temperature to the development of cotton wilt. *Arkansas Agr. Exp. Sta. Bull.* (226).

7

Infection and Disease in Plants

The **infection** stage of pathogenesis is the association of the pathogen with the tissues of the suscept; the pathogen may undergo replication or growth, but there is little or no injury to the tissues. When there is a definite injurious response on the part of the attacked plant cells, **disease** has begun. Disease continues until the cells of the plant no longer respond to the activities of the pathogen.

The term "incubation period" has been used to describe different stages of pathogenesis. Whetzel (1929) considered it the period from deposit of the inoculum in the infection court until initiation of disease in the suscept. Others have used the term to designate the time that elapses between inoculation of the plant and the appearance of symp-oms; still others prefer to use it to designate the time that elapses between penetration of the plant and the development of symptoms of disease, or signs of the pathogen, or both.

Pathological attacks by plant pathogens, the responses of attacked plants, and the factors influencing these responses are summarized in this chapter.

MODES OF ATTACK BY PLANT PATHOGENS

Production of Enzymes

There is no doubt that the attack of green plants by pathogens is partly a matter of enzymatic action (Box 7-1). Certainly, enzymatic activity accounts for the ability of the parasite to establish a food relationship with its host. Enzymatic activity may also account for the establishment of the disease relationship between certain pathogens and their suscepts. The fact that pectolytic enzymes are produced by the soft-rot bacteria was established early in the twentieth century (Jones, 1905; 1909). Apparently, the first pathologic response of attacked fleshy tissue, maceration, is a result of the hydrolysis of pectic materials in the middle lamellae by pectolytic enzymes secreted by the pathogen. The timber-decaying fungi also attack by enzymatic action; the white rots begin with the enzymatic breakdown of lignin, whereas the brown rots begin with destruction of cellulose (Hubert, 1924). The wilt-inducing *Fusarium* spp. (Gothoskar et al., 1955; Waggoner and Dimond, 1955) and *Verticillium* spp. (Scheffer et al., 1956) produce pectolytic enzymes that contribute to the final plugging of the xylem of infected plants. The production of appropriate pectolytic enzymes, however, may not be the only requirement for the pathogenicity of a wilt-inducing fungus to a particular species of plant. Unless the fungus can establish a food relationship with the plant and maintain itself within the plant, no serious pathologic wilting can occur.

Production of Toxins

Many facultative parasites kill cells of the plants they infect at some distance from the actual focus of infection; this was reported first in 1886 by deBary, who studied the attack made by *Sclerotinia sclerotiorum*. The wildfire bacterium, *Pseudomonas tabaci*, produces a toxin that kills the cells of tobacco leaves (Johnson and Murwin, 1925; Clayton, 1934). This toxin is a structural analog of methionine, and apparently exerts its toxicity as an antimetabolite of that essential amino acid (Braun, 1955). *Alternaria solani*, the causal agent of early blight of potato and tomato produces a toxin, alternaric acid, that causes chlorosis and necrosis of susceptible tissues (Brian et al., 1949; Pound and Stahmann, 1951). Toxins probably account for the necroses produced in *Botrytis* infections of broad bean (Ainsworth et al., 1938) and onion (Segall, 1954).

The toxin production by wilt-inducing fungi has received much attention in connection with efforts to provide an explanation of pathologic wilting in plants. It has long been known that wilt-inducing *Fusarium* spp. produce, in culture, toxic compounds capable of inducing wilting of the foliage of shoots placed in solutions of those compounds. Results of extensive investigations in this area led Gäumann and Jaag (1947) to postulate that pathologic wilting in plants results from the toxic action of substances produced by pathogens. It was suggested that pathologic wilting does not result so much from water loss as from toxin-induced disturbance of the osmotic system as a consequence of destruction of the differential permeability of the plasma membranes. Although the toxins produced by *Fusarium* spp. may well play an important role in the attack, recent investigations strongly support the hypothesis that pectolytic enzymes are also involved. These enzymes break down pectic materials in the walls of cells in the xylem and xylem parenchyma. As a consequence of this enzymatic activity, a gel forms in the lumen of invaded vessels that impedes the transport of water in the xylem. There is good evidence that the xylem becomes plugged (Ludwig, 1952), but not by the fungal mycelium alone (Waggoner and Dimond, 1954). This does not disprove the hypothesis that toxins play a part in pathologic wilting. Until further evidence is obtained, it seems reasonable to attribute the success of wilt-inducing fungi to the combined effects of toxins and pectolytic enzymes. Moreover, auxins, cellulolytic enzymes, and oxidizing enzymes contribute to the complexity of pathological wilting in plants.

Pectolytic enzymes do not play an important role in the pathological wilting caused by *Pseudomonas solanacearum*, however. Extracellular slime formed around bacteria in the vessels of infected plants increases the viscosity of the liquid there and reduces the rate of flow of water; tyloses and collapse of some vessels also impedes the movement of water in diseased plants (Buddenhagen and Kelman, 1964).

Toxin development in some plant-parasitic fungi is so advanced that the symptoms of the diseases they cause are incited almost solely by the toxins, each of which acts upon only one host plant (Wheeler and Luke, 1963; Pringle and Sheffer, 1964). Host-specific toxins are produced by at least three plant parasitic fungi, those that cause the Victoria blight of oats (Meehan and Murphy, 1947), the milo disease of grain sorghum (Scheffer and Pringle, 1961), and the black-spot disease of Japanese pears (Tanaka, 1933).

BOX 7-1
ENZYMES AND PLANT DISEASE

Although many plant pathogens cause diseases in susceptible plants by producing toxic substances, they must first infect the plant—that is, they must grow within the invaded plant. With the exception of the plant viruses, plant pathogens invariably derive energy for their growth through enzymatic breakdown of foodstuffs within the infected plant tissues. Doubtless, a certain amount of proteolysis occurs, and certainly some plant pathogens (for example, the pathogen of apple scab, *Venturia inaequalis*) are able to hydrolyze lipids. Also, certain wood-destroying fungi enzymatically degrade lignin (Kirk, 1971), thereby obtaining soluble carbon-containing nutrients while causing the white rots of lumber and timber. We suspect, however, that most plant pathogens obtain their energy for growth from soluble minerals and amino acids and from hydrolyzed carbohydrates, notably cellulosic and pectic substances.

Cellulose, the principal constituent of parenchymatous cell walls, is a network of cross-linked linear polymers of β-1,4-linked D-glucose. It can be degraded by cellulolytic enzymes—in step-wise fashion—to linear chains, to soluble saccharides of low molecular weight, and, finally, to cellobiose. Then, catalyzed by glucosidases, cellobiose is hydrolyzed to glucose, which can be metabolized by a tissue-invading plant-pathogenic bacterium or fungus. Important as cellulolytic enzymes might be in plant pathogenesis, their significance has not been so well established as that of the pectolytic enzymes, at least among the facultatively parasitic plant pathogens.

Pectic substances, which make up the middle lamellae that cement cells together and constitute the matrix in which cutin is imbedded in the cuticle, are polymers of galacturonic acid.

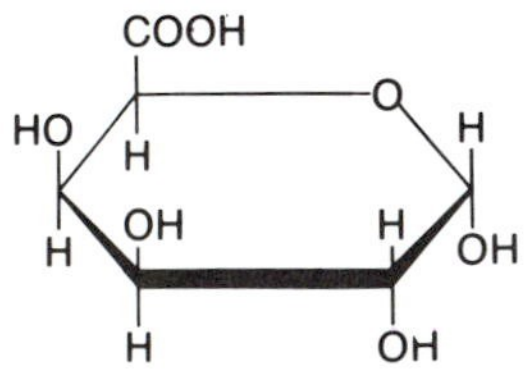

galacturonic acid

Nonmethylated chains are termed pectic acid, chains with fewer than 75% of the galacturonic acid units methylated are termed pectinic acid, and those with 75% or more of the units methylated are termed pectin, whose structural formula follows:

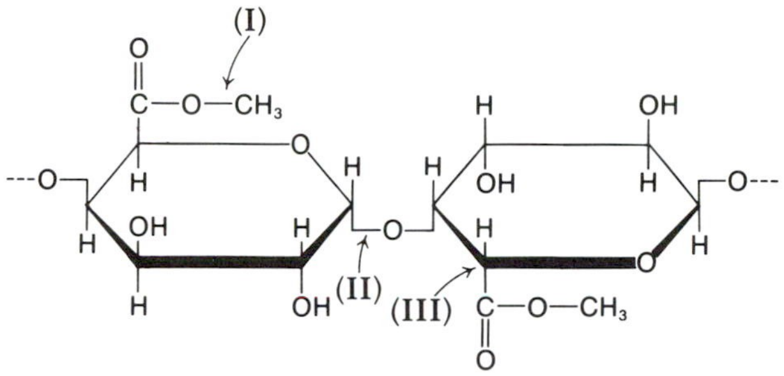

pectin

Two groups of pectolytic enzymes are produced by plant pathogens, particularly those that macerate infected tissue. The first, pectinesterase—or pectin methylesterase (PME)—breaks the ester bond—labeled (I) in the diagram—and removes methyl groups from pectin or pectinic acid to yield pectic acid and methanol. The other group of pectolytic enzymes catalyze reactions that break the 1,4-glycosidic bonds—labeled (II) in the diagram—between the subunits. The bonds may be broken by hydrolytic activity or by the action of a transeliminase. Pectic acid chains are hydrolyzed by polygalacturonase (PG), and chains of pectin or pectinic acid, by pectin methylgalacturonase (PMG). Pectin transeliminase (PTE) also cleaves 1,4-glycosidic linkages and simultaneously removes hydrogen from the fifth carbon—labeled (III) in the diagram—thereby unsaturating the ring between the fourth and fifth carbon atoms.

The chain-splitting pectolytic enzymes, whether hydrolytic or transeliminative, may occur as endoenzymes or exoenzymes. The endo- forms catalyze reactions that split the polymer at any 1,4-glycosidic linkage, whereas the exo- forms attack only the terminal linkages. The endo- forms, because they are unrestricted as to points of attack, are more effective than the exo- forms in tissue maceration. Depolymerase (DP), the term often used for macerating enzymes, is synonymous with the endo- form of polygalacturonase (endo-PG).

Although it has been clearly demonstrated that a number of plant pathogens produce pectolytic enzymes, there is yet no proof that this ability actually determines their pathogenicity (Bateman and Millar, 1966). We can only presume a cause-and-effect relationship between pectolytic enzyme production and pathogenicity.

Establishment of Food Relations Between Parasite and Host

The success of the attack depends ultimately upon the ability of the pathogen to become a parasite—to establish and maintain a food relation with the living plant, its host. The cabbage-wilt *Fusarium* produces toxins and enzymes capable of inducing wilting in tomato shoots, but it cannot actually infect tomato plants. Apparently, this is because it cannot establish a food relation with the tomato plant.

Often, the individual tissues of a plant may be pathogen-specific. Only the xylem and adjacent parenchymatous cells are invaded by wilt-inducing bacteria and such fungi as *Verticillium albo-atrum* (Rudolph, 1931), *Fusarium* spp., and *Ceratocystis ulmi.* Bacteria and fungi that cause rots of plant tissues invade the cells of the pith or cortex, and seem to be able to establish food relations most easily in parenchymatous tissues. All but one or two leafhopper-transmitted plant viruses seem to be restricted to the phloem, whereas mechanically transmissible stylet-borne plant viruses, although translocated in the phloem, occur in high concentrations in parenchymatous tissues.

It is not clear how viruses move in plants. Certainly, long-distance movement—translocation—usually occurs in the phloem, although a few viruses move in the xylem. Virus may be translocated in the phloem along with elaborated food materials (Bennett, 1937), but there is evidence that some mechanically transmissible viruses are able to move long distances against the direction of flow of elaborated food materials (Roberts, 1952). Cell-to-cell movement of plant viruses probably occurs in plasmodesmata (Esau, 1941, 1967; Kassanis et al., 1958).

Bacteria invade plant tissues intercellularly, probably along pathways created by enzymatic digestion of middle lamellae. The bacterial cells increase in numbers, and the masses of bacteria occupy the spaces between plant cells.

A number of plant-pathogenic fungi invade susceptible tissues intercellularly. When this occurs, there is probably some enzymatic digestion of middle lamellae, although the mechanical pressure exerted by the growing hyphae may be sufficient to enable the hyphae to force themselves between plant cells. Certain plant-pathogenic fungi are intracellular, and are able to grow directly through cell walls. From the evidence available, it is likely that fungal penetration of cell walls can be accomplished through mechanical pressure. This seems true for *Pythium debaryanum* (Hawkins and Harvey, 1919), but the pos-

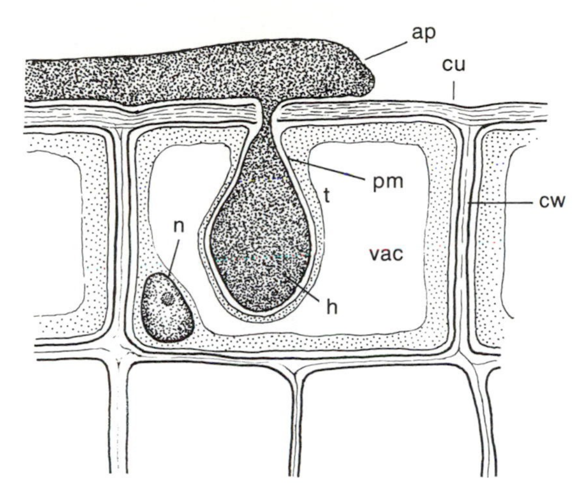

FIGURE 7-1
A haustorium (*h*) within an epidermal cell: the penetration peg from the appressorium (*ap*) pierces the cuticle (*cu*) and cell wall (*cw*), but invaginates the plasma membrane (*pm*) and the tonoplast (*t*) surrounding the vacuole (*vac*). Haustoria often penetrate close to nuclei (*n*).

sibility that chemical actions are involved cannot be completely discounted.

The invading hyphae of obligately parasitic fungi and of certain facultative saprophytes grow intercellularly, but push specialized hyphal branches into plant cells (Rice, 1935, 1945). These thin-walled branches often enlarge inside the plant cell and sometimes take on characteristic shapes. These structures are feeding organs and are termed **haustoria** (Figure 7-1). The cell wall is penetrated, but the enlarging haustorium invaginates the cytoplasm without actually penetrating into the cytoplast or the vacuole (Rice, 1927). Thus, a large absorbing surface of the parasite contacts a large cytoplasmic surface in the host cell, obviously favoring the acquisition of food materials from the host by the parasite. The haustorium of the stem-rust pathogen of wheat is surrounded by a capsule that resembles the sheath around the haustorium of the powdery-mildew fungus of grasses. In the latter fungus, the body of the haustorium is enclosed within a plasma membrane and wall; this, in turn, is surrounded by the sheath, which is presumably of fungal origin. The sheath has inner and outer boundaries (membranes) that are in contact with, respectively, the wall of the haustorium and the plasma membrane of the host cell. The sheath is thought to function in the establishment and maintenance of the infection processes (Ehrlich and Ehrlich, 1963). Anatomical features at host-parasite interfaces have been described by Ehrlich and Ehrlich (1971).

Although the spacial relations between bacterial and fungal parasites and their hosts have been determined, little is known about the way in which food is obtained at the expense of the host. Certainly, the parasite absorbs and accumulates materials; it is probable that many of these materials are breakdown products of the host's tissues, produced by the enzymatic activities of the parasite. The differential permeability of the membranes of attacked and attacking cells undoubtedly plays an important role in parasitism (Thatcher, 1943). A difference between the osmotic concentration of attacking fungal cells and that of attacked plant cells could account only for the movement of water across the membranes; osmotic concentration *per se* is probably not as important in parasitism as is the nature of the permeability to solutes of the various membranes.

RESPONSES OF ATTACKED PLANTS

The responses of attacked plants lead ultimately to the production of symptoms; infected plants respond necrotically, hypoplastically, or hyperplastically to the activities of the pathogen. Toxins and hydrolytic enzymes produced by pathogens aften account for the necrotic reactions of attacked plant cells. Hypoplastic effects, such as those brought about by most plant viruses, probably result from pathologic diversion of essential materials from use by the plant to use by the pathogen. Further, pathogens might check certain metabolic reactions that are necessary for optimum plant growth. Those plant pathogens that stimulate attacked plants to react hyperplastically exert a profound influence on the growth-regulating mechanisms of the plant. The production of tumors in plants infected by the crown gall bacterium proceeds in two phases: in the first phase (conditioning), attacked cells acquire the potentiality for autonomous growth; in the second phase (tumor induction), there is continuous unregulated hyperplasia. Conditioning occurs only as the result of wounding; wounded cells exposed to the tumor-inducing principle are permanently altered and continue to divide, even in the absence of the pathogen (Braun, 1954).

Not all responses to pathogens are harmful to the plants. Certain pathogens incite defensive reactions in some of the plants that they attack. A **defensive reaction** is a reaction of the suscept to the pathogenic activities of an attacking entity (Gäumann, 1946). The defense may be a physiological property of the protoplast itself, in which case the reaction may or may not be necrotic. The nature of the protoplasmic defense in the absence of necrosis is not at all understood. Such a defense renders the plant refractive to the activities of the pathogens, and is the basis for the control of certain plant diseases by the use of resistant varieties. Some parasitic pathogens induce severe local necrosis in certain plants; this necrosis may be so sudden and so complete in the immediate vicinity of the invading pathogen that the pathogen, being unable to establish food relationships except in living cells, fails to survive. This kind of necrotic reaction is termed **hypersensitivity** (Gäumann, 1946), and results in the "field immunity" of certain plants from diseases caused by such obligately parasitic pathogens as viruses and the powdery-mildew and rust fungi.

The **phytoalexin theory** was proposed by Müller and Börger (1940) to explain the hypersensitive defensive reaction in plants. This theory, which has been restated by Müller (1961) and reviewed by Cruikshank (1963), states that, in living cells of hypersensitive hosts, substances formed in response to parasitic attack inhibit further development of the parasite. Aside from stimulating the production of phytoalexins, pathogens increase the activity of such enzymes as polyphenol oxidase within attacked cells. Products of phenol oxidation accumulate in diseased cells, which turn brown or black and then die. The rapid darkening and death of hypersensitive tissue could contribute to the isolation of the pathogen within the affected area. The subjects of hypersensitivity and of pathological darkening of plant tissue have been reviewed recently (Müller, 1959; Rubin and Artsikhovskaya, 1964).

A hypersensitive defensive reaction in a viral infection has been described by Ross (1961a, 1961b). A high degree of resistance to several viruses developed in a narrow zone surrounding local necrotic lesions induced by tobacco mosaic virus. Moreover, resistance also developed in uninoculated leaves; this resistance could be detected two or three days after inoculation, was most pronounced seven days after inoculation, and persisted for at least twenty days. Possibly, the phytoalexin theory can be expanded to encompass this acquired resistance.

Certain other defensive mechanisms are characterized by the formation of cellular or gummous demarcations in response to pathogenic attacks. Such demarcations limit the spread of pathogens and restrict the movement of solutes from other parts of the plant into the invaded area. A histological demarcation is formed in peach leaves attacked by the shot-hole fungus (Higgins, 1914) and in leaves of other plants attacked by various fungi. Cunningham (1928) termed such demarcations **cicatrices,** and defined a cicatrix as a band of healing tissue about the lesion or wound, the cells of which arise by division and enlargement of mesophyll cells. The cicatrix forms in some plants in response to mechanical injury alone, and some pathogenic fungi are able to prevent cicatrix formation. Some cells of wheat roots produce elongated protuberances, called **lignitubers,** in front, and around the tips, of attacking hypha of *Ophiobolus graminis,* the pathogen of the root-rot (take-all) disease (Fellows, 1928); lignitubers, however, do not prevent the advance of the hyphae, which finally penetrate the lignitubers and spread between the cells and through the cortex.

An unusual defense reaction was reported by Wingard (1928), who observed that tobacco plants infected by tobacco-ringspot virus underwent a severe necrotic disease that was followed by almost complete symptomlessness in leaves at the tops of diseased plants. The leaves that recover from severe disease contain virus, but in concentrations lower than in those leaves showing necrotic symptoms, and are protected against further inoculations by the same virus. The plant acquires an immunity of the active, nonsterile type, but acquires it at the expense of continuously suffering from a mild disease. A number of plants may exhibit this tolerance after they have been attacked by ringspot and related viruses. It is generally agreed that symptomlessness follows early invasion of meristematic tissues, as opposed to tissues with rapidly expanding cells, but there is no satisfactory explanation for the mechanism by which infected plants recover from severe symptoms of viral disease.

Factors Affecting Pathological Responses of Plants

The Presence of Other Pathogens. When a plant is attacked simultaneously by two pathogens, it may respond in one of three ways: its pathological reactions may be no different from those expected on the basis of its reactions to each pathogen alone; it may have an anergistic reaction, in which the disease is less severe than expected; or it may have a synergistic reaction, in which the disease is more severe than expected. Anergistic (White and McCulloch, 1934) and synergistic (Burkholder and Guterman, 1932) reactions of plants to infections by bacteria have been reported. Synergistic reactions of plants to double infections by viruses are of economic importance in the streak disease of tomatoes and rugose mosaic of potatoes. The tomato disease results from simultaneous infection by tobacco-mosaic virus and potato virus X, and the potato disease is caused by a simultaneous infection by potato virus X and potato virus Y. Results of quantitative studies of the viral content of doubly infected plants that react synergistically (Bennett, 1949; Rochow and Ross, 1955) indicate that there is an increase in concentration of the milder of the two viruses, whereas the concentration of the more severe virus is little different from the concentration attained in singly infected plants. Rochow and Ross (1955) suggested that, in double infections, the synthesis of potato virus Y somehow increases the synthesis of potato virus X.

The reaction of a plant to simultaneous attack by two pathogens may be governed, to a large extent, by the closeness of the taxonomic relationship of the pathogens. Holmes (1956) pointed out that mixed infections by two strains within a single pathogenic species seldom evoke anergistic or synergistic reactions, and postulated that the occurrence of a synergistic reaction in double infections may indicate that the two pathogens involved are not closely related.

Environment. When environmental factors operative during the infection stage of pathogenesis are such that the inherent susceptibility of the attacked plant is increased, those factors are said to *predispose* the plant to disease. For example, the susceptibility of tomato plants to *Fusarium* wilt is increased in dry soil, by short day length, low light intensity, and by low *p*H of the soil solution (Foster and Walker, 1947). Conditions under which the plants are grown before transplanting may well determine the severity of disease that develops after the plants have been transplanted into infested soil. The literature on mineral nutrition as it predisposes plants to disease is voluminous, but there is still much controversy over the influence of particular nutrients on susceptibility. One point seems clear, however: excessive applications of nitrogen predispose plants to diseases caused by obligately parasitic pathogens (Gothoskar et al., 1955).

The soaking of leaves that occurs during rain storms favors development of certain diseases (Johnson, 1947). In the wildfire disease of tobacco, the flooding of intercellular spaces facilitates the spread of bacteria within the leaf tissues after invasion (Clayton, 1936). Soaking alone is injurious to tobacco leaves, however, and it is difficult to distinguish between the damage done directly by water and that done by bacteria in watersoaked tissue (Valleau et al., 1944). By contrast, infiltration with water actually increased the resistance of bean leaves to rust (Cohen, 1951).

Literature Cited

Ainsworth, G. C., E. Oyler, and W. H. Read, 1938. Observations on the spotting of tomato fruits by *Botrytis cinerea* Pers. *Ann. Appl. Biol.* 25:308–321.

Bateman, D. F., and R. L. Millar, 1966. Pectic enzymes in tissue degradation. *Annu. Rev. Phytopathol.* 4:119–146.

Bennett, C. W., 1937. Correlation between movement of the curly-top virus and translocation of food in tobacco and sugar beet. *J. Agr. Res.* 54: 479–502.

———, 1949. Recovery of plants from doddar latent mosaic. *Phytopathology* 39:637–646.

Braun, A. C., 1954. The physiology of plant tumors. *Annu. Rev. Plant Physiol.* 5:133–162.

———, 1955. A study on the mode of action of the wildfire toxin. *Phytopathology* 45:659–664.

Brian, P. W., P. J. Curtis, H. G. Hemming, C. H. Unwin, and J. M. Wright, 1949. Alternaric acid, a biologically active metabolic produce of the fungus *Alternaria solani. Nature* 164:534–535.

Buddenhagen, I., and A. Kelman, 1964. Biological and physiological aspects of bacterial wilt caused by *Pseudomonas solanacearum. Annu. Rev. Phytopathol.* 2:203–230.

Burkholder, W. H., and C. E. F. Guterman, 1932. Synergism in a bacterial disease of *Hedera helix. Phytopathology* 22:781–784.

Butler, E. J., and S. G. Jones, 1949. *Plant Pathology.* London: Macmillan.

Clayton, E. E., 1934. Toxin produced by *Bacterium tabacum* and its relation to host range. *J. Agr. Res.* 48:411–426.

———, 1936. Water soaking of leaves in relation to development of the wildfire disease of tobacco. *J. Agr. Res.* 52:239–269.

Cohen, M., 1951. Increased resistance to bean rust associated with water infiltration. *Phytopathology* 41:937. (Abstr.)

Cruikshank, I. A. M., 1963. Phytoalexins. *Annu. Rev. Phytopathol.* 1:351–374.

Cunningham, H. S., 1928. A study of the histologic changes induced in leaves by certain leaf-spotting fungi. *Phytopathology* 18:717–751.

deBary, A., 1886. Über einige Sclerotineen und Scleroteenkrankheiten. *Bot. Ztg.* 44:337–387, 393–404, 409–426, 433–441, 449–461, 465–474.

Ehrlich, H. G., and M. A. Ehrlich, 1963. Electron microscopy of the sheath surrounding the haustorium of *Erysiphe graminis. Phytopathology* 53: 1378–1380.

Ehrlich, M. A., and H. G. Ehrlich, 1971. Fine structure of the host-parasite interfaces in mycoparasitism. *Annu. Rev. Phytopathol.* 9:155–184.

Esau, K., 1941. Inclusions in guard cells of tobacco affected with mosaic. *Hilgardia* 13:428–430.

———, 1967. Anatomy of plant virus infections. *Annu. Rev. Phytopathol.* 5:45–76.

Fellows, H., 1928. Some chemical and morphological phenomena attending infection of the wheat plant by *Ophiobolus graminis. J. Agr. Res.* 37: 647–661.

Foster, R. E., and J. C. Walker, 1947. Predisposition of tomato to *Fusarium* wilt. *J. Agr. Res.* 74:165–185.

Gäumann, E., 1946. Types of defensive reactions in plants. *Phytopathology* 36:624–633.

———, and O. Jaag, 1947. Die physiologischen Grundlagen des parasitogenen Welkens. *Ber. Schweiz. Bot. Ges.* 57:3–34, 132–148, 227–241.

Gothoskar, S. S., R. P. Scheffer, J. C. Walker, and M. A. Stahmann, 1955. The role of enzymes in the development of *Fusarium* wilt of tomato. *Phytopathology* 45:381–387.

Hawkins, L. A., and R. B. Harvey, 1919. Physiological study of the parasitism of *Pythium debaryanum* Hesse on the potato tuber. *J. Agr. Res.* 18:275–297.

Higgins, B. B., 1914. Contribution to the life history and physiology of *Cylindrosporium* on stone fruits. *Amer. J. Bot.* 1:145–173.

Holmes, F. O., 1956. A simultaneous-infection test for viral interrelationships as applied to aspermy and other viruses. *Virology* 2:611–617.

Hubert, E. E., 1924. The diagnosis of decay in wood. *J. Agr. Res.* 29:523–567.

Johnson, J., 1947. Water-congestion in plants in relation to disease. *Wis. Agr. Exp. Sta. Res. Bull.* (160).

———, and H. F. Murwin, 1925. Experiments on the control of wildfire of tobacco. *Wis. Agr. Exp. Sta. Res. Bull.* (62)

Jones, L. R., 1905. The cytolytic enzymes produced by *Bacillus carotovorus* and certain other soft rot bacteria. *Zentralbl. Bakteriol.* (2)14:257–272.

———, 1909. Pectinase, the cytolytic enzyme produced by *Bacillus carotovorus* and certain other soft rot organisms. *N. Y. Agr. Exp. Sta. Geneva Tech. Bull.* (11):291–368.

Kassanis, B., T. W. Tinsley, and F. Quak, 1958. The inoculation of tobacco callus tissue with tobacco mosaic virus. *Ann. Appl. Biol.* 46:11–19.

Kirk, T. K., 1971. Effects of microorganisms on lignin. *Annu. Rev. Phytopathol.* 9:185–210.

Ludwig, R. A., 1952. Studies on the physiology of hydromycotic wilting in the tomato plant. *McGill Univ. Macdonald Coll. Agr. Tech. Bull.* (20).

Meehan, F., and H. C. Murphy, 1947. Differential phytotoxicity of metabolic by-products of *Helminthosporium victoriae*. *Science* 106:270–271.

Müller, K. O., 1959. Hyersensitivity. *In* J. G. Horsfall and A. E. Dimond, eds., *Plant Pathology, an Advanced Treatise,* vol. I (The diseased plant), pp. 469–519. New York: Academic Press.

———, 1961. The phytoalexin concept and its methodological significance. *Recent Advan. Bot.* 1:396–400.

———, and H. Börger, 1940. Experimentelle Utersuchungen über der *Phytophthora*-Resistenz der Kartoffel. *Landwirt. Jahrb. Berlin* 87:189–231.

Pound, G. S., and M. A. Stahmann, 1951. The production of a toxic material by *Alternaria solani* and its relation to the early blight disease of tomato. *Phytopathology* 41:1104–1111.

Pringle, R. B., and R. P. Scheffer, 1964. Host-specific plant toxins. *Annu. Rev. Phytopathol.* 2:133–156.

Rice, M. A., 1927. The haustoria of certain rusts and the relation between host and pathogene. *Bull. Torrey Bot. Club* 54:63–153.

———, 1935. The cytology of host-parasite relations. *Bot. Rev.* 1:327–354.

———, 1945. The cytology of host-parasite relations, II. *Bot. Rev.* 11:288–298.

Roberts, D. A., 1952. Independent translocation of sap-transmissible viruses. *Phytopathology* 42:381–387.

Rochow, W. F., and A. F. Ross, 1955. Virus multiplication in plants doubly infected by potato viruses X and Y. *Virology* 1:10–27.

Ross, A. F., 1961a. Localized acquired resistance to plant virus infection in hypersensitive hosts. *Virology* 14:329–339.

———, 1961b. Systemic acquired resistance induced by localized virus infections in plants. *Virology* 14:340–358.

Rubin, B. A., and E. V. Artsikhovskaya, 1964. Biochemistry of pathological darkening of plant tissues. *Annu. Rev. Phytopathol.* 2:157–178.

Rudolph, B. A., 1931. *Verticillium* hadromycosis. *Hilgardia* 5:197–360.

Scheffer, R. P., S. S. Gothoskar, C. F. Pierson, and R. P. Collins, 1956. Physiological aspects of *Verticillium* wilt. *Phytopathology* 46:73–87.

———, and R. B. Pringle, 1961. A selective toxin produced by *Periconia circinata*. *Nature* 191:912–913.

Segall, R. H., 1954. Onion blast or leaf spotting caused by species of *Botrytis*. Ph.D. Thesis, Cornell University.

Tanaka, S., 1933. Studies on black spot disease of Japanese pears (*Pyrus serotina*). *Mem. Coll. Agr. Kyoto Univ.* 28:1–31.

Thatcher, F. S., 1943. Cellular changes in relation to rust resistance. *Can. J. Res.* (C)21:151–172.

Valleau, W. D., E. M. Johnson, and S. Diachun, 1944. Root infection of crop plants and weeds by tobacco leaf-spot bacteria. *Phytopathology* 34:163–174.

Waggoner, P. E., and A. E. Dimond, 1954. Reduction in water flow by mycelium in vessels. *Amer. J. Bot.* 41:637–640.

———, and ———, 1955. Production and role of extracellular pectic enzymes of *Fusarium oxysporum* f. *lycopersici*. *Phytopathology* 45: 79–87.

Wheeler, H., and H. H. Luke, 1963. Microbial toxins in plant disease. *Annu. Rev. Microbiol.* 17:223–242.

Whetzel, H. H., 1929. The terminology of phytopathology. *Int. Congr. Plant Sci. Proc.* 2:1204–1215.

White, R. P., and L. McCulloch, 1934. A bacterial disease of *Hedera helix*. *J. Agr. Res.* 48:807–815.

Wingard, S. A., 1928. Hosts and symptoms of ring spot, a virus disease of plants. *J. Agr. Res.* 37:127–153.

8

The Epidemiology of Plant Diseases

Strictly speaking, an epidemic disease is one that affects, at the same time, many people in a community; such diseases in plants are termed epiphytotic, and those in animals are said to be epizootic. The term epidemic, however, is commonly used to describe widespread outbreaks of any disease, irrespective of whether it occurs in people, other animals, or plants. This usage of the term, of course, is no more illogical than using the word population (from the Latin *populus*, people) when referring to a group of plants or animals other than man. Moreover, it permits the construction of the noun *epidemiology*, which means, with no ambiguity, the science of epidemic diseases. What should be the corresponding term derived from epiphytotic? Epiphytology? No, for this word actually denotes the study of epiphytic plants ("air plants"). Epiphytoticology or epiphytotiology? These words are decidedly awkward. Epidemiology, clearly, is the most useful term.

The conditions that must be fulfilled before widespread plant disease can occur have already been stated in qualitative terms (Chapter 1). To reiterate, a widespread outbreak of a disease can occur in plants only when there is a large number of susceptible plants, an abundance of virulent inoculum, and a favorable environment that persists over a relatively long time. The ways in which these conditions were met in 1970 to make possible the corn-blight epidemic are discussed in Box 8-1. These conditions are considered in earlier and subsequent chapters as factors that influence the initiation and development of disease in plants. In this chapter, the nature of epidemics themselves and the application of epidemiology will be considered. The subject has been reviewed by van der Plank (1959, 1960, 1963, 1965).

THE NATURE OF EPIDEMICS

The science of epidemiology deals with the increase of disease in a population of plant suscepts. Of prime importance are a knowledge of the rate of increase and a knowledge of the factors affecting that rate. As van der Plank (1965) has pointed out, there is no need to distinguish between the increase in disease and the increase of the abundance of pathogens in and on susceptible plants.

The late blight disease of potatoes dramatically illustrates the dynamics of an epidemic disease of plants. The pathogen overwinters as fungal mycelium in diseased tubers, which may be discarded in cull piles or even cut up and planted as seed pieces. Shoots that develop from such tubers in the following spring are invaded by the fungus, which later produces spores that are dispersed by wind or splashing rain as inocula for the primary cycles of the pathogen. If, in a susceptible variety of potato, there is a single diseased sprout per 250 acres, the pathogen must increase itself by a factor of approximately one billion in order to destroy all the fields of potatoes (van der Zaag, 1956). This phenomenal increase can be accomplished, under favorable environmental conditions, in less than three months (van der Plank, 1965). Although other plant diseases may develop more slowly and under different circumstances than late blight, the fact remains that all epidemics have one common denominator: they all "have in common a dynamic process of increase of the pathogen—of the fungus, bacterium, or virus that causes the disease" (van der Plank, 1965).

Unchecked, an epidemic proceeds geometrically and occurs in three distinct stages. In the first, or **lag stage,** the rate of spread is slow. Then, in the second, or **logarithmic stage,** the rate is rapid. Finally, the rate decreases in the third, or **postlogarithmic stage.** Toward the end of an epidemic, the rate of spread of the pathogen is clearly limited by a paucity of susceptible tissues. There is a more subtle reason for the slow rate of spread that characterizes the early stage of an epidemic.

BOX 8-1
SOUTHERN LEAF BLIGHT OF CORN

The corn crop of the United States was threatened in 1970 by an epidemic of southern leaf blight, caused by the fungus *Helminthosporium maydis*. Yields were drastically reduced in the southern states, where losses of 50–75% were average and total losses were common. As it turned out, the disease was less damaging in the northern "corn belt," so that yield loss, nationwide, was finally put at approximately 15%. Even so, the epidemic had far-reaching implications, and it reminded us of just how vulnerable our major food crops are to unexpected outbreaks of a devastating plant disease.

Why was leaf blight so severe in 1970? Would it be worse in 1971? What are the prospects for the ensuing years? These questions were uppermost in the minds of agriculturists in 1970 as they watched the blight spread through the corn fields of the southern and midwestern United States.

Strains of *H. maydis* had long since become established in the United States, where they had always caused some leaf spotting in corn; but the disease was never very destructive, at first, because the varieties of corn commonly grown there were somewhat resistant. Recently, however, increasing amounts of susceptible hybrid varieties have been grown. These hybrids were developed from inbred lines whose female parents were male-sterile—a desirable characteristic, because it made it unnecessary to detassel the female plants by hand in the seed-production fields. Male sterility is inherited through the cytoplasm, and the most popular varieties of corn were derived from parents with the so-called Texas male-sterile, or T-type male-sterile (Tms), cytoplasm. Despite the known susceptibility of these varieties to *H. maydis,* they were used successfully for five to ten years in the United States, where virulent strains of *H. maydis* were not prevalent.

By 1970, almost all the varieties of corn planted in the United States had the Tms cytoplasm. Moreover, there was (at least in the southern states) sufficient inoculum of the T-race of *H. maydis* to initiate an epidemic. Because *H. maydis* is a "high birth-rate pathogen," only a small amount of initial inoculum can, if the environment is favorable, lead to an epidemic. Unfortunately, warm, moist weather persisted during the 1970 growing season, and secondary cycles of the southern leaf-blight fungus developed regularly and rapidly. In the northern states, initial inoculum that may have survived the winter there was supplemented by windblown conidia from farther south. All conditions for a full-scale epidemic were fulfilled in 1970: there was an abundance of susceptible plants, an adequate supply of initial inoculum of a virulent strain of the pathogenic fungus, and a set of environmental factors highly favorable to development of the disease.

Although sufficient inoculum survived to initiate an epidemic in 1971, the disease was significantly less damaging than it was the preceding year. Seed producers grew crops of resistant corn during the winter of 1970–71 in tropical and subtropical regions. These enterprises provided seed of resistant varieties (containing the normal male-fertile, or N-type, cytoplasm) for about one-third of the commercial acreage in 1971. Moreover, initial inoculum was reduced in corn-growing areas by plowing under the stubble of infected corn plants following the 1970 season; *H. maydis* survives best in standing stubble or in crop refuse lying on the top of the ground. Finally, the warm, moist environment so favorable to secondary cycles of southern leaf blight of corn did not persist throughout the 1971 growing season. An epidemic was averted in 1971—partly because of increased use of resistant varieties, partly because initial inoculum had been reduced by fall plowing, and partly because the environment was less favorable to disease development during the growing season.

Infected tissue does not contribute to further spread of the pathogen until it has grown in those tissues and produced inoculum for subsequent cycles. The time required for infected tissue to become "infectious," that is, to produce inoculum, is the **period of latency** (van der Plank, 1965). The earliest stage of an epidemic is characterized by a high proportion of the relatively few infected plants in the period of latency. The rate of increase of the disease becomes rapid as relatively fewer of the infected plants are in the period of latency and relatively more of them become "infectious."

Although infections characterized by local lesions have received the most attention from students of epidemiology, systemic infections of plants often result in the greatest crop losses after an epidemic has run its course. Epidemics of systemic infections develop much more slowly than do those of local infections, but they can ultimately cause extensive losses because each infection causes great damage. Epidemics may proceed over many years in perennial plants, and vegetative propagation of diseased plant parts perpetuates the pathogen (van der Plank, 1959).

Whereas the lesions of the potato late-blight disease can increase in number by a factor of one billion in a single year, and those of wheat-leaf rust by a factor of about thirty million in the same time (Chester, 1946), systemic infections, such as viral infections of trees, increase only about tenfold annually. Some systemic infections of herbaceous plants may increase by a factor of ten thousand in a year's time (van der Plank, 1959). In general, systemic diseases that spread slowly are more amenable than others to prevention by sanitation, the removal of infected tissues. Conversely, sanitary measures seldom control diseases characterized by local lesions; when caused by fungi, such diseases are usually prevented by protective spraying or dusting with a fungicide or by the use of resistant varieties of crop plants.

Knowledge of epidemiology is often used in determining the necessity for certain measures of plant-disease prevention. When epidemics can be predicted, control measures can be recommended in time to prevent the epidemic. The severity of some plant diseases can be forecast with considerable accuracy, and the subject of forecasting illustrates the applied aspects of epidemiology.

THE FORECASTING OF EPIDEMICS

A forecast is a prophecy or prediction of a future event or condition. The forecasting of epidemics is, therefore, simply a prognostication, or foretelling. The practical implications of forecasting have been incorporated in the definition proposed by Miller and O'Brien (1952), who state:

> Forecasting involves all the activity in ascertaining and notifying the growers of a community that conditions are sufficiently favorable for certain diseases, that application of control measures will result in economic gain or, on the other hand, and just as important, that the amount of disease expected is unlikely to be enough to justify the expenditure of time, energy and money for control.

Miller and O'Brien thus include the concept of a warning service for plant-disease prevention in their definition of forecasting.

Plant-Disease Warning Services

Sometimes one or two of the factors that influence the development of an epidemic are so important that their effects override those of all the other factors combined. When the magnitude of a few factors can be foretold with some accuracy, it is possible to develop a warning service against a particular disease. Several warning services for specific plant diseases are discussed below. They are outstanding examples of "quantitative epidemiology applied with courage" (Waggoner, 1950). Additional information on plant disease forecasting can be found in reviews by Miller and O'Brien (1952, 1957), Miller (1959), and Waggoner (1960).

The Decay of Cranberries in Storage. N. E. Stevens made what is probably the first formal warning to growers on the likely severity of a plant disease. His first forecast of the keeping quality of cranberries was published in *The Courier*, a newspaper in Wareham, Massachusetts, during September of 1923. Stevens (1943) found that the fungi that cause rots of stored cranberries are most destructive when high temperatures prevail in May and June, the two months before blossoming. Decay in storage is also positively correlated with high amounts of rainfall in July and August and with unusually high yields of cranberries. Growers have been advised on the growing, handling, and marketing of cranberries on the basis of forecasts of the keeping quality of the product.

Curly Top of Sugar Beet. The severity of the curly-top disease of sugar beet can be predicted if the conditions affecting the overwintering and subsequent migration of the leafhopper vectors of the causative virus are determined (Carter, 1930). Epidemics of curly top are likely if large numbers of leafhoppers survive the winter and if their early spring migrations coincide with the seedling stage of beets in the field. Cool, wet summers favor a build-up of the leafhopper population because the growth of hosts for summer breeding and fall breeding is favored. Mild winter weather favors survival of the insects, and warm weather early in the spring favors early migration of the insects. An epidemic is likely after such conditions, provided, of course, a large number of the surviving leafhoppers are viruliferous. An impending epidemic can often be averted by the simple expedient of delaying the planting of beet seed so that seedlings are not abundant when the first insect migration occurs. Under extreme conditions, much of the acreage normally planted to sugar beets can be temporarily devoted to other crops, in order to minimize economic loss.

Bacterial Wilt of Sweet Corn. The knowledge that the bacterium that causes Stewart's wilt of sweet corn survives the winter in adult flea beetles was the basis for a simple and highly effective forecasting and warning service. Warm winter weather favors survival of a high percentage of flea beetles and therefore permits the survival of enough bacteria to provide sufficient inoculum for an epidemic during the subsequent spring and summer. Again, it was N. E. Stevens (1934) who first associated severe outbreaks of Stewart's wilt of corn with the relatively high temperatures of the preceding winter. Stevens and Haensler (1941) developed a warning service based on a single index, the sum of the monthly mean temperatures (in degrees Fahrenheit) for December, January, and February. Little or no disease occurs if the index is below 90, but an epidemic is likely if it exceeds 100. A moderate amount of disease usually occurs when the index is between 90 and 100. On the basis of this simple index, growers of sweet corn can determine, as much as two months in advance, whether it will be profitable to use an insecticide on corn in the early season to control the flea beetles and, thus, the wilt disease caused by the bacteria they harbor.

Leaf Rust of Wheat. K. S. Chester developed an accurate method for forecasting epidemics of leaf rust of wheat in Oklahoma. His forecasts were so accurate that they "have been decisive in determining whether to allow crops of borderline condition to go to maturity or, alternatively, to abandon the wheat in favor of spring-planted summer crops" (Chester, 1950). Successful forecasting of leaf rust of wheat in Oklahoma is based on the knowledge that the weather after April is seldom critical to the subsequent development of rust, that overwintered infections are the sources of the inoculum, that spring renewal of rust is determined by the weather conditions from December through March, and that the level of rust on April 1 is the chief factor determining rust damage before harvest in June.

Apple Scab. Venturia inaequalis, the pathogen of apple scab, usually overwinters as perithecia in fallen leaves. Ascospores that are discharged early in the spring initiate the primary cycle of the pathogen. It is essential that fungicidal sprays be applied at the proper time to prevent infection during the blossoming period, for failure to do so cannot be corrected later in the summer, even by as many as three or four applications of fungicide (Miller and O'Brien, 1952).

For the purpose of developing warning services, workers in many parts of the world have studied the relationships among the fungus, the apple tree, and the meteorological environment. The primary cycle of *Venturia inaequalis* can be foretold on the basis of knowledge of the maturity of perithecia, the temperature and moisture conditions, and the stage of growth of the apple tree in the spring. For example, a severe outbreak of scab can be expected within one week if blossom buds are swelling and beginning to open, if the temperature is near 20°C, and if the blossoms remain wet for approximately 18 hours. Under such conditions, it is important that the first protective fungicidal application be made before or during the first several hours of the period of wetting.

The literature describing apple-scab warnings in the United States, Canada, Europe, New Zealand, and South Africa has been reviewed by Miller and O'Brien (1952).

Downy Mildew of Grape. Warning services in Europe have long been of service to growers of grapes, which are subject to serious crop losses due to the downy-mildew fungus *Plasmopara viticola.* These services not only provide the warnings that fungicidal applications must be made, but also predict, with accuracy, when applications can be safely omitted. The services thus assist in control of the disease when an epidemic threatens and also enable growers to save on production costs by omitting sprays when they would serve no useful purpose.

One of the most reliable systems of forecasting vine mildew was developed in Germany by Müller, who proposed an "incubation calendar" as early as 1913. The calendar is a guide to determining when protective spraying should be used to prevent the disease. Its success is based on conditions of temperature and moisture and how these factors influence the rate of disease development (Müller, 1936). The "incubation calendar" is reliable throughout Europe. In areas where there is much fog or dew, however, its applicability is restricted (Miller and O'Brien, 1952).

Late Blight of Potato and Tomato. Late blight of potato has probably received more attention from students of epidemiology than any other disease. Perhaps this has stemmed from the drama attached to the disease—the potato famine in Ireland in 1845 and subsequent years. But, aside from that, late blight of potato is particularly well suited to disease forecasting and warning services; moreover, it is a devastating disease of one of our major food crops.

Assuming an abundance of virulent inoculum and a large number of susceptible plants in the area, late blight will become epidemic if temperature and moisture conditions are favorable. Most potatoes are produced in regions where the temperature during the last two months of the growing season is within the favorable range of 16–21°C. Within these regions, then, the frequency of "blight years" varies with moisture conditions. As a rule, the severity of the disease is not dependent upon average rainfall over a long time, but is closely correlated with humidity at leaf surfaces. Late blight is thus most frequent in maritime regions and at high altitudes in the tropics, where there is an abundance of clouds, fog, and dew (Cox and Large, 1960). For example, severe blight is expected during roughly eight of every ten years in Penzance, in southwestern England, but only three of every ten years in Warsaw, yet both regions report annual temperatures favorable to blight and both receive approximately three inches of rain during the last two months of the growing season. The moist maritime air of Penzance, however, results in many more hours of high relative humidity than does the drier air of continental Warsaw.

Warning services for late blight have been developed rapidly ever since van Everdingen (1926) published his rules for forecasting severe outbreaks of the disease: (1) there must be dew during at least four hours at night; (2) the minimum temperature must be 10°C or higher; (3) mean cloudiness the following day must be 0.8 or more; and (4) measurable rainfall during the next 24 hours must be 0.1 mm or more. These rules were used successfully in southwest England and the Netherlands, and, as the result of a long-term study, Beaumont (1947) was able to simplify them: outbreaks of late blight can be expected 14–21 days after 48 consecutive hours when the relative humidity is 75% or more and the temperature is at least 16°C. Humidity and temperature can also be used to make accurate forecasts of late blight in the midwestern United States. Wallin (1951) made accurate experimental forecasts with the knowledge that favorable conditions for sporulation, spore germination, and infection were temperature below 24°C and relative humidity of at least 95% for ten or more hours in succession.

A reliable warning service for late blight in eastern Virginia (Cook, 1949) is based on temperature below 24°C and cumulative rainfall above the "critical rainfall line." Plots were made of cumulative rainfall against time (for the period May 8 through July 30) for blight years and for years when late blight was not severe. The critical rainfall line was "a straight median line . . . constructed between the lines for the blight and nonblight years" (Cook, 1949).

The seven diseases cited in the foregoing section are by no means the only ones for which reliable forecasts can possibly be made. Considerable research has been completed and much progress has been made in the prediction of epidemics of cereal rusts, rice blast, ergot of rye, the sugar-beet nematode, beet yellows, southern blight, blue mold of tobacco, bacterial wilt of alfalfa, Texas root-rot of cotton, and wood decay in stands of timber (see, for example, Miller and O'Brien, 1952). Clearly, the practical application of epidemiology has progressed to a point where substantial benefits have accrued to the growers of plants. Yet much remains to be done. Chester's observation that "plant pathologists are unduly timid about disease forecasting" (Chester, 1950), made almost two decades ago, remains apt today.

Perhaps this persistent timidity will yield to confidence as the plant pathologist comes better to understand the nature of epidemics in plants. The complexities of the subject are enormous, for factors relating to pathogen, suscepts, and their environment interact qualitatively and quantitatively. They affect the latent period, the duration of the infectious period, the virulence of the pathogens, and the susceptibility of the plants they attack. These effects must be comprehended if the science of plant pathology is to be understood. Van der Plank (1963) stated that "within a few years it will be taken for granted that it is as essential for a plant pathologist to be trained in epidemiology as it is for him to be taught mycology, virology, or genetics."

Literature Cited

Beaumont, A., 1947. The dependence on the weather of the dates of outbreak of potato blight epidemics. *Trans. Brit. Mycol. Soc.* 31:45–53.

Carter, W., 1930. Ecological studies of the beet leaf hopper. *U. S. Dept. Agr. Tech. Bull.* (206).

Chester, K. S., 1950. Validity and value of plant disease forecasting. *Plant Dis. Rep.* 190(Suppl.):5–8.

Cook, H. T., 1949. Forecasting late blight epiphytotics of potatoes and tomatoes. *J. Agr. Res.* 78:545–563.

Cox, A. E., and E. C. Large, 1960. Potato blight epidemics throughout the world. *U. S. Dep. Agr. Agr. Handb.* (174).

Miller, P. R., 1959. Plant disease forecasting. *In* C. S. Holton, et al., eds., *Plant Pathology—Problems and Progress, 1908–1958,* pp. 557–565. Madison, Wis.: University of Wisconsin Press.

———, and M. O'Brien, 1952. Plant disease forecasting. *Bot. Rev.* 18:547–601.

———, and ———, 1957. Prediction of plant disease epidemics. *Annu. Rev. Microbiol.* 11:77–110.

Müller, K., 1936. Die biologischen Grundlagen fur die Peronosporabekampfung nach der Inkubationskalender-Methode. *Z. Pflanzenkr.* (*Pflanzenpathol.*) *Pflanzenschutz* 46:104–108.

Stevens, N. E., 1934. Stewart's disease in relation to winter temperatures. *Plant Dis. Rep.* 18:141–149.

———, 1943. Relation of weather to the keeping quality of Massachusetts cranberries. *Mass. Agr. Exp. Sta. Bull.* (402):68–83.

van der Plank, J. E., 1959. Some epidemiological consequences of systemic infection. *In* C. S. Holton, et al., eds., *Plant Pathology—Problems and Progress, 1908–1958,* pp. 566–573. Madison, Wis.: University of Wisconsin Press.

———, 1960. Analysis of epidemics. *In* J. G. Horsfall and A. E. Dimond, eds., *Plant Pathology, an Advanced Treatise,* vol. 3 (The diseased population, epidemics and control), pp. 229–289. New York: Academic Press.

———, 1963. *Plant Diseases: Epidemics and Control.* New York: Academic Press.

———, 1965. Dynamics of epidemics of plant disease. *Science* 147:120–124.

van der Zaag, D. E., 1956. Overwintering en epidemiologie van *Phytophthora infestans,* tevens enige nieuwe bestrijdingsmogelijkheden. *Tijdschr. Plantenziekten* 62:89–156.

van Everdingen, E., 1926. Het verband tussen de weersgesteldheid en de aardappelziekte (*Phytophthora infestans*). *Tijdschr. Plantenziekten* 32:129–140.

Waggoner, P. E., 1960. Forecasting epidemics. *In* J. G. Horsfall and A. E. Dimond, eds., *Plant Pathology, an Advanced Treatise,* vol. 3 (The diseased population, epidemics and control), pp. 291–312. New York: Academic Press.

Wallin, J. R., 1951. Forecasting tomato and potato late blight in the north-central region. *Phytopathology* 41:37–38. (Abstr.)

9

The Prevention of Disease in Plants

There are four principles of plant disease control: (1) exclusion of plant pathogens, (2) eradication of plant pathogens, (3) protection of suscepts, and (4) development of resistance in suscepts. These principles were defined earlier and their interrelationship with the principles of pathogenesis have been demonstrated. It is the purpose of this chapter to present some of the facts that are the basis for the applied science of plant pathology, whose ultimate aim is the prevention of plant disease.

EXCLUSION OF PLANT PATHOGENS

The successful exclusion of a pathogen from a given geographical area depends mainly upon what knowledge may be gained of its agents of distribution. Obviously, a pathogen that is widely distributed by wind (*Puccinia graminis,* for example) cannot be excluded. Exclusion can be eminently successful, however, in the control of plant diseases whose pathogens may be distributed with the propagative parts of susceptible plants, because these parts can be made subject to treatment and inspection.

Treatment of Propagative Parts of Plants

Seed and other propagative parts may be treated with chemicals or heat to kill pathogens associated with them. Such treatments eradicate pathogens from the plant parts, but it is logical to consider them also as exclusionary measures, because they prevent the pathogens from invading new areas. Moreover, some treatments of propagative parts are also protective measures: seed dressings may protect seedlings during the short time that they are susceptible to attack by the soil-borne damping-off fungi.

Propagative parts of plants may be either infested or infected by plant pathogens. Chemical treatments of seed and other propagative parts are usually successful only as disinfestants, although organic mercury compounds kill some fungi borne within infected seed. Hot-water treatments are most successful in the control of diseases in which the seed is infected.

Seed was treated with chemicals to control certain plant diseases as early as the seventeenth century. In 1807, Prévost provided the basis for the control of bunt of wheat by developing seed treatments with copper sulfate. The copper sulfate treatment was effective against the smut fungus, but was toxic to the wheat seeds and seedlings as well. Later, formaldehyde was used to disinfest seeds (Bolley, 1891; Geuther, 1895), but it, too, was injurious to seedlings. The copper carbonate treatment of seeds to control bunt was used by von Tubeuf (1902) and later by Darnell-Smith (1917); copper carbonate soon became the most widely used seed-treatment chemical for the control of wheat bunt.

Liquid preparations of organic mercury compounds were used successfully by Riehm (1913), and organic mercury dusts became increasingly popular. Nonmetallic organic compounds, the dithiocarbamates, were patented by Tisdale and Williams (1934), and since that time many similar compounds have gained acceptance as seed dressings—for example, captan, chloranil, dexon, dichlone, and thiram. In recent years, use of some chemicals, notably the mercury compounds, has been prohibited because of environmental pollution and possible toxicity to humans and animals.

Seed dressings of many compounds fail to control those diseases in which seed is actually infected by pathogenic organisms, because the organism is inside the tissues and, thus, protected from chemical action. Jensen (1888) described a hot-water treatment for the seeds of small grains. The pathogens of loose smut of wheat and loose smut of barley were killed when the seeds were immersed for 15 minutes in water at 132°F. Special hot-water treatments have been devised to control black rot of cabbage (Clayton, 1929) and virus diseases of sugar cane.

A major breakthrough in the control of seed-borne pathogens came with the discovery by von Schmeling and Kulka (1966) that derivatives of oxathiin (5,6-dihydro-2-methyl-1,4-oxathiin-3-carboxanalide) are effective for seed treatment (Erwin, 1970). These chemicals (Vitavax, carboxin, DMOC) act systemically—that is, they are absorbed by the seed and translocated to the point of infection—and are particularly effective against the seed-borne cereal smuts.

Inspection of Propagative Materials and Elimination of Plant Pathogens

Certification of Propagative Materials. The voluntary inspection of fields in which seed and other propagative materials are produced for commercial use has led to programs of certification, which insure the marketing of materials that do not harbor plant pathogens. Inspection and certification programs have been particularly successful in the control of viral diseases of potatoes and citrus, smut diseases of small grains, nematode-induced diseases of woody ornamentals and fruits, and viral, bacterial, and fungal diseases of beans.

Quarantines. Plant **quarantine** is the legally forced stoppage of shipments against the possibility that the materials being shipped are carrying pathogens that are potentially dangerous in the area of import. Inspections at ports of entry are theoretically effective against plant pathogens that may be transported in or on plants or plant parts or their containers. In general, quarantines have been most effective against the introduction of viral, bacterial, and nematode-induced diseases. The subject of quarantine as an exclusionary measure for the control of plant disease has been reviewed by McCubbin (1946), and Hunt (1946) has discussed important plant diseases that have not yet been found in the United States.

ERADICATION OF PLANT PATHOGENS

Plant pathogens may be eradicated from areas in which they have become established by the eradication of susceptible plants, by crop rotation with insusceptible species, or by heat or chemical treatments of the soil or of the susceptible plants themselves.

Eradication of Susceptible Plants

The eradication of certain plant pathogens can be accomplished through the eradication of the susceptible plant of chief economic importance or through the eradication of alternative or alternate hosts. An **alternative host** is a plant species, other than the one of chief economic importance, in which a given parasite may live, whereas an **alternate host** is a species that is indispensible to the parasite (specifically, to a heteroecious rust parasite) for completion of its life cycle: the pycnial and aecial stages of heteroecious rust fungi develop in alternate hosts. The eradication from Florida of *Xanthomonas citri*, the cause of citrus canker, is a striking example of plant-disease control through the eradication of economically important suscepts. Apparently, all diseased trees were destroyed, beginning in 1915, and subsequent quarantine measures have proved successful in preventing the reintroduction of the pathogen. A number of viral diseases, notably cucumber mosaic (Doolittle and Walker, 1926) and southern celery mosaic (Wellman, 1937), are controlled by eradication of alternative hosts. The control of rust diseases through the eradication of alternate hosts has been discussed by Fulling (1943). The elimination of the uredial and telial host would insure the eradication of a rust pathogen, because the uredial stage is the repeating one and the telial stage is often the surviving one. Unfortunately, however, many important crop plants are hosts for these two stages. In the control of *Puccinia graminis*, the elimination of barberry is only partially successful, because the uredia overwinter in the southern United States and Mexico and provide inoculum for primary cycles in wheat the following spring. The eradication of the pycnial and aecial hosts of rust fungi sometimes gives a degree of disease control in the uredial and telial host, and it results in a reduction of the number of pathologic races that develop in a given period of time.

Crop Rotation

In a sense, crop rotation is eradication of the chief and alternative hosts. Many plant pathogens survive for only a short time in the absence of susceptible weeds or crop plants. Some of these pathogens can survive for a time as saprophytes in decaying crop residues, but most are unable to live indefinitely in the soil. Such soil invaders are subject to eradication by starvation, which is best accomplished through crop rotation. True soil inhabitants, however, are not easily controlled by such measures.

Crop rotation is not always feasible, however, because annual cash income from a single crop may often be essential to the economic well-being of certain farmers. This is true not only for wheat farmers in the United States but for rice growers in Southeast Asia as well. However, a single annual plant species, grown year after year in the same soil, is vulnerable to periodic outbreaks of devastating plant diseases, particularly those caused by nematodes or those caused by root-invading fungi—specifically, those facultative saprophytes that cause root rots and vascular wilts. Such diseases may be controlled in monocultures by planting resistant varieties and, sometimes, by chemically treating the infested soil; but when resistant varieties are not available or prove to be only mildly resistant, and when chemical treatment of the soil is not feasible, crop rotation remains the best possible control measure.

Heat and Chemical Treatments

Treatments of Soil and Other Sources of Inoculum. Plant pathogens can be eradicated from soil by dry heat, steam or hot water, flooding, and fumigation with chemicals. Heat and chemical treatments have been used with much success in the eradication of plant pathogens from greenhouse soils and soils in plant beds; treatments of soils in large fields have not been so effective. It is possible to eradicate nematodes in the field by fumigation with a dichloropropane-dichloropropene (DD) mixture and with other nematicides. The treatment of soil in the furrow with pentachloronitrobenzene (PCNB) at planting time has proved effective in the eradication of certain sclerotia-producing soil-borne pathogens. Formaldehyde or captan applied in the row are effective in controlling the onion-smut disease. The literature on the subject of soil disinfestation has been reviewed by Newhall (1955).

Sources of primary inoculum other than soil may be treated to eradicate plant pathogens. For example, *Venturia inaequalis* can be eradicated from fallen apple leaves by eradicative spraying (Keitt et al., 1941), and *Taphrina deformans* can be eradicated from dormant peach trees by autumn or spring applications of an eradicative fungicide (Wallace and Whetzel, 1910).

Treatment of Diseased Plants During the Growing Season. Some plant viruses can be eradicated from growing plants by heat treatments (Bawden, 1964), and chemical treatments of growing plants are known to eliminate certain plant pathogens. The apple-scab fungus can be destroyed after infection occurs by applying an eradicative fungicide (Keitt and Palmiter, 1937), and the crown-gall bacterium can be killed *in vivo* by applying an eradicative fungicide containing sodium 4,6-dinitro-*o*-cresoxide (Ark, 1941). Many modern fungicides used for the control of diseases—for example, dodine for apple scab—are able to "burn out" infected leaf areas as well as to protect healthy foliage from attack by the pathogen (Weaver, 1958).

Eradicative fungicides must be applied in the correct proportions, because they may be phytotoxic in concentrations high enough to kill the pathogen within plant tissues. The chemical treatment of infections is termed **chemotherapy** (Horsfall, 1956), and it has been used successfully in the control of certain local infections (including certain leafspot, powdery-mildew, and loose-smut diseases) and in the control of wilt diseases in which the pathogen develops within the vascular system of the plant. A great deal of attention is currently being given to a group of chemicals that are distributed systemically in plants grown from treated seed, in plants grown in treated soil, or in plants whose foliage is sprayed directly (Erwin, 1970). These chemicals—benomyl, for example—may destroy pathogens already established in the plant, or provide protection against invasion by pathogens over an extended period of time.

Biological Eradication of Plant Pathogens

It has long been known that the presence of one organism in the soil may strikingly influence the growth of another. Soil-borne organisms compete for substrate, and plant-pathogenic soil invaders are at a disadvantage when forced to compete with aggressively saprophytic soil inhabitants. It is possible that the control of plant diseases that results from green manuring is a consequence of having improved the environment for saprophytic soil inhabitants, thereby making them better able to compete with soil invaders. Soil inhabitants not only compete with soil invaders for substrate, but some, such as *Trichoderma lignorum* (Weindling, 1946), actually parasitize other fungi. Further, certain soil-borne organisms produce antibiotics that impede the rate of growth of, or even kill, nearby plant pathogens. Some fungi are known that trap and feed upon nematodes. Although a considerable amount of biological eradication occurs in nature, little success has been achieved in this area of plant-disease control on a commercial scale. The possibilities for biological control richly deserve more attention (Wood and Tveit, 1955).

PROTECTION OF SUSCEPTIBLE PLANTS

Protective Spraying and Dusting

Spraying or dusting plants as a protective measure is a familiar method of plant-disease control, and has received much attention. It should be kept clearly in mind that protective spraying or dusting is effective only when circumstances permit the deposition of fungicidal or bactericidal chemicals in the infection courts without doing harm to the plants under treatment. This requires that the fungicidal deposit be relative insoluble in water so that its phytotoxic components will not rapidly enter plant tissues. Good protective fungicides are slightly soluble in the water in the infection court, which makes their active principles available to pathogenic fungi during the prepenetration stage. Thus, the fungus can be destroyed in the infection court prior to ingress. Many of the most important plant diseases (McCallan, 1946) are best controlled by protective spraying or dusting, and millions of dollars are spent annually to protect crop plants by this method. The safety of fungicides to man is discussed in Box 9-1.

Most protective fungicides are either inorganic copper or sulfur compounds, or organic compounds containing chlorine, the carbamate radical, quinones and other ketones, or heterocyclic groups. Although the fungitoxicity of copper was recognized in the eighteenth century, it was not until 1885, when Millardet published his findings on Bordeaux mixture, that a satisfactory inorganic copper-containing protective fungicide was developed. When copper sulfate, slaked lime, and water are mixed, a copper complex of low solubility is formed. When deposited on foliage, it adheres after drying, and cupric ions are freed slowly when water (from dew, rain, and so forth) comes in contact with the leaves. Because Bordeaux mixture must be solubilized before it can release fungitoxic ions, it is most effective in protecting against plant pathogens whose spores require free water for germination. This holds true for most protective fungicides. Sulfur fungicides volatilize, however, and are therefore not dependent upon solubilization by water; because of their volatility, they are effective in the control of diseases caused by powdery mildew and other fungi with spores that germinate well in the absence of free water.

Although copper- and sulfur-containing fungicides are still widely used in programs of protective spraying, they are being replaced, in many of their former applications, by organic fungicides. Some of the more popular ones are captan, zineb, nabam, ferbam, maneb, and

BOX 9-1
FUNGICIDES AND THE ENVIRONMENT

The fact and fiction of man's polluting influence on his environment are so confused with each other that it is often hard to tell where sense leaves off and nonsense begins. Certainly, man has polluted his soil, his air, and his streams, lakes, and oceans; but he has, however belatedly, shown the good sense to recognize and to rectify at least some of his past mistakes.

We comment in Chapter 23 on the direct effects of certain industrial and agricultural pollutants upon crop plants, but it is beyond the scope of this book to treat the subject of man's use and abuse of his environment. Rather, we wish to focus attention here to a single component born of agricultural technology: chemicals, mostly fungicides, that are used to control plant diseases. Assuming their effectiveness for their intended purposes, we are concerned here with their safety to the environment, which means, in the end, their safety to mankind.

Fortunately, most chemicals used to control plant disease are relatively safe to use, but a few (for example, mercury compounds and halogenated hydrocarbons) are highly toxic to man. Moreover, many of the relatively nontoxic fungicides are mixed with highly toxic insecticides before they are applied to plants. Consequently, we urge everyone to treat all disease-control chemicals as if they were deadly poisons. Don't spill them on the skin, don't inhale their fumes, and, above all, don't swallow them. People who use agricultural chemicals should handle them carefully, measure them accurately, and apply them correctly.

Chemicals for the control of plant diseases are either eradicative or protective, and some—the newly developed systemic fungicides —may act eradicatively as well as protectively. Those used to eradicate plant pathogens from the soil or from seed are usually more toxic to man and other animals than those used in protective applications. The generally acknowledged toxicities of several fungicides, bactericides, and nematicides to rats are listed in the table below, along with comparable figures for several familiar insecticides and herbicides. The toxicity of each compound is expressed as the median lethal dose (LD_{50})—that is, the number of milligrams of toxicant per kilogram of body weight required to kill 50% of the test animals (laboratory rats).

Pesticide	Use	Toxicity to rats (LD_{50}, mg/kg body weight)	
		Oral	Dermal
Benomyl	Fungicide	9590	—[a]
Captan	Fungicide	9000–15,000	—[a]
Fixed copper	Fungicide	3000–6000	—[a]
Maneb	Fungicide	6750–7500	4444
Plantvax	Fungicide	3200–3820	8000
Streptomycin	Bactericide	9000	600
Sulfur	Fungicide	17,000	—[b]
D-D mixture	Nematicide	140	2100
2,4-D	Herbicide	375–800	800–1500
Dieldrin	Insecticide	46	60–90
Methoxychlor	Insecticide	5000	76,000
Parathion	Insecticide	4–13	7–21
Rotenone	Insecticide	50–75	7940

Source: Pesticide Conference, Ithaca, N. Y., Nov. 8–11, 1971.
[a] Nontoxic at concentrations tested.
[b] Slightly toxic.

Inherent mammalian toxicity, however, is not the only criterion for the safety of an agricultural chemical to man and the environment. No matter how low the level of toxicity, a chemical may endanger the environment if it persists and accumulates. For example, whatever danger there might be in the continued widespread use of the insecticide DDT derives from the fact that it is persistent and it accumulates, at least to a certain level, in adipose tissues of animals. Whether or not the accumulation of such chemicals has any direct adverse effects upon the health of humans remains a controversial question.

We believe that scientists have a responsibility beyond proving the agricultural effectiveness of chemical compounds—a responsibility for learning, to the limits of their talents and techniques, the facts about the impact of agricultural chemicals upon man and his environment. By assuming this basic responsibility, scientists can pave the way for the controlled use of effective agricultural chemicals, and can prevent the necessity of banning, through ignorance, first-rate pesticides of great value to man in his struggle against hunger.

dichlone. Until recently, fungicides for plant-disease control have been successful only in topical applications. The oxathiins, however, can be used as systemic fungicides to control certain cereal smuts and other plant diseases (von Schmeling and Kulka, 1966). The chemical names of the active ingredients of various protective and eradicative fungicides have been listed by McCallan et al. (1955) and Sharvelle (1961).

Fungicides kill spores by altering the normal metabolism of the fungi. Much remains to be learned about the modes of fungicidal action, although there is a wealth of information in the literature, as reviewed by Horsfall (1956), that provides a basis for further study.

Factors affecting the efficiency of a protective fungicide include its inherent toxicity, its ability to permeate fungal spores, its tenacity, its spreading and covering ability, and its ability to be redistributed while weathering on the protected plant (Horsfall, 1956). It is axiomatic that, in order to protect growing plants, spraying with fungicides must be repeated frequently enough to insure coverage of new growth before it is attacked.

Some plant diseases are so markedly influenced by weather that it is possible to forecast epidemics. Conversely, disease-free periods can sometimes be forecast. When this is possible, the number of protective application required for control of leaf diseases may be reduced with considerable saving to the grower.

Protection Through Cultural Practices

Plants may be protected from disease through cultural practices that neither exclude nor eradicate plant pathogens but that so affect the growth of the plant that its escape from disease is facilitated. For example, because plants are susceptible to damping-off only in the seedling stage, any treatment that will induce early emergence and early maturation of stem tissue will protect the young plants. The seedling blights of wheat and corn are caused by the same fungus, but the disease in corn is favored by cool weather, whereas its development in wheat is favored by warm weather (Dickson, 1923); weather conditions that favor disease are those that are unfavorable for early maturation stem tissues. Seeds of most crops should be planted as shallowly as possible, after the soil has become warm in the spring, and fertilizer rich in phosphorus should be applied to hasten the maturation of tissues. Such treatments are rarely sufficient to insure satisfactory control, but they ameliorate the severity of disease and make for increased effectiveness of other control measures.

Those diseases caused by viruses, bacteria, and fungi that have insects as their agents of inoculation are often controlled along with control of the insects. Stewart's wilt of sweet corn is controlled by applications of an insecticide that reduces the population of flea beetles, and some persistent viruses are eliminated upon the destruction of their vectors.

DEVELOPMENT OF RESISTANCE IN PLANTS

Not all plants within a species are equally vulnerable to attack by a given pathogen, and the susceptibility or resistance possessed by a plant is heritable. These facts form the basis for the development of resistance, and for the use of immune or resistant varieties of crop plants.

Breeding and Selection of Resistant Varieties

Although there is evidence that resistance of plants to disease may be enhanced by ionizing radiation (Konzak, 1955) or by internal chemical treatments that alter metabolism (Horsfall, 1956), all resistant varieties now used in commercial agriculture have been developed through selection and breeding for resistance. Biffen (1905) provided the genetic basis for resistance to disease in plants when he reported that the inheritance in wheat of susceptibility to attack by *Puccinia glumarum* follows a simple Mendelian ratio for a single gene factor for dominance of susceptibility. The widespread use of disease-resistant food crops offers considerable hope for success in man's continuing war against hunger in the most populous regions of the world (see Box 9-2).

Morphological Resistance. This kind of resistance usually implies resistance to invasion, and results from the occurrence in the plant of some structure that prevents the ingress of a pathogen. For some plants, the thickness of the skin of fruit or tubers is positively correlated with resistance. In potato tubers of certain resistant strains, the scab pathogen is unable to establish itself because of the early formation of wound periderm, which walls off the wounded area containing the bacteria. Scab resistance, apparently, is correlated with the degree of corking of the lenticel cambium, and the ability to form cork cells varies from one variety of potato to another (Longree, 1931). The seed of certain varieties of flax split more readily than do those of others, and are thus more susceptible to seed decay; the tendency for flax seeds to split is heritable (Kommedahl and Culbertson, 1954). Mature plants are more resistant to certain diseases than young plants, presumably because of some morphological feature that protects older plants from invasion. Mature wheat plants are somewhat resistant to

BOX 9-2
MIRACLE FOOD GRAINS

The challenge of world hunger is being met, however slowly, on many fronts—economic, sociological, industrial, and agricultural. The plant scientist's role is one of breeding and selecting high-yielding, disease-resistant, nutritious varieties of food crops that are well adapted to the soils and climates of those populous regions in which food production is presently inadequate to the needs of the people. There are two programs that deserve special acclaim—the programs that have led to the development of the so-called miracle wheat and miracle rice varieties.

High-yielding, stiff-strawed varieties of wheat have recently been released by the Rockefeller Foundation as a result of its long-term program of agricultural research in Mexico. For his work as project director, Dr. Norman E. Borlaug, agronomic plant breeder and pathologist, was awarded the Nobel Prize for Peace in 1970. Thus far, the new varieties of wheat have been grown successfully in several places outside of Mexico, notably in India.

The International Rice Research Institute, located in the Republic of the Philippines, has recently released varieties of high-yielding rice that are resistant to several diseases. These varieties were developed through the cooperative efforts of geneticists, plant physiologists, plant pathologists, entomologists, soil scientists, and agronomists.

Existing economic and social problems of world-wide poverty and hunger lie far beyond the scope of this book. Moreover, we cannot predict to what extent the hunger problem might be alleviated, nor can we foresee in what ways miracle food grains might create unexpected problems. From the standpoint of plant pathology, however, a word of caution seems advisable.

rust (Hart, 1931), and mature corn plants are more resistant than young ones to bacterial wilt (Elliott, 1942).

Functional Resistance. The functioning of plants may confer resistance to invasion, at least in one plant disease: those varieties of wheat in which the stomata open late in the morning and close early in the

Until newly developed disease-resistant varieties have stood the test of time in a given region, we cannot be certain that they will survive the onslaught of plant pathogens. There are two explanations, both of them based on the differences between the populations of plant pathogens in the regions in which the new varieties were developed and those in the regions in which such varieties are planted: First, certain races of pathogens from outside the regions of development may be virulent to the new varieties. Second, agents of entirely different diseases—diseases to which the new varieties could be susceptible—may be present in those regions into which the new varieties are introduced. For example, rust-resistant 'Victoria' oats, developed in Australia, could not be grown in the United States because of Victoria blight, a disease caused by the fungus *Helminthosporium victoriae,* to which 'Victoria' oats had not been exposed until grown in the United States.

We do not intend to imply that "miracle crops" should have their distribution restricted to regions in which field-testing has been completed. Instead, we urge their widespread use for as long as possible, and encourage the authorities to expend all energies to support the establishment and maintenance of agricultural research and testing stations in all regions of the world. The challenge of world poverty and hunger, in the long run, must be met head-on by the people of the impoverished countries themselves (Myrdal, 1970). The goal of the agricultural scientists of affluent nations, it seems, should be to assist the people of less fortunate countries in their efforts to solve their own problems for themselves. Certainly, the recently released miracle food grains are a giant step in this direction.

afternoon resist stem rust (Hart, 1929). Apparently, the germination tube of the fungus enters closed stomata only with difficulty; further, the dry conditions in the infection court during the middle of a summer day are not favorable for development of the inoculum.

Protoplasmic Resistance. A plant possessing resistance of this kind is able to ward off infection by virtue of some property of the protoplast that prevents a pathogen from establishing a food relation with its tissues. Immune plants are entered by pathogens as easily as susceptible ones, but the protoplast of an immune plant is refractive to the activities of the pathogen. The biochemical nature of protoplasmic resistance is not understood. Many attempts have been made to correlate the resistance of plants with the occurrence of specific chemicals in them, but virtually all have been unsuccessful. The one notable exception was the discovery that varieties of onions with brown outer scales resist the smudge disease, owing to the presence of protocatechuic acid (Link et al., 1929) and catechol (Link and Walker, 1933).

Whatever the chemistry of protoplasmic resistance, it is certain that such resistance is heritable. The inheritance of resistance may be monogenic or polygenic, depending on the host and the parasite, and the resistance may be dominant or recessive. As a general rule, monogenic inheritance of resistance frequently occurs when there is a highly specialized or obligate food relation between parasite and host, whereas polygenic inheritance of resistance seems to occur most often when the attacking organism is a faculative saprophyte. Van der Plank (1963, 1968) uses the terms vertical and horizontal resistance. In **vertical resistance,** a variety is resistant to certain races of a pathogen, but susceptible to others. In **horizontal resistance,** the variety is resistant to most or all races of the pathogen. The genetics of resistance to disease in plants has been reviewed by Wingard (1941).

Pathogenic Variations Within Species of Plant Pathogens

As green plants of a given species vary in susceptibility, so plant pathogens vary in virulence. Such variation in pathogens remains undetected as long as the plants under attack are highly susceptible. Resistant plants in the population, however, are subject to attack only by those pathogens that are more virulent than the common ones of their species. Thus, when disease-resistant varieties are grown, a substrate for exceptionally virulent members of the pathogen population is provided. Over the years, as a greater number of resistant crop plants were grown, a greater number of virulent pathogen types became evident, and—at least in some instances—the use of resistant varieties in the control of plant disease has become a race between the plant breeders and the plant pathogens.

A pure line of an organism is a **strain** (Christensen and Rodenhiser, 1940; Reed, 1935), and when a strain of pathogen differs from other strains of the same species only in its virulence, it is either a special form or a pathologic race. **Special forms** are strains that differ from others in their pathogenicity to different *species* of susceptible plants; **pathologic** (or pathogenic) **races** are strains that differ in pathogenicity to different *varieties* within a susceptible plant species. Special forms are given latinized names; races, though sometimes designated by letters, are usually númbered.

Strains of Plant Viruses. Strains of plant viruses were discussed briefly in Chapter 3. Viral strains arise as mutants of "parent" strains; strains of a given virus are serologically and immunologically related, and differ among themselves in their host range, in the symptoms they induce in host plants, and their vector specificity. Thus, the criteria for determining strains of plant viruses are similar to, but not exactly the same as, those for determining special forms and pathogenic races of plant-parasitic bacteria and fungi. Because of the highly specialized metabolic relationship between plant viruses and their hosts, varietal resistance, when it occurs, if of the highest order; complete immunity is not uncommon

Races of Plant-Pathogenic Bacteria. Strains of many plant-pathogenic bacteria differ among themselves in their virulence to different species of susceptible plants and in their virulence to different varieties within a susceptible species (Burkholder, 1948). Variations in pathogenicity are evidence that races of bacteria also differ biochemically; the biochemical differences thus far noted, however, are not sufficient to provide an explanation of the observed differences in pathogenicity. Similarly, serological differences among bacterial strains are not necessarily correlated with pathogenic differences. *Xanthomonas barbarae,* which infects winter cress, and *X. campestris,* which infects cabbage, are serologically alike, but neither infects the host for the other (Elrod and Braun, 1947). *X. phaseoli* and *X. vignicola* differ serologically, but induce the same symptom syndrome in beans; *X. vignicola,* however, produces cankers in stems of cowpeas, whereas *X. phaseoli* does not (Burkholder, 1948). Much more sensitive methods of making chemical

BOX 9-3
THE GENE-FOR-GENE HYPOTHESIS

Disease resistance in plants may be monogenic (under the control of a single gene) or it may be polygenic (conferred by the interactions of the effects of several genes). Whereas polygenic resistance is often expressed in reduced numbers of disease lesions and may be influenced by environmental factors, monogenic resistance usually results in decreased size of lesions and is relatively unaffected by the environment. Monogenic resistance is thus absolute, at least under normal field conditions, and is frequently expressed by the hypersensitive reaction, a reaction of the plant to the pathogen that results in little or no spread of the pathogen in the tissues and, therefore, only a minor spotting or flecking.

The gene-for-gene hypothesis, based on Flor's (1946) classical experiments with flax rust, states that, for every gene for virulence in the pathogen, there is a corresponding gene for susceptibility in the susceptible plant species (Flor, 1971). Experience with diseases other than flax rust suggests the general truth of the hypothesis, at least for parasitic pathogens and monogenically resistant plant varieties that react hypersensitively.

The hypothesis, in its simplest form, is illustrated by the following table, in which *A* represents the dominant gene for avirulence (nonpathogenicity) and *a* the recessive gene for virulence (pathogenicity) in the pathogen, and where *R* represents the dominant gene for resistance and *r* the recessive gene for susceptibility in the attacked plant.

Genes of the pathogen	Genes in the plant: *R*	Genes in the plant: *r*
A	*A–R* (Resistant)	*A–r* (Susceptible)
a	*a–R* (Susceptible)	*a–r* (Susceptible)

According to the gene-for-gene hypothesis, only one of the possible gene combinations above, *A–R*, would result in resistance (hypersensitivity). In all other gene combinations, the susceptible reaction would occur because the host plant is susceptible (*r*), the parasite is virulent (*a*), or both conditions (*a–r*) are fulfilled in the same host–parasite interaction.

The interactions between corresponding pairs of genes (the A genes of the parasite and the R genes of the host) are highly specific. This significant aspect of the gene-for-gene concept can be illustrated only by considering interactions in which the corresponding pairs of genes occur at two or more different loci on the chromosomes. With two loci, four different gene combinations are possible. Notations for genes at two loci in the parasite would be A_1A_2, A_1a_2, a_1A_2, and a_1a_2. For corresponding genes at two loci in the host, the notations would be R_1R_2, R_1r_2, r_1R_2, and r_1r_2. All possible interactions between corresponding pairs of genes, along with the disease reaction that would result from each interaction, are set forth in the table below.

Disease reactions following interactions between corresponding genes at two different loci

Genes of the parasitic pathogen	Genes of the host plant	Disease reaction
A_1A_2	R_1R_2	Resistant
A_1A_2	R_1r_2	Resistant
A_1A_2	r_1R_2	Resistant
A_1A_2	r_1r_2	Susceptible
A_1a_2	R_1R_2	Resistant
A_1a_2	R_1r_2	Resistant
A_1a_2	r_1R_2	Susceptible
A_1a_2	r_1r_2	Susceptible
a_1A_2	R_1R_2	Resistant
a_1A_2	R_1r_2	Susceptible
a_1A_2	r_1R_2	Resistant
a_1A_2	r_1r_2	Susceptible
a_1a_2	R_1R_2	Susceptible
a_1a_2	R_1r_2	Susceptible
a_1a_2	r_1R_2	Susceptible
a_1a_2	r_1r_2	Susceptible

Because of the specificity of interaction, resistance (hypersensitivity) is expressed only when combinations A_1–R_1 or A_2–R_2 occur in the same parasite–host system. That is, A_1 "recognizes" only R_1, and A_2 "recognizes" only R_2.

Finally, it is significant that hypersensitivity ensues when but a single gene for avirulence (A) at a given locus (x) in the parasite interacts with a single gene for resistance (R) at the corresponding locus (x) in the host. The A_x–R_x interaction thus overrides any reaction between corresponding genes at other loci.

and serological determinations are required before variations in pathogenicity among bacteria can be correlated with biochemical and serological differences.

Races of Plant-Pathogenic Fungi. Pathogenic variation within a species of fungus was first demonstrated by Eriksson (1894), who found that strains of *Puccinia graminis* from one host species did not usually infect other related species of small grains. Stakman (1914) proved later that each of the special forms of *P. graminis* described by Eriksson consists of several races, each of which is specialized in its pathogenicity to a variety of host species. The existence of pathogenic races, as races are presently defined, was also demonstrated by Barrus (1911), who designated strains within the bean-anthracnose pathogen. Since that time, many plant pathogens—bacteria, fungi, nematodes, and viruses—have been shown to exist in several pathogenic races.

The determination of pathogenic races is based on the reactions to attack, by different races of a parasite, of a number of arbitrarily selected host varieties. With *Puccinia graminis,* for example, twelve different varieties of wheat are inoculated with an isolate of the fungus, and six disease reactions (ranging in intensity from complete immunity to complete susceptibility) are possible in each variety. Reaction by a variety is always specific for a given race; therefore, identification of the unknown race may be made after observing all of the wheat varieties inoculated. Roughly 250 races of the wheat-attacking form of *P. graminis* have been identified.

From results of extensive studies of the flax-rust pathogen, *Melampsora lini,* Flor (1946) concluded that for every pair of genetic factors governing the pathologic reaction of the suscept, there is a corresponding pair of factors in the pathogen that governs its virulence (see Box 9-3). Genetic differences among pathogenic races are certainly responsible for biochemical differences that can be correlated with virulence, but there is, as yet, no adequate proof of this. The various races of *Phytophthora infestans* differ in their ability to use organic nitrogen as substrates (French, 1953); the enzymatic reactions controlling their ability to utilize such substances as aspartic acid and asparagine may or may not be governed by the same genes that govern their virulence.

Organisms that reproduce sexually develop new races as a result of segregation and recombination during **hybridization** (Christensen and Rodenhiser, 1940). The majority of races of rust and other fungi probably have arisen as a result of hybridization, but races can also develop, particularly in imperfect fungi, by **mutations,** which are sudden, permanent changes in the genetic material of an organism. [New races of viruses (Bawden, 1964) and bacteria (McNew, 1948) arise as mutants, but the possibility of genetic segregation and recombination in these pathogens deserves further attention.] Reddick and Mills (1938) reported that the virulence of *Phytophthora infestans* could be increased by serial passage of the fungus through resistant varieties of potato. This is evidence for the origin of races by **adaptation**; this controversial subject has been reviewed by Christensen and DeVay (1955). In certain imperfect fungi, races arise through **heterocaryosis,** in which individual spores or cells of a fungus contain at least two distinct kinds of nuclei (Buxton, 1956; Hansen, 1938). The reassortment of kinds of nuclei in *Botrytis cinerea* occurs through hyphal anastomoses or through unequal cell divisions (Hansen and Smith, 1932). A final method by which new races may arise is **parasexualism,** in which sexual fusion and meiosis occurs in heterocaryotic, undifferentiated hyphae (Buxton, 1956; Pontecorvo and Sermonti, 1954; Pontecorvo et al., 1953).

Literature Cited

Ark, P. A., 1941. Chemical eradication of crown gall on almond trees. *Phytopathology* 31:956–957.

Bawden, F. C., 1964. *Plant Viruses and Virus Diseases* (4th ed.). New York: Ronald Press.

Barrus, M. F., 1911. Variations of varieties of beans in their susceptibility to anthracnose. *Phytopathology* 1:190–195.

Biffen, R. H., 1905. Mendel's laws of inheritance and wheat breeding. *J. Agr. Sci.* 1:4–48.

Bolley, H. L., 1891. Grain smuts. *N. Dak. Agr. Exp. Sta. Bull.* (1).

Burkholder, W. H., 1948. Bacteria as plant pathogens. *Annu. Rev. Microbiol.* 1948:389–412.

Buxton, E. W., 1956. Heterokaryosis and parasexual recombination in pathogenic strains of *Fusarium oxysporum*. *J. Gen. Microbiol.* 15:133–139.

Christensen, J. J., and J. E. DeVay, 1955. Adaptation of plant pathogen to host. *Annu. Rev. Plant Physiol.* 6:367–392.

———, and H. A. Rodenhiser, 1940. Physiologic specialization and genetics of the smut fungi. *Bot. Rev.* 6:389–425.

Clayton, E. E., 1929. Studies of the black-rot or blight disease of cauliflower. *N. Y. Agr. Exp. Sta. Geneva Bull.* (576).

Darnell-Smith, G. P., 1917. The prevention of bunt. Experiments with various fungicides. *Agr. Gaz. N. S. W.* 28:185–189.

Dickson, J. G., 1923. Influence of soil temperature and moisture on the development of the seedling-blight of wheat and corn caused by *Gibberella saubinetii. J. Agr. Res.* 23:837–870.

Doolittle, S. P., and M. N. Walker, 1926. Control of cucumber mosaic by eradication of wild host plants. *U. S. Dep. Agr. Bull.* (1461).

Elliott, C., 1942. Bacterial wilt of dent corn inbreds. *Phytopathology* 32: 262–265.

Elrod, R. P., and A. C. Braun, 1947. Serological studies of the genus *Xanthomonas,* III. The *Xanthomonas vascularum* and *Xanthomonas phaseoli* groups; the intermediate position of *Xanthomonas campestris. J. Bacteriol.* 54:349–357.

Eriksson, J., 1894. Über die Specialisirung des Parasitismus bei den Getreiderostpilzen. *Ber. Deut. Bot. Ges.* 12:292–391.

Erwin, D. C., 1970. Progress in the development of systemic fungitoxic chemicals for control of plant diseases. *FAO* (*Food Agr. Organ. U. N.*) *Plant Prot. Bull.* 18:73–82.

Flor, H. H., 1946. Genetics of pathogenicity in *Melampsora lini. J. Agr. Res.* 73:335–357.

———, 1971. Current status of the gene-for-gene concept. *Annu. Rev. Phytopathol.* 9:275–296.

French, A. M., 1953. Physiologic differences between two physiologic races of *Phytophthora infestans. Phytopathology* 43:513–516.

Fulling, E. H., 1943. Plant life and the law of man, IV. Barberry, currant and gooseberry, and cedar control. *Bot. Rev.* 9:483–592.

Geuther, T., 1895. Über die Einwirkung des Formaldehyde-lösungen auf Getreidebrand. *Ber. Pharmakol. Ges.* 5:325–329.

Hansen, H. N., 1938. The dual phenomenon in imperfect fungi. *Mycologia* 30:442–455.

———, and R. E. Smith, 1932. The mechanism of variation in imperfect fungi: *Botrytis cinerea. Phytopathology* 22:953–964.

Hart, H., 1929. Relation of stomatal behavior to stem-rust resistance in wheat. *J. Agr. Res.* 39:929–948.

———, 1931. Morphologic and physiologic studies on stem-rust resistance in cereals. *U. S. Dep. Agr. Tech. Bull.* (266).

Horsfall, J. G., 1956. *Principles of Fungicidal Action.* Waltham, Mass.: Chronica Botanica.

Hunt, N. R., 1946. Destructive plant diseases not yet established in North America. *Bot. Rev.* 12:593–627.

Jensen, J. L., 1888. The propagation and prevention of smut in oats and barley. *J. Roy. Agr. Soc.* (2)24:397–415.

Keitt, G. W., C. N. Clayton, and M. H. Langford, 1941. Experiments with eradicant fungicides for combating apple scab. *Phytopathology* 31: 296–322.

———, and L. K. Jones, 1926. Studies of the epidemiology and control of apple scab. *Wis. Agr. Exp. Sta. Res. Bull.* (73).

———, and D. H. Palmiter, 1937. Potentialities of eradicant fungicides for combating apple scab and some other plant disease *J. Agr. Res.* 55:397–437.

Kommedahl, T., and J. O. Culbertson, 1954. Comparative seed damage within paired isogenic lines of flax. *Phytopathology* 44:495. (Abstr.)

Konzak, C. F., 1954. Stem rust resistance in oats induced by nuclear radiation. *Agron. J.* 46:538–540.

Large, E. E., 1953. Potato blight forecasting investigations in England and Wales, 1950–52. *Plant Pathol.* 2:1–15.

Link, K. P., H. R. Angell, and J. C. Walker, 1929. The isolation of protocatechuic acid from pigmented onion scales and its significance in relation to disease resistance in onion. *J. Biol. Chem.* 81:369–375.

———, and J. C. Walker, 1933. The isolation of catechol from pigmented onion scales and its significance in relation to disease resistance in onions. *J. Biol. Chem.* 100:379–383.

Longree, K., 1931. Untersuchungen über die Ursache des verschiedenen verhaltens Kartoffelsorten gegen Schorf. *Arb. Biol. Reichanst. Land-Forstwirt.* 19:285–336.

McCallan, S. E. A., 1946. Outstanding diseases of agricultural crops and uses of fungicides in the United States. *Contrib. Boyce Thompson Inst. Plant Res.* 14:105–115.

———, L. P. Miller, and M. A. Magill, 1955. Chemical names for active ingredients of fungicides. *Phytopathology* 45:295–302.

McCubbin, W. A., 1946. Preventing plant disease introduction. *Bot. Rev.* 12:101–139.

McNew, G. L., 1938. The relation of nitrogen nutrition to virulence in *Phytomonas stewarti. Phytopathology* 28:769–787.

Millardet, A., 1885. De l'action des melanges de sulfate de cuivre et de chaux sur le mildiou. *C. R. Hebd. Seances Acad. Sci.* (*Paris*) 101:929–932.

Myrdal, G., 1970. *The Challenge of World Poverty: A World Anti-Poverty Program in Outline.* New York: Pantheon.

Newhall, A. G., 1955. Disinfestation of soil by heat, flooding, and fumigation. *Bot. Rev.* 21:189–250.

Pesticide Conference, Ithaca, N. Y., Nov. 8–11, 1971. Toxicity ratings of recommended pesticides. In *New York State Insecticide, Fungicide, and Herbicide Recommendations,* pp. 3–7. [Thirty-third Annual Report]. Ithaca, N. Y.: Cornell University.

Pontecorvo, G., J. A. Roper, and E. Forbes, 1953. Genetic recombination without sexual reproduction in *Aspergillus niger*. *J. Gen. Microbiol.* 8:198–210.

———, and G. Sermonti, 1954. Parasexual recombination in *Penicillium chrysogenum*. *J. Gen. Microbiol.* 11:94–104.

Reddick, D., and W. R. Mills, 1938. Building up virulence in *Phytophthora infestans*. *Amer. Potato J.* 15:29–34.

Reed, G. M., 1935. Physiologic specialization of the parasitic fungi. *Bot. Rev.* 1:119–137.

Riehm, E., 1913. Prufung einiger Mittel zur Bekämpfung des Steinbrandes. *Mitt. Kaiserl. Biol. Anst. Land- Forstwirt.* 14:8–9.

Sharvelle, E. G., 1961. *The Nature and Uses of Modern Fungicides.* Minneapolis, Minn.: Burgess.

Stakman, E. C., 1914. A study in cereal rusts: physiological races. *Minn. Agr. Exp. Sta. Bull.* (138).

Tisdale, W. H., and I. Williams, 1934. Disinfectant. U. S. Pat. 1,972,961. U. S. Patent Office, Washington, D. C.

van der Plank, J. E., 1963. *Plant Diseases: Epidemics and Control.* New York: Academic Press.

———, 1968. *Disease Resistance in Plants.* New York: Academic Press.

von Schmeling, B., and M. Kulka, 1966. Systemic fungicidal activity of 1,4-oxathiin derivatives. *Science* 152:659–660.

von Tubeuf, C. F., 1902. Weitere Beiträge zur Kenntniss der Brandkrankheiten des Getreides und ihre Bekämpfung. *Arb. Kaiserl. Gesundheitsamte (Biol.) Berlin* 2:437–467.

Wallace, E., and H. H. Whetzel, 1910. Peach leaf curl. *Cornell Univ. Agr. Exp. Sta. Bull.* (218):137–161.

Weaver, L. O., 1958. Cyprex, an eradicant for apple scab. *Del. Agr. Exp. Sta. Bull.* 48:49.

Weindling, R., 1946. Microbial antagonism and disease control. *Soil Sci.* 61:23–30.

Wellman, F. L., 1937. Control of southern celery mosaic in Florida by removing weeds that serve as sources of mosaic infection. *U. S. Dep. Agr. Tech. Bull.* (548).

Wingard, S. A., 1941. The nature of disease resistance in plants, I. *Bot. Rev.* 7:59–109.

Wood, R. K. S., and M. Tveit, 1955. Control of plant disease by use of antagonistic organisms. *Bot. Rev.* 21:441–492.

II

PLANT PATHOLOGY IN PRACTICE

10

Diseases Affecting Storage of Food Materials—Rots of Fleshy Tissues

Rots of the fleshy parts of plants—among the most familiar symptoms of disease—develop as tissues are disintegrated by the action of microorganisms. Rots may be soft and wet, in which case they may often be followed by putrefaction, or they may be hard and dry, especially when disease develops in storage tissues kept in a relatively dry atmosphere.

Soil-infesting bacteria or fungi that cause rots of fleshy tissues typically infect plants at the time of harvest or somewhat before; infection may occur, however, during postharvest handling or in storage. Ubiquitous air-borne molds, such as *Penicillium* spp., may gain ingress into susceptible tissues and cause rots during packing and shipping operations. Regardless of the source of inoculum, most rot-inducing plant pathogens are unable to enter fleshy tissues except through open wounds. This fact is the basis for an effective control measure: the prevention of wounds through careful handling of harvested plants and plant parts. Moreover, if wounding cannot be entirely prevented, the harvested material may be placed in an environment favorable to the rapid healing of wounds, thus preventing development of secondary cycles of the pathogen in storage.

All important rot pathogens initiate disease in the same general way: they all produce extracellular enzymes that start the degenerative processes in advance of the bacterial cells or fungal hyphae of the attacking pathogens. These organisms, then, are all saprophytic pathogens and obtain nourishment from dead or dying cells. On the basis of their food relations with affected tissue, the rot pathogens are classed as facultative parasites. The fact that there is an unspecialized food relation between rot pathogens and suscept tissues suggests that protoplasmic resistance to rot does not occur. Those horticultural varieties of a crop species that resist rot are able to do so because they possess a form of morphological resistance: they may, for example, produce fruit, tubers, or roots with tough skins that resist wounding.

The development of most rots is favored by a moist, warm environment, and effective disease control can be had by storing susceptible plant parts in a cool, dry place, or under refrigeration.

The generalizations mentioned here apply to moldiness of grain and to the rots of timber and lumber, as well as to those of fleshy tissues, some of which will be discussed below.

BACTERIAL SOFT ROT

Bacterial soft rot of succulent plant tissues is of world-wide occurrence. It is one of the most important diseases of vegetables and ornamental plants in transit and storage, and is among the most important of the bacterial diseases of growing plants. Affected tissues of such fleshy organs as tubers, roots, and bulbs become hydrotic and soft, turn brown, and often emit offensive odors. There is a sharp demarcation between intact, healthy tissues and the macerated, diseased tissues (see Figure 2-1). Masses of bacteria, mixed with plant fluids, often ooze out through breaks in the surface of badly diseased plant parts. The diagnostic symptoms of the disease are maceration, hydrosis, and soft rot; the diagnostic sign of the pathogen is the bacterial ooze.

Etiology

The causal agent, *Erwinia carotovora* (Jones) Holland, is a peritrichously flagellate rod, $0.7\,\mu \times 1.5\,\mu$, that forms white, somewhat roughened colonies in culture. In nature, the bacterium survives from season to season in the soil, deriving nourishment from plant debris. The primary cycle of the bacterium is initiated when bacteria from the soil are deposited in wounds in susceptible plant parts, usually at the time of harvest. *E. carotovora* also causes rot of the fleshy roots of radish and stems of cabbage in the field. The cabbage maggot moves the bacteria in the soil, and makes the wounds through which they enter cabbage stems. The cabbage maggot and other insects may also serve as agents of dispersal and inoculation.

Upon reaching the infection court, the bacteria are nourished by nutrients that exude from the wounded tissues. Infection begins with the multiplication of the bacteria, and the attack is then made upon healthy suscept cells. Disease begins with their first pathological responses.

The disease is characterized by three phenomena—hydrosis, maceration, and then necrosis. *E. carotovora* produces extracellular pectolytic enzymes that digest middle lamellae. The resultant maceration is followed by the death of cells in the affected area, possibly as a consequence of toxin production by the pathogen. The bacteria exert enzymatic and toxic influences well in advance of themselves. As they multiply, daughter cells occupy the intercellular spaces in the macerated tissue. The disease process usually continues until the attacked plant part has rotted completely. Bacteria ooze out of diseased tissues, from which they may be spread to other plants or storage organs, in which the secondary cycles proceed, usually in transit or in storage.

Epidemiology

Moisture in the infection court is essential to development of the bacteria during the infection stage. Sufficient moisture is supplied at first by the wounded cells of the infection court. A moist environment prevents drying of the surface of the infection court and favors the development of the bacteria. The bacteria grow rapidly at 24–27°C, the optimum temperature for disease development.

Prevention of Disease

Wounds (often made by insects) are required for ingress by the bacteria, and soft rot is favored by a warm, moist environment; therefore, the disease in transit and storage can be controlled effectively by such protective measures as insect control, careful handling to avoid wounding the product, and the use of cool (or cold), well-ventilated storage places. Other control measures may also be effective, and are used when feasible: Diseased vegetative plant parts for propagation can be excluded from an area. The bacteria may be eradicated from soil by crop rotation or by sterilizing soil to be used in greenhouses. When tubers or roots are allowed to dry in the sun for a few hours at harvest time, the bacteria will be eradicated from their surfaces. Before grading and storing, potatoes may be kept for a week at 20–27°C in a moist environment to favor the development of wound periderm, which effectively walls off incipient infections. Almost all carbohydrate-rich fleshy tissues of plants are susceptible; therefore, protoplasmically resistant varieties of the suscept species are not known. Some varieties of potato, however, are morphologically resistant because they have tough skins that are resistant to wounding.

BROWN ROT OF STONE FRUIT

Brown rot occurs throughout the world, wherever the suscepts are grown, and is the most serious disease of stone fruit. Crop losses of 25–75% are common as a result both of fruit rot, particularly in transit and on the market, and of blossom and twig blight caused by the same organism.

Infected blossoms and young twigs are blighted by the disease, and cankers form on older twigs and branches. Rot of the fruit is first visible as a small brown spot. This enlarges rapidly, and most of the flesh of the fruit soon becomes affected. Diseased fruits that fall to the ground are completely destroyed by the soft, mushy, brown rot. When diseased fruits cling to the tree, their water is rapidly lost, and a dry rot completely destroys the flesh of the fruit. The skin of diseased fruit wrinkles and becomes dry and leathery; diseased fruits tend to persist, and hang in the tree as "mummies." Gray to tan tufts of sporodochia form on them, often in concentric rings (Figure 10-1); catenulate ovoid conidia develop on branched conidiophores that arise from the sporodochia. Apothecia (Figure 10-2) develop from mummified fruits in which the fungus has been dormant after they have fallen to the ground.

Etiology

Brown rot is caused by one of three fungi. *Monilinia fructicola* (Winter) Honey [= *Sclerotinia fructicola* (Winter) Rehm] is most prevalent in American-grown peaches, whereas *Monilinia fructigena* is most common in Europe. Brown rot of nectarines and other suscepts in parts of Europe and in a few orchards in the United States is caused by *Monilinia laxa.*

M. fructicola survives between growing seasons in mummified fruit

FIGURE 10-1
Rings of sporodochia on cherries affected by brown rot.

and in twig cankers. Sporodochia are formed in twig cankers and on the surfaces of mummified fruit in the tree; they produce conidia that may initiate primary cycles of the fungus in flowers, young twigs, and even leaves, where it may produce a leaf blotch. Conidia are probably the chief inocula for primary cycles in warm-temperate regions; in colder parts of temperate regions, however, the fungus survives in fallen mummified fruit, from which apothecia develop in the spring (Figure 10-3).

In a moist environment, as the weather becomes warm in the spring, apothecial fundaments and microconidia form on the surfaces of fallen mummified fruit. Apparently, the microconidia act as spermatia, germinating to produce hyphae from which antheridia branch. Antheridia contact archegonia in the apothecial fundaments, plasmogamy occurs, and branches of dicaryotic hyphae develop from the archegonium. These grow upward, surrounded by developing vegetative hyphae that organize to form the stalk and the enlarged, saucer-shaped apex of the apothecium. Asci form at the ends of the dicaryotic hyphae that occur in the upper surface of the top of the apothecium. Individual asci develop by crozier formation.

In an appropriate environment, ascospores are forcibly ejected from

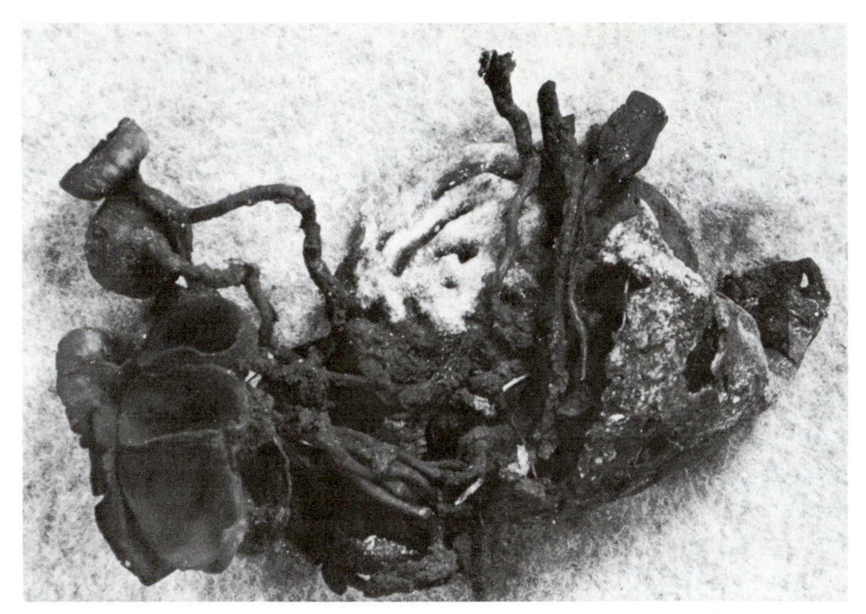

FIGURE 10-2
Apothecia from pseudosclerotia within mummified fruit infected by *Monilinia fructicola.*

the asci by a puffing mechanism. In the mature apothecium, water is imbibed by asci and by filamentous sterile cells (paraphyses), creating considerable hydrostatic pressure. The apex of the ascus contains a plug of starch or glycogen, in contrast to the cellulose in the remainder of the wall. In the presence of imbibed water, hydrolytic enzymes are activated and the starch plug is digested, making a weak point in the wall of the ascus. When external pressure is suddenly reduced, as when air moves across the apothecium, the pressure from within expels ascospores upward in a cloud from the apothecium. They are carried by wind to such infection courts as flowers, young shoots, and leaves, where primary cycles of the pathogen are induced.

Primary cycles of the fungus may also be initiated by conidia from twig cankers or from mummified fruit. They are splashed, blown, or carried by insects to infection courts. Ingress by hyphae from other ascospores or conidia may be by direct penetration of floral parts, of young leaves, or, later on, of ripening fruit. Young fruits are usually penetrated through wounds, particularly those made by the plum curculio. Cells that make up the hair sockets may be penetrated directly. Conidia are produced on infected plant parts, and these initiate sec-

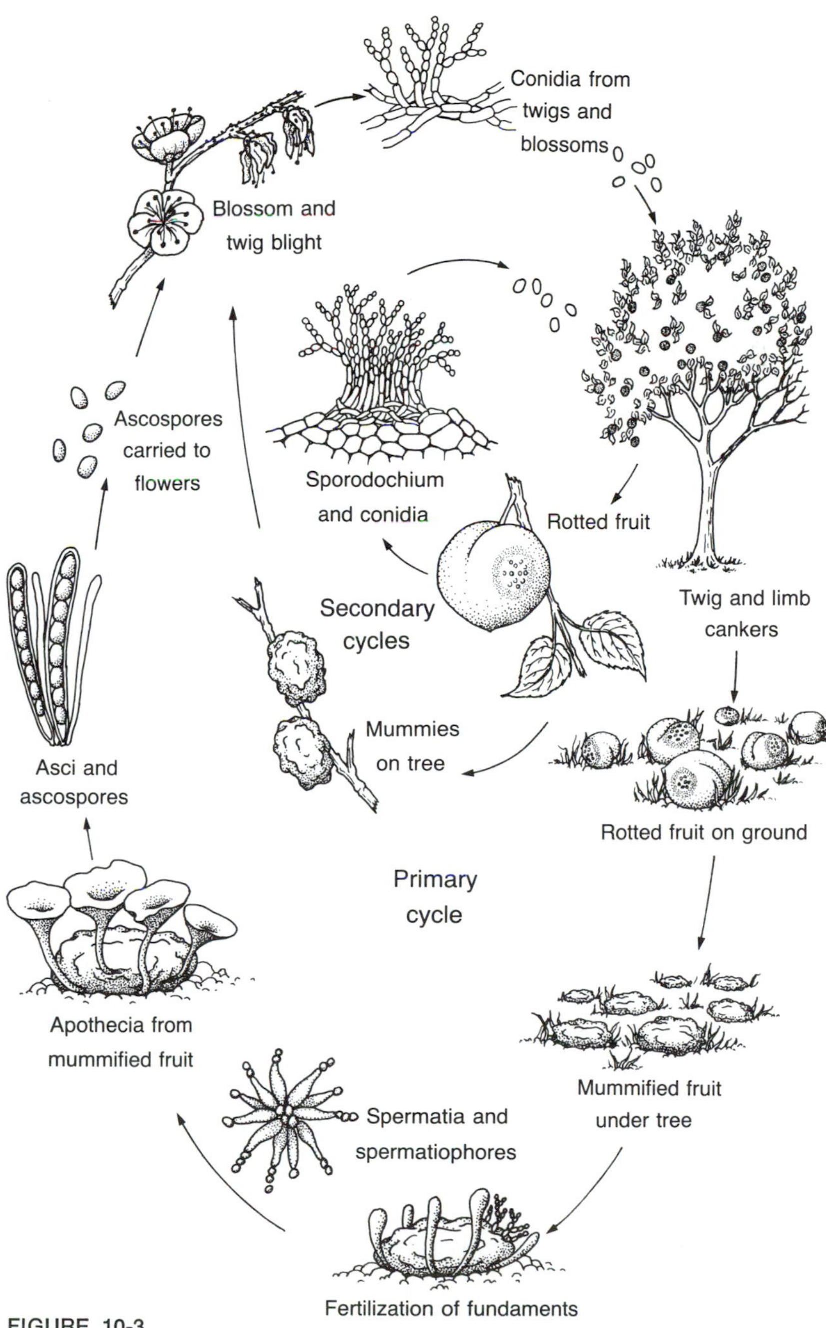

FIGURE 10-3
Brown rot of stone fruit.

ondary cycles of the fungus in ripening fruit during the growing season. They are also responsible for the secondary cycles that occur in transit and in the market.

In diseased fruit, the fungus produces pectolytic enzymes that diffuse into the tissue in advance of the fungal mycelia. The middle lamellae are digested, and maceration ensues. Affected tissues are hydrotic at first, but if water is evaporated rapidly, and if the affected tissue is not invaded by another microorganism, the rot may be dry rather than wet. Affected tissues become necrotic and are characteristically brown. Evidently, there is no toxic action by *M. fructicola*.

Primary and secondary cycles occur only in wet weather; indeed, free water in the infection court is required for spore germination, germ tube growth, and ingress. Primary cycles develop most rapidly between 15°C and 25°C, but may occur between 5°C and 30°C.

Prevention of Disease

Exclusionary control measures are not practicable, and, although modern varieties of peach are more resistant than old ones, resistance is not normally considered a control measure. Instead, eradicative and protective measures are relied upon to control brown rot.

Fallen mummified fruit may be buried by plowing, or disturbed by disking, so that primary inoculum will not form. Eradicative sprays may be used to reduce the numbers of conidia that might be formed on sporodochia in cankers and in mummified fruit on the tree, but this practice has not proved very effective. Because the fungus may survive in wild plums, their eradication is indicated.

Protective sprays of such organic compounds as Botran, captan, dichlone, and thiram, or wettable sulfur, should be applied at intervals throughout blossoming and again just prior to harvest. Post-harvest dips of the fruit in Botran, captan, or a mixture of both, are recommended. During the growing season, control of the plum curculio will prevent infection of ripening fruit. Precooling the fruit and the rail cars in which the fruit is to be shipped will lessen the development of rot in transit. It is axiomatic that fruit should be handled carefully during and after harvest to prevent wounds.

COTTONY ROT OF VEGETABLES

Cottony rot is perhaps the most devastating disease of cultivated plants in cool-temperate climates and in subtropical regions of the

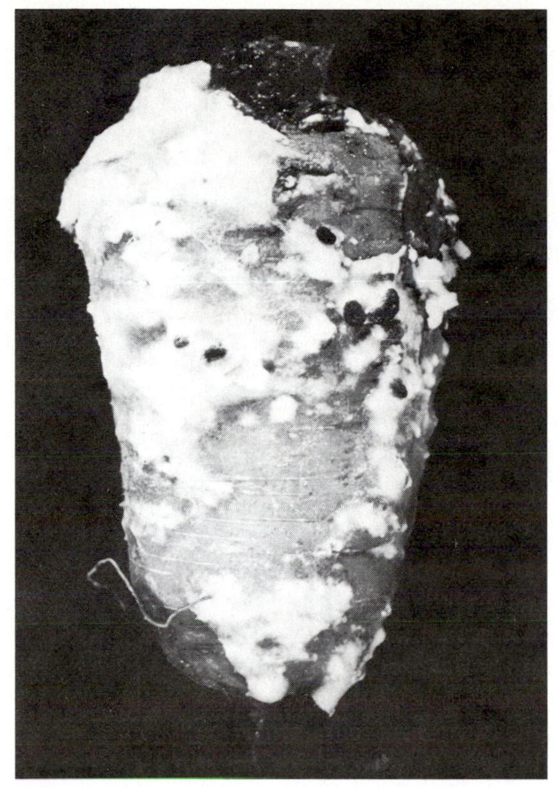

FIGURE 10-4
Cottony mycelium and sclerotia of *Sclerotinia sclerotiorum* growing on a diseased carrot root.

world. The disease is most severe in herbaceous dicotyledonous plants and in a few woody plants. The disease has many common names, including crown rot (of legumes), pink rot (of celery), drop (of lettuce), stalk rot (of potato), timber rot (of tomato), watery soft rot, and sclerotiniose. The disease is important in the field as well as in transit and in storage.

Watery soft rot is the diagnostic symptom of diseased fleshy organs in storage, or of diseased stems, petioles, flowers, and fruit of plants in the field. The invasion of seedling stems results in damping-off, the toppling of seedlings. The diagnostic sign of the pathogen is the mass of white mycelium that envelops diseased tissue and gives the appearance of a mass of cotton (Figure 10-4). Later, black sclerotia develop within the cottony weft of mycelium. The sclerotia are irregularly shaped, somewhat elongated, and often as large as bean seed.

Etiology

Cottony rot is caused by the ascomycete *Sclerotinia sclerotiorum* (Libert) deBary. The pathogen survives between crop seasons as soil-borne sclerotia. The cortex of carbohydrate-rich fungal cells is surrounded by a protective layer of thick-walled suberized cells. In a moist, cool (12–20°C) environment, the sclerotia germinate to produce inoculum for the primary cycle. The cycle may be initiated by mycelium or by ascospores.

When the sclerotium germinates directly, strands of hyphae that develop from it may contact stems of susceptible plants near the soil line. It is presumed that strands of hyphae grow along the surface of the infection court, to which the hyphal tips become appressed. Branches from these hyphae penetrate directly, and infection ensues.

When the sclerotium germinates indirectly, apothecia develop and mature ascospores are expelled from the layer of asci produced on the upper surface of the apothecium. Ascospores may be blown or splashed onto infection courts. Apothecial production requires moisture, and proceeds most rapidly at about 18°C, some three to six weeks being required for the formation of apothecia. After the sclerotium has imbibed water, the rind cracks, and masses of cells derived from cortical cells develop and protrude through the broken rind. These masses are fundaments of the apothecium. At the same time, microconidia are produced from other masses of cells. Apparently, the microconidia serve as spermatia, and plasmogamy is accomplished when antheridia derived from germinated microconidia fuse with archegonia in the apothecial fundaments. The mycelium that develops thereafter is dicaryotic, and grows in the central part of the stalk of the developing apothecium. In the top of the apothecium, the dicaryotic mycelium gives rise to asci. The supporting tissue of the apothecium is composed of haploid mycelium, and takes the shape of a saucer at the top of a stalk. Mature ascospores are forcibly discharged from the asci.

Apothecia are formed in the autumn and spring, and primary cycles may be initiated by spores blown to susceptible plants in the area. After inoculation, and in the presence of a film of water in the infection court, the ascospores germinate, producing germ tubes that branch and develop into small thalli. The tips of young hyphae enlarge and are held fast to the surface of the infection court, perhaps by a gelatinous secretion. From these swollen hyphal tips, which are termed appressoria, slender penetration pegs form and bore through the cuticle. Once ingress has occurred, the invading hyphae enlarge and ramify between the suscept's cells. The progress of the invading hyphae is facilitated by extracellular pectolytic enzymes that partially digest the middle lamellae between suscept's cells. The cells are killed in advance of the fungus, which absorbs nutrients from cells killed by the action of enzymes and toxins that it produces.

Pathogenesis proceeds in essentially the same way, whether the inoculum is hyphae from ascospores or hyphae from sclerotia that germinate directly, and whether the infection court is in a plant growing in the field or in fleshy plant parts in transit or storage. The disease develops most rapidly in moist, cool (12–18°C) weather.

Prevention of Disease

The pathogen may be excluded from areas in which it has not become established through the use of seed that are not contaminated with sclerotia. It is difficult to eradicate sclerotia from infested soil. Small areas can be sterilized with methyl bromide, and soil to be used in greenhouses may be sterilized with chemicals or by steam heat. In certain areas of Florida, sclerotia can be killed by flooding the soil for four to six weeks during the summer. Water need not stand on the field, but the soil should be kept at 100% of its moisture-holding capacity. In some areas, susceptible crops can be rotated with rice, the flooding required in rice culture being sufficient to bring about the decay of most sclerotia. Sclerotia may perish if susceptible crops are rotated with grass or if cyanamid is used to treat the soil before planting. Fleshy plant parts in transit or in the market can be protected by refrigeration and ventilation. The produce should be kept at the lowest temperature that will not impair its palatability. Resistant varieties are not likely to be found, because the carbohydrate-rich storage tissue of many suscepts is a suitable medium for the growth of *S. sclerotiorum*.

Selected References

BACTERIAL SOFT ROT

Echandi, E., S. D. van Gundy, and J. C. Walker, 1957. Pectolytic enzymes secreted by soft rot bacteria. *Phytopathology* 47:549–552.

Elliott, C., 1951. *Manual of Bacterial Plant Pathogens* (2nd ed.). Waltham, Mass.: Chronica Botanica.

Johnson, D. E., 1930. The relation of the cabbage maggot and other insects to the spread and development of soft rot of Cruciferae. *Phytopathology* 20:857–872.

Jones, L. R., 1909. Pectinase, the cytolytic enzyme produced by *Bacillus carotovorus* and certain other soft-rot organisms. *Vt. Agr. Exp. Sta. Tech. Bull.* (147):281–360.

Ruehle, G. D., 1940. Bacterial soft rot of potatoes in southern Florida. *Fla. Agr. Exp. Sta. Res. Bull.* (348).

Smith, W. L., Jr., 1955. Streptomycin sulfate for the reduction of bacterial soft rot of packaged spinach. *Phytopathology* 45:88–90.

Wehlburg, C., and R. W. Meyer, 1966. Bacterial soft rot of Iceburg (Great Lakes) lettuce in the Florida Everglades. *Plant Dis. Rep.* 50:938–941.

Winfree, J. P., R. S. Cox, and D. S. Harrison, 1958. Influence of bacterial soft rot, depth to water table, source of nitrogen, and soil fumigation on production of lettuce in the Everglades. *Phytopathology* 48:311–316.

BROWN ROT OF STONE FRUIT

Chandler, W. A., 1968. Preharvest fungicides for peach brown rot control. *Plant Dis. Rep.* 52:695–697.

Heuberger, J. W., 1934. Fruit-rotting Sclerotinias, IV. A cytological study of *Sclerotinia fructicola* (Wint.) Rehm. *Md. Agr. Exp. Sta. Bull.* (371):167–189.

Ogawa, J. M., J. L. Sandeno, and J. H. Mathre, 1963. Comparisons in development and chemical control of decay causing organisms on mechanical and hand harvested stone fruits. *Plant Dis. Rep.* 47:129–133.

Roberts, E. W., and J. G. Dunegan, 1932. Peach brown rot. *U. S. Dep. Agr. Tech. Bull.* (328).

Smith, W. L., and W. H. Redit, 1968. Post harvest decay of peaches as affected by hot water treatments, cooling methods, and sanitation. *U. S. Dep. Agr. Agr. Res. Serv.* (*Rep.*) (807).

Wormald, H., 1954. The brown rot diseases of fruit trees. *Minist. Agr.* (*London*) *Tech. Bull.* (3).

COTTONY ROT OF VEGETABLES

Boyle, C., 1921. Studies in the physiology of parasitism, VI. Infection by *Sclerotinia libertiana. Ann. Bot.* (*London*) 35:337–347.

Brown, J. G., and K. D. Butler, 1936. Sclerotiniose of lettuce in Arizona. *Ariz. Agr. Exp. Sta. Bull.* (63):475–506.

Moore, W. D., R. A. Conover, and D. L. Stoddard, 1949. The sclerotiniose disease of vegetable crops in Florida. *Fla. Agr. Exp. Sta. Bull.* (457).

Natti, J. J., 1967. Bean white mold control with foliar sprays. *Farm Res.* 33:10–11.

11

Diseases Affecting Breakdown and Utilization of Stored Food Materials —Seedling Blights

Any one of numerous fungi may be capable of causing seedling blight (damping-off), but two species, *Pythium debaryanum* and *Rhizoctonia solani,* are responsible for most losses in commercial plantings. Their ability to live saprophytically in soil and to compete successfully with the myriads of nonpathogenic saprophytes there make *P. debaryanum* and *R. solani* serious threats to crop production. Moreover, both fungi are often involved in root or stem rots of mature plants.

Regardless of which fungus is the causal agent, a seedling blight results from the decay of seedling stems that are invaded shortly after germination of the seed. Ingress is by direct penetration, and maturing stems have natural morphological resistance to invasion. Thus, any treatment that will reduce the chances of infection during the first one or two weeks of plant growth will result in a degree of disease control. One such treatment is chemical seed dressing; the fungicide eradicates the pathogen in an area immediately surrounding the seed, and it also provides a protective barrier around the emerging hypocotyl. A fair degree of control is sometimes achieved through manipulation of cultural practices. Planting seed at a time favorable to early maturation of stem tissue (usually late in the spring) is a common practice.

The damping-off fungi are soil-inhabiting; therefore, sterilization by heat or with chemicals results in good control in greenhouse-grown crops. Field-scale sterilization of soil is not feasible, but chemicals placed in the row at planting time may help to check the severity of stem rots caused by *R. solani* and by *Sclerotium rolfsii*.

The important soil-inhabiting fungi that cause seedling blights are saprophytic pathogens, killing suscept cells by enzyme action (and possibly also by toxic action) well in advance of their hyphal tips. They have unspecialized food relations with the plants they attack, and, as might be expected, have wide suscept ranges. Thus, damping-off usually cannot be prevented by crop rotations or through the development of protoplasmically resistant varieties of crop plants.

PYTHIUM-INDUCED DAMPING-OFF OF SEEDLINGS

Damping-off may occur in herbaceous and woody plant seedlings of almost all plant species except species of grasses. The disease is worldwide, occurring in the temperate zones and the tropics.

The pathogen may attack seed, seedling stems, or roots, causing, respectively, seed decay, damping-off, or root rot. The first two kinds of disease will be considered here. Seed sown in infested soil may be attacked about the time of germination, whereupon the seed themselves are rotted. Commonly, however, the seedling stem is the infection court. Hydrosis occurs at the soil line, and this is followed by necrosis of the juvenile stem, loss of turgor, and toppling of the entire seedling (Figure 11-1). Necrotic areas of the stem become bleached and shriveled.

Etiology

Several fungi, notably *Rhizoctonia solani* Kühn and *Pythium debaryanum* Hesse, cause damping-off of seedlings; the *Pythium*-induced disease (Figure 11-2) is the subject of this section.

The life cycle of *P. debaryanum* is typical the family Pythiaceae (class Phycomycetes). Oospores form at the end of, or after, the infection stage of pathogenesis. A short branch of a hypha swells at the tip to form a spherical cell, the oogonium, which is separated from the stalk by a crosswall. A branch from the oogonial stalk elongates and partly encircles the oogonium. This branch, the antheridium, is

FIGURE 11-1
Damping-off of tomato seedlings infected by *Pythium debaryanum.*

also abjointed from the base. The oogonium and the antheridium are multinucleate; each contains nuclei of at least two mating types, one of which is arbitrarily designated +, the other −. The tip of the slender, fingerlike antheridium is held fast to the thickening wall of the oogonium by a gelatinous secretion. A small penetration branch develops from the appressed tip of the antheridium and grows through the oogonial wall. Nuclei flow from the antheridium into the oogonium, where plasmogamy occurs as each nucleus pairs with another of the opposite mating type. All dicaryons except one migrate to the periphery of the oogonium and abort; the remaining dicaryon stays intact in the center of the oogonium. The rupture in the oogonial wall heals and the outer wall becomes sclerotized; the former oogonium is now termed the oosphere. Later, when its wall has matured and when the two haploid nuclei fuse (caryogamy), the oosphere becomes the oospore, which is adapted for dormant survival because of the thick outer wall that surrounds the membranous inner wall of the spore.

In an appropriate environment, the oospore germinates directly, by

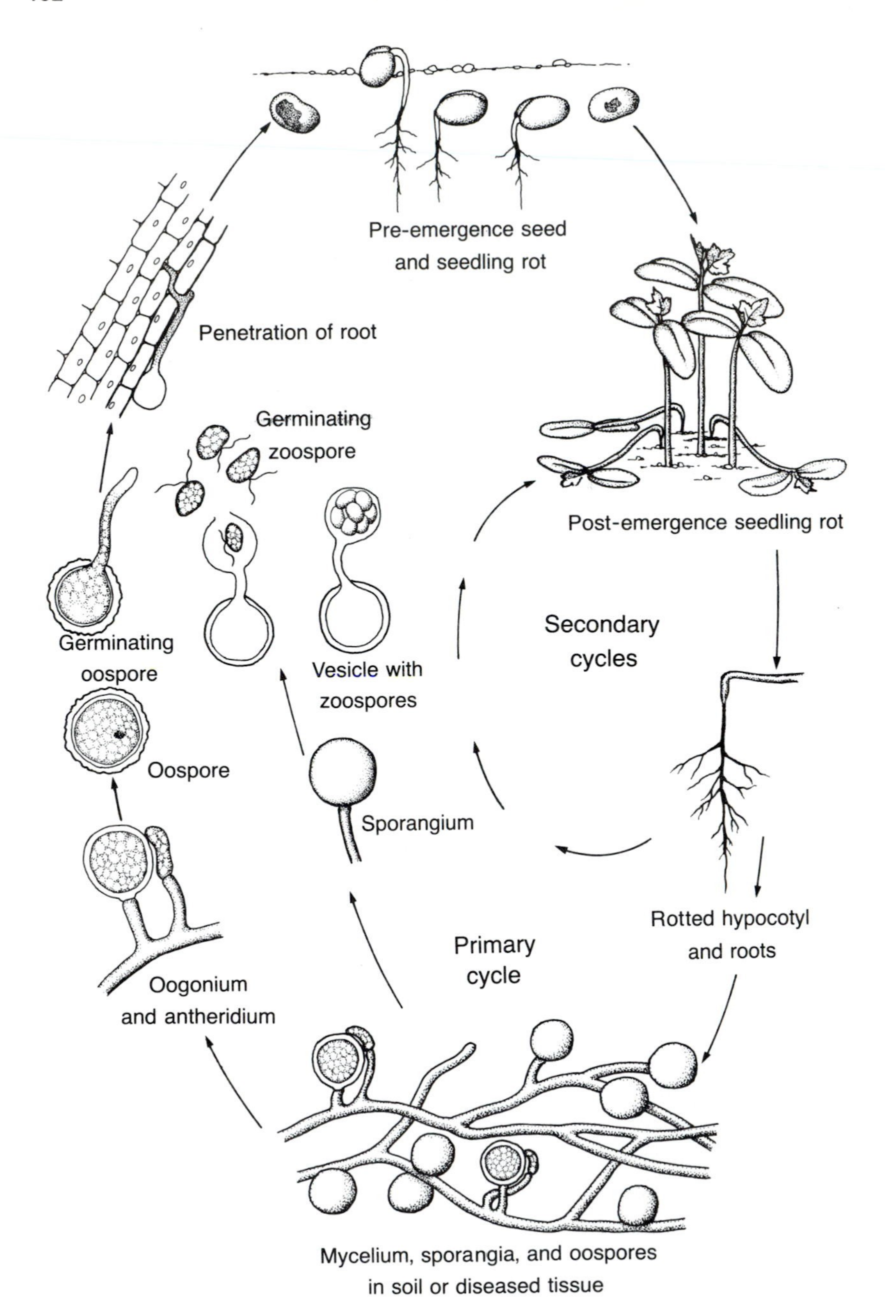

FIGURE 11-2
Pythium-induced damping-off of seedlings.

means of a germ tube, after meiosis has occurred. The elongating tube develops to form coenocytic hyphae, which may come into contact with juvenile tissue of a seedling stem that can be attacked. The oospore may also germinate indirectly, forming asexual, haploid, biflagellate zoospores (swarmspores), each of which may give rise to hyphae. When indirect germination occurs, meiosis is followed by the growth of a short tube that enlarges at the tip to form a spore sac (sporangium), which is separated from the stalk by a septum. Cytoplasm aggregates around each of the nuclei, and membranous walls develop around each mass of protoplasm. Two flagella form on each cell (zoospore). The sporangium germinates by producing a swollen sac (vesicle) that is attached to the sporangium by a short tube. Zoospores move from the sporangium into the vesicle, whose wall ruptures to release the zoospores. Sporangia are also formed on the mycelium of the fungus during infection. These may survive as dormant structures, but, being thin-walled, sporangia are not nearly as important in survival as oospores or chlamydospores.

Chlamydospores are vegetative spores formed directly on the mycelium. The tips of some branches swell, become abstricted by a septum, and form a spherical structure whose walls thicken and become hardened. They break away from the mycelium and lie dormant. Germination results in the production of a germ tube that may invade seedling stems.

From the foregoing account, it is clear that *P. debaryanum* may survive in the absence of susceptible plants by means of oospores, chlamydospores, and, possibly, sporangia. Another means of survival, by means of mycelium growing saprophytically in the soil, deserves special attention: whereas many plant pathogens are also highly specialized parasites, *P. debaryanum*, like the rot-inducing molds and such other soil-borne fungi as *Rhizoctonia solani*, is a saprophytic pathogen. Thus, *P. debaryanum* is not dependent upon living green plants for food materials. The fungus grows well as a saprophyte in the soil, competing successfully with nonpathogenic soil-borne organisms. The tips of actively growing hyphae may come into contact with seedling stems and subsequently induce damping-off.

Irrespective of the manner of its survival in the soil, inoculation of emerging hypocotyls or seedling stems is accomplished when the infection court is contacted by the tip of a growing hypha or by a viable zoospore. When inoculation is by a hypha, the prepenetration stage includes the growth of the hyphal strand along the surface of the hypocotyl and the adhesion of its tip to the surface. Entrance occurs by means of direct penetration of the hypocotyl by a hyphal branch.

When a zoospore is the inoculum, it contacts and adheres to the surface of the infection court. The flagella are withdrawn, the spore rounds off, and the germ tube that emerges grows directly into the epidermal cell beneath the spore. Extracellular enzymes and toxins produced by the invading fungus evoke necrotic responses from suscept cells in advance of the invading fungus. The fungus absorbs nutrients from previously killed cells and grows saprophytically through and between the dead and dying suscept cells. Within 24–28 hours after the start of infection, the invaded hypocotyl or stem dies.

Although the fungus continues to grow in and away from killed seedlings, relatively few secondary cycles of the damping-off pathogen occur. This is because only the uncutinized stem of juvenile plants can be penetrated by *P. debaryanum*. By the time the pathogen has killed one plant and has grown through the soil to another, the second plant has usually matured to such an extent that its stem can no longer be penetrated. Most damping-off occurs, then, as a consequence of many primary cycles of the pathogen.

Epidemiology

Pythium debaryanum thrives in moist soils, growing best when the water content of a soil is at least 50% of its water-holding capacity. Relatively warm (20–25°C) and slightly acid or alkaline soils also favor activity of the pathogen. As might be expected, *Pythium*-induced damping-off is serious during the rainy season in tropical America, and during wet weather in the spring, as the soils become warm, in the humid northeastern United States. *Pythium*-induced damping-off is not so severe in the warm soils of the humid southeastern United States, probably because the sandy soils do not remain saturated to a point above 50% of their water-holding capacity for very long, even after a heavy rain. In such areas, damping-off is most likely to be caused by *Rhizoctonia solani*.

Prevention of Disease

The survival of *Pythium debaryanum* in the soil is the basis for the success of certain eradicative control measures. Soil to be used in greenhouse can be sterilized by heat (120°C at 15 lb/in^2 for 3 hours) or by chemicals (methyl bromide at the rate of 25 lb/1000 ft^3, or methyl bromide plus chloropicrin). Soil in outdoor beds or in fields to be planted is usually treated with such chemicals as Vapam or Vorlex.

Under greenhouse conditions, soil sterilization alone is insufficient to insure control of damping-off. All containers or benches should also be treated to kill the fungus that might have survived in bits of soil from a previous planting. Benches and containers may be treated with steam, volatile chemicals, or copper sulfate solution (1 oz/gal); wooden flats or metal containers may also be sterilized by heat. Once soil and containers have been sterilized, special precautions must be taken to prevent recontamination. The usual source of recontamination is infested soil on the floor of the greenhouse. Bits of such soil may cling to water hoses and be washed into sterilized soil on the bench during routine irrigation practices.

The fact that mature stems cannot be penetrated by *Pythium debaryanum* suggests two additional protective measures. One, the treatment of the surface of seed with a fungicide, provides a chemical barrier around the seed and the emerging hypocotyl and radicle for several days. The chemical should also kill any of the fungus in the immediate vicinity of the seed. The chemical does not persist in fungitoxic concentration for more than one or two weeks, but by the time the chemical has disappeared, the mature stem is structurally resistant to penetration by the fungus. Three commonly used seed dressings are captan, dichlone, and thiram (2–6 oz/100 lb seed, depending upon type of seed to be treated).

The second protective measure is the manipulation of the environment in such a way that the susceptible seedlings escape infection. Any treatment that hastens maturation of the stem will lessen the severity of disease. Seed should be planted shallowly in warm soil, and irrigation should be done in the mornings so there is rapid evaporation of surface moisture. It has also been suggested that the nitrogen and potassium in the fertilizer should be balanced with adequate phosphorus to hasten maturation. If excessive amounts of nitrogen are applied, seedlings will *grow* rapidly, but will remain succulent and susceptible longer than if they *mature* rapidly. Commercial fertilizers used on plant beds should contain nitrogen, phosphorus, and potassium in the ratio of 1:2:1 to insure rapid maturation of seedling stems. Although shallow planting in warm soil that has been properly fertilized will reduce the incidence of damping-off, the disease cannot be properly controlled except through seed dressings or soil sterilization.

RHIZOCTONIA-INDUCED DISEASES

Black scurf, a *Rhizoctonia*-induced disease, has long been recognized as a common disease of potatoes throughout the world. The causal

FIGURE 11-3
Sclerotia of *Rhizoctonia solani* on the surface of a potato tuber.

agent *Rhizoctonia solani* Kühn, also attacks hundreds of plants other than potato, usually causing the damping-off of seedlings.

To the potato industry, black scurf ranks in importance with late blight, scab, and the viral diseases. Losses attributed to the *Rhizoctonia* disease average less than 5%, but, even today, the yields of some potato fields may be reduced by as much as 25–50%.

In potatoes, the disease is first apparent early in the season, when it manifests itself as "missing hills." When sprouts are attacked before emergence, the tips become brown and die; when sprouts are attacked at or soon after emergence, reddish-brown necrotic lesions develop in the young stems. When such shoots are girdled, they die and topple over. Maturing stems may be attacked near the soil line, whereupon a rot of the cortical tissues ensues. Aerial tubers often form in leaf axils above stem lesions, and a "nest" of small underground tubers usually forms on shortened stolons of plants whose stems are partially rotted. Brown lesions may occur on stolons. The russetting of the skin of tubers has been attributed to *Rhizoctonia.* Moreover, the fungus sometimes causes production of cankers in lenticels of tubers. The diagnostic sign of the pathogen, its sclerotia, occurs as slightly raised, rough, irregularly shaped black bodies that adhere to the surfaces of affected tubers (Figure 11-3). The sclerotia constitute the "black scurf" or the "dirt that won't wash off."

Stem cankers frequently form in other suscepts, particularly in tomatoes. Tomato fruit that touch the ground may also be invaded; a dark brown to black soft rot occurs in affected fruit. *Rhizoctonia solani* also causes typical damping-off of seedlings, the symptoms of which are usually indistinguishable from those caused by *Pythium;* in some instances, however, seedling stems infected by *Rhizoctonia* are darker brown than the straw-colored stems of *Pythium*-infected plants. Also, *Pythium* is more likely than *Rhizoctonia* to cause root necroses in seedlings.

Etiology

The causal agent forms no spores during its vegetative growth, and was first classified in the order Mycelia Sterilia of the Fungi Imperfecti. The perfect (sexual) stage of the fungus has been observed, however, on the surfaces of diseased stems during prolonged periods of warm, wet weather. The sexual stage is *Thanatephorus cucumeris* (Frank) Donk [= *Pellicularia filamentosa* (Patouillard) Rogers], a basidiomycete. The hyphae that are destined to bear the sexual spores grow along the surface of the stem and form a pellicle. Haploid hyphae containing nuclei of different mating types may anastomose, and the nucleus from one hypha migrates into the other hypha, thus accomplishing plasmogamy. Hyphae that develop from the dicaryotic cell remain binucleate, and conjugate division of the nuclei in each cell is accompanied by the "clamp-connection phenomenon." A short branch grows laterally from the apical cell, and one nucleus migrates into the protuberance, which then curls back to the surface of the hyphal cell wall. The two nuclei divide, and an angled septum is formed across the hypha in the region of the protuberance (clamp). The septum separates daughter and parent nuclei. Finally, tip cells of dicaryotic hyphae swell to form basidia, and the two nuclei in each fuse, accomplishing caryogamy. Meiosis occurs, giving rise to four haploid nuclei. At this time, four apical stalks (sterigmata) develop on each basidium, and each nucleus migrates into a stalk. The apex of each stalk swells, is entered by the nucleus, and the swollen part is cut off from the stalk by a septum. Thus, basidiospores are formed. The sexual stage is unimportant pathologically, although basidiospores may germinate to produce mycelium that may live saprophytically in the soil and attack the seedling stems in much the same way that *Pythium* does.

Rhizoctonia solani survives between crop seasons in the form of mycelium or sclerotia in the soil or as sclerotia on the surface of tubers.

In a moist environment, when weather becomes warm in the spring, the sclerotia germinate, producing mycelium. This grows in the soil and contaminates seedling stems or sprouts of potatoes. When potato sprouts are attacked, the prepenetration stage consists of the growth of strands of mycelium ("marching hyphae") along the surface of the sprout. Branches from the attacking hyphae penetrate the tip of the sprout directly, and grow intercellularly to the region of differentiation, where cells are killed by extracellular toxins and enzymes produced by the fungus. When seedling stems are attacked, toxins from the hyphae diffuse through the uncuticularized cells of the juvenile stem and evoke a necrotic response even before the fungal hyphae have penetrated. This is an unusual example of the initiation of disease prior to ingress by the pathogen.

Rhizoctonia solani, like *Pythium debaryanum,* is a saprophytic pathogen. It ramifies through diseased tissue, obtaining nourishment from cells that have been killed in advance of the mycelium.

Mature stems of potato, tomato, and other suscepts may be invaded through wounds or lenticels. Infection of mature stems results in a rot of cortical tissues, described above.

Epidemiology

Although *Rhizoctonia solani* grows most rapidly in culture at about 30°C, the diseases it causes in seedling stems or potato sprouts are most severe in moist soils at about 18°C. The fungus grows fairly well at 15–20°C, and most seedling stems mature slowly at this temperature. Thus, the period of susceptibility is longer than at higher temperatures, when seedlings not only grow faster, but also mature faster.

Prevention of Disease

Although sclerotia may be killed by chemical treatment of seed tubers, this exclusionary measure is rarely undertaken. The small amount of inoculum derived from tuber-borne sclerotia seems unimportant in comparison to the abundance of mycelium and sclerotia present in some soils.

The fungus may be eradicated—usually by steam sterilization, but also by such broad spectrum chemicals as methyl bromide and Vapam —from soil to be used in greenhouses. In the field, it can be eradicated from the immediate vicinity of the seed by pentachloronitrobenzene (75%) and captan (50%) in a 1:1 mixture by weight at the rate of

8–10 lb/acre. A three- to five-year rotation with small grains may be a helpful eradicative control measure.

Chemical seed dressings, as recommended for control of *Pythium*-induced damping-off, will also control damping-off by *Rhizoctonia*. Late, shallow planting is another excellent protective measure. If potato seed tubers are allowed to sprout in the light, stubby green sprouts that resist ingress by *R. solani* will form in a week. Little sprout rot occurs when tubers have been "green sprouted" just before cutting and planting.

Selected References

PYTHIUM-INDUCED DAMPING-OFF OF SEEDLINGS

Alexander, L. J., H. C. Young, and C. M. Kiger, 1931. The causes and control of damping-off of tomato seedlings. *Ohio Agr. Exp. Sta. Bull.* (496).

Anderson, P. J., 1934. *Pythium* damping-off and root rot in the seed bed. *Conn. Agr. Exp. Sta. New Haven Bull.* (359):336–354.

Jones, J. P., 1961. Comparative effects of soil fungicide treatments on soil rot and damping off of cucumber. *Plant Dis. Rep.* 45:376–379.

Miyake, K., 1901. The fertilization of *Pythium debaryanum. Ann. Bot. (London)* 15:653–667.

Roth, L. F., and A. J. Riker, 1943. Life history and distribution of *Pythium* and *Rhizoctonia* in relation to damping-off of red pine seedlings. *J. Agr. Res.* 67:129–148.

Wood, R. K. S., and S. C. Gupta, 1958. Studies in the physiology of parasitism, XXV. Some properties of the pectic enzymes secreted by *Pythium debaryanum. Ann. Bot. (London)* (N.S.)22:309–319.

RHIZOCTONIA-INDUCED DISEASE

Blair, I. D., 1943. Behavior of the fungus *Rhizoctonia solani* Kühn in the soil. *Ann. Appl. Biol.* 30:118–127.

Boosalis, M. G., 1950. Studies on the parasitism of *Rhizoctonia solani* Kuehn on soybeans. *Phytopathology* 40:820–831.

Duggar, B. M., and F. C. Stewart, 1901. The sterile fungus *Rhizoctonia* as a cause of plant diseases in America. *N. Y. Agr. Exp. Sta. Geneva Bull.* (186):50–76.

Kerr, A., and N. T. Flentje, 1957. Host infection of *Pellicularia filamentosa* controlled by chemical stimuli. *Nature* 179:204–205.

Müller, K. O., 1924. Untersuchungen zur Entwicklungsgeschichte und Biologie von *Hypochnus solani* P. u. D. (*Rhizoctonia solani* K.). *Arb. Biol. Reichanst. Land- Forstwirt.* 13:198–262.

Parmeter, J. R., ed., 1970. Rhizoctonia solani: *Biology and Pathology.* [Based on a symposium on *Rhizoctonia solani* held in Miami, Florida, in October, 1965.] Berkeley and Los Angeles: University of California Press.

12

Diseases Affecting Absorption and Accumulation—Cortical Rots of Roots and Stems

Rots of cortices of roots and stems are all caused by fungi that are associated with the soil. Some (*Rhizoctonia solani,* for example) are soil inhabitors, whereas others (*Thielaviopsis basicola*) are soil invaders. Soil invaders may produce long-lasting structures, such as chlamydospores, that enable them to persist for as much as ten years. Moreover, they may well live in the mycelial stage, growing saprophytically on decaying organic matter in the soil, for surprisingly long periods.

Whereas the rot-inducing bacteria and fungi can attack almost any fleshy, parenchymatous plant tissues, and the damping-off fungi can attack almost any seedling stem, the root-rotting fungi are specialized to the extent that only cortical tissues provide the appropriate milieu for their activities. Moreover, a number of the root-rotting fungi have relatively narrow suscept ranges.

Because these organisms are soil-borne, soil sterilization is an effective control measure for greenhouse-grown plants. Seed treatments are ineffective because the root-rotting fungi can attack a plant throughout the growing season, and the chemicals used as seed dressings are dissipated within a few days of planting the treated seed.

The somewhat specialized food relations between a root-rotting fungus and its suscept suggest two control measures. One, crop rotation, is successful when the pathogen does not have a wide suscept range. The other, the use of resistant varieties, has been used in the successful control of some—but not all—root rots.

Special methods are used to combat root rots of forest and fruit trees because rotation of such crop plants is virtually impossible and breeding and selection for resistance is often a lifetime project. Diseased trees can be removed, the soil can be treated, and young trees can be planted in the treated soil; moreover, infested areas can be isolated by ditches from uninfested areas. Root-rot control in orchards is largely a matter of prevention, best accomplished by planting young trees only in soil that is not infested by a root-rotting pathogen.

Various unrelated pathogens are capable of causing root and stem rots; these pathogens differ considerably in their modes of attack and their ability to survive in the soil.

PHYTOPHTHORA ROOT ROTS OF WOODY PLANTS

Phytophthora root rots are serious fungal diseases of many woody plants, and occur commonly in the tropics and the warm parts of the temperate zone. Although several species of the genus *Phytophthora* may cause root rots, only the one caused by *Phytophthora cinnamomi* will be considered here. Approximately 150 plants have been listed as being susceptible, but the suscept range probably includes many other species.

As the name of the fungus implies, the disease was discovered in cinnamon trees; it is also one of the most important root diseases of pine and chestnut in the southeastern United States, and of avocado in the United States, Puerto Rico, and South Africa. Affected trees may react in one of two ways: rapid decay of large roots brings about yellowing of foliage and the sudden collapse and death of the tree, whereas less severe root damage brings about a slow decline characterized by die-back of branches, partly in the top of the affected tree. In pine, needles become yellow and short, and the rate of shoot growth decreases. Below ground, the disease is characterized by cankers at the juncture of root and stem. The cortical and cambial tissues become necrotic, with the affected areas becoming dark brown or black. Such symptoms are characteristic of pines that decline quickly; the roots of pines that decline slowly have infections limited to small rootlets and a few of the larger roots. The mycelium of the fungus can be observed as a sparse mat between the bark and wood of older roots.

Etiology

The causal agent, *Phytophthora cinnamomi* Rands, is a member of the family Pythiaceae, and is, therefore, closely related to *Pythium* and to the water molds. Three kinds of spores—chlamydospores, oospores, and zoospores—are produced. The chlamydospores probably serve as the most important organ of survival in the soil, but oospores are also adapted for survival. Although they are produced in culture, oospores have not been observed regularly in nature, and their function in the life history of the pathogen under field conditions is not known. The fungus may also survive in soil as saprophytic mycelium.

In warm, moist soils, the chlamydospores germinate and produce sporangia from which are released as many as sixteen zoospores. These motile spores may then germinate in the vicinity of the roots of avocado or other suscepts. The zoospore produces a short germ tube, which penetrates directly the tender tissues of feeder roots. The mycelium develops within the young roots and grows into older ones, where the cortical tissues are destroyed. Evidently, the fungus grows intercellularly and absorbs nutrients directly without producing haustoria.

Epidemiology

The disease is invariably severe in infested soils in which there is excessive soil moisture around the roots of susceptible plants. The disease develops rapidly in trees growing under humid conditions, especially in those growing in poorly drained soils. The disease has never been particularly severe in well drained, sandy soils; a closely related fungus causes serious root rot of citrus growing in sandy soils in Florida, but the disease in that state is most serious in those areas where the water table is high. The optimum temperature for the development of the fungus in the soil is approximately 30°C.

Prevention of Disease

The fungus can be spread by means of infected nursery stock; inspection and certification of nursery stock, therefore, are effective exclusionary control measures. The fungus can be eradicated from mildly diseased nursery stock by a hot-water treatment of the roots. The diseased roots should be submerged for 15 minutes in water at 50°C. No satisfactory method of eradicating the fungus from the soil has been recorded, but Vapam and nabam have been used with some success to treat the soil around the roots of trees in orchards.

Orchards can be protected by intermittent irrigation, which prevents saturation of the soil. Although an extensive search for resistant rootstocks has been made, no commercially acceptable avocado has been found to be resistant. Seedlings of *Persea baconia* are highly resistant, but are not graft-compatible with the commercial avocado. The extensive suscept range of the fungus indicates that its food relation with the attacked plant is somewhat primitive, and the possibility of finding resistant varieties seems remote.

THIELAVIOPSIS BLACK ROOT ROT

Black root rot, first observed in tobacco, affects at least twenty-nine other species in the United States and Europe. It has been reported that 75–100% of the plants in an affected tobacco field may be destroyed by the disease.

Occasionally, the causative fungus may be responsible for the damping-off of seedlings, but it is usually restricted to the roots of established plants. In mature plants, infected cortical tissues of the tap root and the lateral roots become necrotic and turn dark brown or black. The stems of field-grown plants pull away easily from the diseased roots. When the tap root is destroyed rapidly, the formation of lateral roots is stimulated; when these roots are not seriously affected, the plant may produce a crop despite the infection. Plants whose roots have been severely affected become stunted, and their foliage turns yellow (Figure 12-1).

Etiology

The causal agent is the imperfect fungus *Thielaviopsis basicola* Kopf. Evidence that it is a form of the ascomycete *Thielavia basicola* is still inconclusive. The imperfect form of the fungus is always associated with black root-rot disease; the perfect stage, if it exists, seems to have nothing to do with disease development.

The mycelium is dark, and is characterized by dichotomous branching. Short chains of dark brown chlamydospores (Figure 12-2) form in

FIGURE 12-1
Resistant healthy tobacco plants (*left*) and susceptible ones with black root rot (*right*). (Courtesy of R. D. Lumsden, ARS–USDA, Beltsville.)

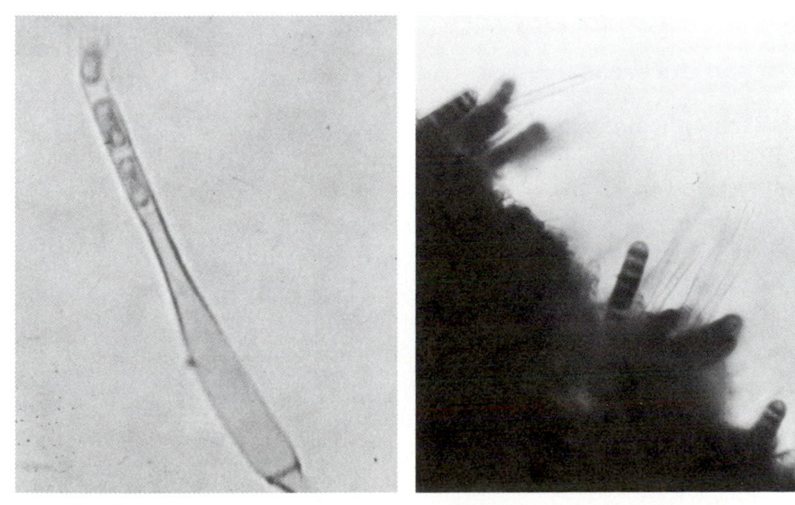

FIGURE 12-2
Endoconidia (*left*) and chlamydospores (*right*) of *Thielaviopsis basicola.*

fascicles on short stalks. Each thick-walled spore has the appearance of a thick disc. Individual spores break apart at maturity. Thin-walled cylindrical endoconidia (Figure 12-2) form at the tips of hyphae. These are released from the "conidiophore" when its tip ruptures at about the time the conidia mature.

As might be expected, the fungus survives between crop seasons as chlamydospores in the soil or in plant debris. Conidia, being thin-walled, can survive neither desiccation nor alternate freezing and thawing. Thus, they do not survive as over-wintering structures.

Surviving chlamydospores germinate in moist soils and inoculation is accomplished when the hyphae touch the roots of susceptible plants. The prepenetration stage is of short duration, and consists of the growing of hyphae along the surface of the roots. Ingress occurs at the juncture of primary and lateral roots. A small wound results when the lateral root, which arises from the pericycle, emerges through the cortex and epidermis of the primary root. If the attacking fungus is present in the area of one of these natural wounds, it grows into the root and initiates disease. The invading fungus ramifies between the cortical cells of the affected root, causing a brown and then black necrosis.

The disease is most severe in moist soils that are rich in humus and have high fertility. The disease develops over a wide range of environmental conditions, and no single environmental factor can be manipulated sufficiently to give control of the disease.

Prevention of Disease

The pathogen is not usually subject to exclusionary measures, but the disease it causes in poinsettia is sometimes controlled in greenhouses through the use of disease-free planting stock. The fungus may be eradicated from soils used in the greenhouse production of susceptible ornamental plants. It may also be eradicated from plant-bed soils, such as those used in the culture of tobacco. Greenhouse soils are usually treated with steam heat, whereas field soils are treated with such chemicals as methyl bromide. The treatment recommended for the sterilization of soil in the control of damping-off is also recommended for the control of the black root-rot disease.

Although the fungus has a wide range of suscepts, it can be eradicated in part from field soils by means of crop rotations that include corn, small grains, and grasses. It is common practice in Canada, for example, to rotate flue-cured tobacco with rye in a two-year rotation.

Most varieties of tobacco are somewhat resistant. Resistant varieties of other crops, however, have not been found satisfactory for commercial use. Physiological races of *Thielaviopsis* exist, and crop varieties resistant in one geographic area may be susceptible in another.

MUSHROOM ROOT ROTS OF TREES

Mushroom root rots, which are of world-wide distribution, cause serious losses in fruit and forest trees and in woody ornamental plants. The mushroom root rot that occurs in relatively heavy soils of the cooler parts of the temperate zones is caused by *Armillaria mellea* (Vahl) Quélet, whereas the one that occurs in well-drained soils in warmer parts of the temperate zone and in the tropics is caused by *Clitocybe tabescens* (Scopoli ex Fries) Bresadola. The two fungi evoke the same symptoms in diseased plants, but the diagnostic signs of the pathogens differ.

Usually, the first evidence of disease is a slow decline and a yellowing of foliage. The disease begins in the root system when lateral roots are girdled by the fungus. Occasionally, in such plants as citrus and cherry, the necrotic zone is limited by production of cankers, and gummosis occurs near the crown. Normally, the necrotic area is restricted to the root system, but sometimes a canker will form at the base of the stem.

Etiology

Whether a plant is infected by *Armillaria* or by *Clitocybe* can be determined when diagnostic signs are viewed beneath the bark of diseased roots. *Armillaria* produces coarse, sclerotized, shoestringlike structures that are called rhizomorphs. These consist of strands of hyphae bundled together and enclosed within suberized cells. *Clitocybe* does not produce rhizomorphs, but white or cream-colored mycelial fans are readily visible in the wood just below the bark of diseased roots. Both fungi are basidiomycetes and belong to the family Agaricaceae. The fruiting bodies are produced in severely diseased roots, and emerge through the soil near the base of the trunk. Both fungi produce mushrooms in groups; *Armillaria* produces mushrooms (Figure 12-3) from the rhizomorphs (Figure 12-4), whereas *Clitocybe* produces mushrooms from the mycelial mats. Moreover, the fruiting body of *Armillaria* has a ring (annulus) around the stalk, whereas the stalk of the mushroom produced by *Clitocybe* is bare.

Inoculation occurs when mycelium comes into contact with the

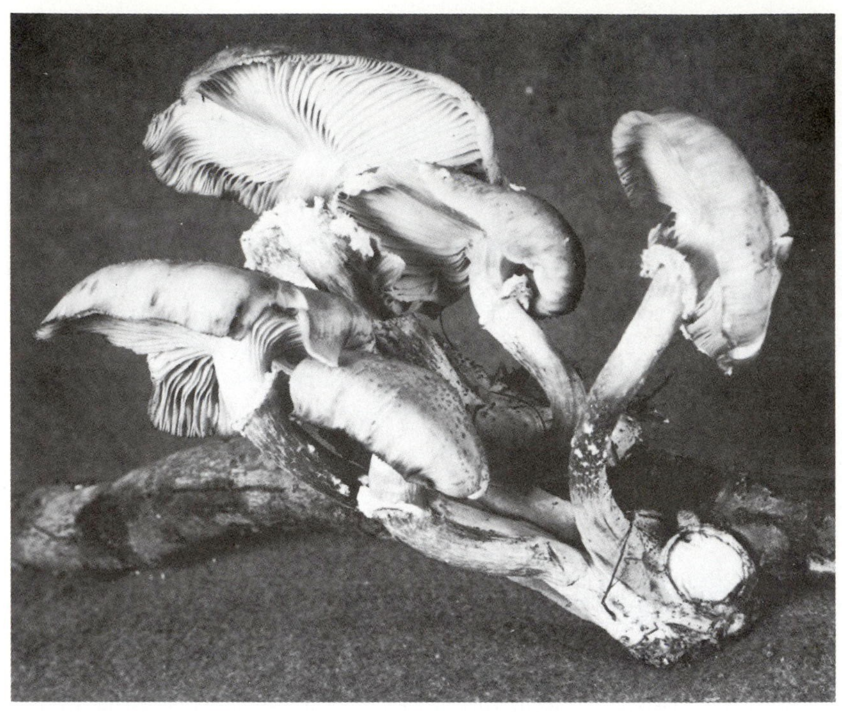

FIGURE 12-3
Mushrooms of *Armillaria mellea* arising from rhizomorphs.

young roots of a susceptible plant; in *Armillaria,* the mycelium is in the form of rhizomorphs. During the prepenetration stage, the strands of mycelium grow along the surface of the root that is to be attacked; these structures adhere to the surface by means of a gelatinous secretion. Ingress is by direct penetration, and the process is essentially the same for both fungi. The young hyphae that are held fast to the surface of the root apparently secrete enzymes that partially digest the cell walls of young roots. Branches from these hyphae then grow into the root and move between the cells. In *Armillaria,* the tips of the rhizomorphs are not suberized, but suberin develops in the hyphal cells of the infection court shortly after ingress has taken place. Both fungi rot the wood by destroying the lignified tissues. Once a plant has been invaded, the fungus continues to ramify through the tissues, even after the plant has died. Thus, both *Armillaria* and *Clitocybe* are capable of surviving for long periods of time in the decayed roots of diseased plants. *Armillaria* may also grow into the trunk of a tree that has been killed by the disease. Sometimes it can be found ten

FIGURE 12-4
Rhizomorphs of *Armillaria mellea* on the base of a tree killed by mushroom root rot.

to twenty feet above the ground in trees that have remained standing two to four years after having been killed (Figure 12-5).

Although the root-rot disease develops most rapidly in warm, moist soils, there is apparently no way in which the disease can be controlled through manipulation of the environment.

Prevention of Disease

Both fungi can be excluded from an area if care is taken to insure that nursery stock imported to that area is disease-free. Once either

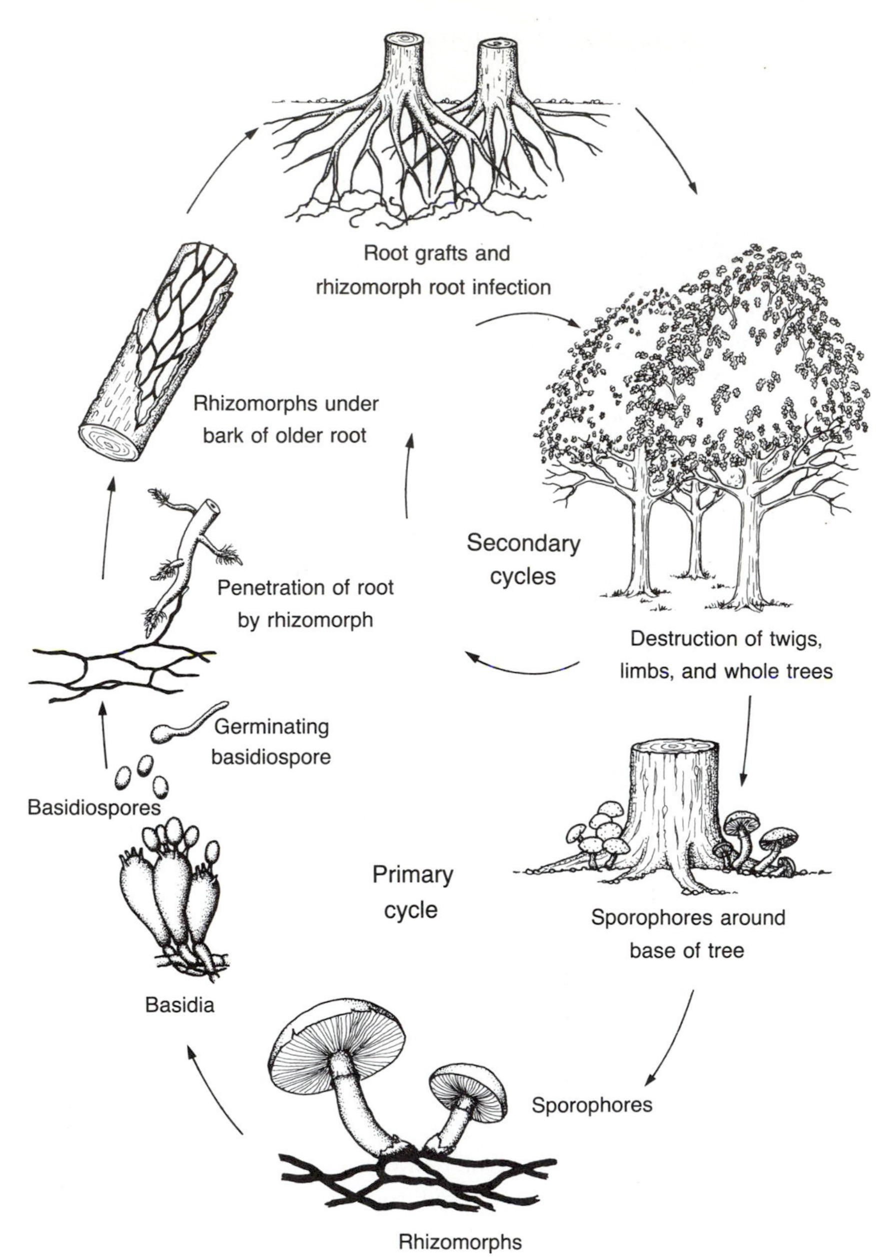

FIGURE 12-5
Shoestring (*Armillaria*) root rot of trees.

fungus has been established in an area, it may be eradicated by removing diseased plants and burning diseased root systems, or by chemical treatment of the soil in small areas. Infested soil may be treated successfully with methyl bromide, formaldehyde, or carbon disulfide. When the disease is discovered early, an individual ornamental plant or fruit tree may be saved by surgery. All diseased roots are removed and destroyed, and the cut surfaces are treated with a wound dressing.

Fruit trees in the vicinity of a diseased tree may be protected if the diseased tree is removed and a trench is maintained around the area in which it grew. If the trench is two or three feet deep, the fungus will be unable to grow through the soil to the root systems of adjacent trees. This practice, however, is usually not feasible, because to maintain an open trench for the 1–2 years required for the decomposition of previously infected roots would require periodic digging to remove the soil that would accumulate from erosion of the walls of the trench.

In citrus, the severity of the disease is reduced by the use of resistant rootstocks, such as sour orange. Grapefruit rootstock is also somewhat more resistant than sweet orange or rough lemon.

In forests, *Armillaria* root rot is less severe in vigorously growing trees. If diseased trees are ringed prior to felling, the stumps that remain after harvest will be low in carbohydrate and will not, therefore, support the rapid growth of the fungus as a saprophyte. Moreover, the stumps of infected trees often become contaminated with fungi that are antagonistic to *Armillaria mellea.* Such fungi (including *Fomes marginatus, Trametes odorata, Polyporus fibrillosus,* and *Coniophora cerebella*) can be grown in culture and then used to infest the stumps of trees infected by *Armillaria mellea.*

Selected References

PHYTOPHTHORA ROOT ROTS OF WOODY PLANTS

Blackwell, E., 1943. The life history of *Phytophthora cactorum* (Leb. and Cohn.) Schroet. *Trans. Brit. Mycol. Soc.* 26:71–89.

Knorr, L. C., R. F. Suit, and E. P. Ducharme, 1957. Handbook of *Citrus* diseases in Florida. *Fla. Agr. Exp. Sta. Bull.* (587).

Stuntz, D. E., and C. E. Seliskar, 1943. A stem canker of dogwood and madrona. *Mycologia* 35:207–221.

Torgeson, D. C., 1954. Root rot of Lawson cypress and other ornamentals caused by *Phytophthora cinnamomi*. *Contrib. Boyce Thompson Inst. Plant Res.* 17:359–373.

Zentmyer, G. A., A. O. Paulus, and R. M. Burns, 1962. Avocado root rot. *Calif. Agr. Exp. Sta. Ext. Serv. Circ.* (511).

THIELAVIOPSIS BLACK ROOT ROT

Clayton, E. E., 1969. Studies of resistance to black root rot disease of tobacco. *Tobacco (New York)* 168:30–37.

Conant, G. H., 1927. Histological studies of resistance in tobacco to *Thielavia basicola*. *Amer. J. Bot.* 14:457–480.

Gayed, S. K., 1969. The relation between tobacco leaf and root necrosis induced by *Thielaviopsis basicola* and its bearing on the nature of tobacco resistance to black root rot. *Phytopathology* 59:1596–1600.

Gilbert, W. W., 1909. The root-rot of tobacco caused by *Thielavia basicola*. *U. S. Dep. Agr. Bur. Plant Ind. Bull.* (158).

Peters, L., 1921. Zur Biologie von *Thielavia basicola* Zopf. *Ber. Biol. Reichanst. Land- Forstwirt.* 16:63–74.

Stover, R. H., 1950. The black root rot disease of tobacco, I. Studies on the causal organism *Thielaviopsis basicola*. *Can. J. Res.* (C)28:445–470.

Todd, F. A., 1967. Control of black root rot of Burley tobacco. *N. C. Agr. Ext. Serv. Ext. Circ.* (477).

MUSHROOM ROOT ROTS OF TREES

Cooley, J. S., 1943. *Armillaria* root rot of fruit trees in the eastern United States. *Phytopathology* 33:812–817.

Kaufman, C. H., 1923. The genus *Armillaria mellea* in the U. S. and its relationships. *Pap. Mich. Acad. Sci. Arts Lett.* 2:53–67.

Rhoads, A. S., 1948. *Clitocybe* root rot of citrus trees in Florida. *Phytopathology* 38:44–61.

———, 1956. The occurrence and destructiveness of *Clitocybe* root rot of wood plants in Florida. *Lloydia* 19:193–240.

Thomas, H. E., 1934. Studies on *Armillaria mellea* (Vahl) Quel., infection, parasitism, and host resistance. *J. Agr. Res.* 48:187–218.

13

Diseases Affecting Meristematic Activity—Bacterial and Fungal Galls

The bacteria and fungi that induce the formation of galls in plants adversely affect growth processes, largely by stimulating mature cells to resume meristematic growth. Many of these microorganisms survive for a time in the soil, and may be eradicated through the practice of crop rotation. Others survive between growing seasons as spores on perennial plants, and are vulnerable to eradication by means of fungicidal sprays applied during the suscept's dormancy period. Control of some gall diseases is achieved through the use of resistant varieties of crop plants.

CROWN GALL

Unlike other plant-pathogenic bacteria, which induce necroses, the crown-gall bacterium induces the diagnostic symptom of tumefaction (Figure 13-1). The galls, which are most injurious to young, rapidly growing plants, may weigh as little as a fraction of an ounce or as much as 50 pounds. In herbaceous plants, the galls are composed of tender parenchymatous tissues that are subject to decay by rot-induc-

FIGURE 13-1
Crown gall of rose.

ing organisms. In woody plants, however, the galls are composed of strong cells that are not likely to decay. Galls are produced as a result of hyperplasia and hypertrophy. The outer portions of the galls crack and the tissues containing the bacteria slough off and fall to the soil.

Etiology

The crown-gall bacterium, *Agrobacterium tumefaciens,* is a short rod with peritrichous flagella; it survives in organic debris in the soil, and is carried to the infection court by insects or by splashing water. The bacterium enters plants only through wounds, including graft unions of susceptible nursery stocks, and grows intercellularly. It secretes a toxic tumor-inducing principle that stimulates adjacent cells to divide, thus causing the hyperplasia. The resulting tumors are not self-limiting, because their cells retain the ability to divide long after the bacteria themselves have ceased to grow.

In some infections, secondary galls may appear at nodes some distance from the original infection; this condition suggests the metastasis of tumors in animals. At one time, these secondary galls were thought to be connected with the primary ones by strands of bacteria; later, it was discovered that primary and secondary galls are not necessarily connected. The size of a gall in a given infection court seems to vary with the size of the original wound through which the bacteria entered. Thus, a small gall develops around a small wound, whereas a large gall forms around a large wound. Apparently, two phenomena—conditioning and induction—are involved in gall formation. The first is accomplished only by wounding, and only those cells predisposed or conditioned by wounding are capable of responding hyperplastically to the "tumor-inducing principle."

Prevention of Disease

The disease is most serious in woody-ornamental and fruit-tree nurseries, where it is controlled by using uninfested soil and by carefully wrapping graft unions with antiseptic tape. It is further controlled through the elimination of soil-infesting insects. In commercial orchards and plantings of woody ornamentals, the disease is controlled by the exclusionary practice of using plants certified to be free of crown gall. The bacteria can be eradicated from gall tissue by painting the gall and a portion of the healthy tissue around it with a mixture of hydrocarbons manufactured under the name Bacticin. This treatment does not kill contiguous healthy tissue of the plant. It is recommended, however, when galls are detected early. If the disease has progressed extensively, the plant may be removed, the soil treated with a chemical such as methyl bromide, and a new, healthy plant set in the disinfected soil.

CLUB ROOT OF CRUCIFERS

Club-root disease may seriously affect cruciferous plants in the cooler regions of the temperate zones. As a result of tumefaction in their root systems, diseased plants are stunted and may never produce marketable products.

FIGURE 13-2
Club root of cabbage.

Although yellowing and dwarfing—the first above-ground symptoms of the disease—may appear in young plants in nursery beds, it is not uncommon to find that transplants in the incipient stages of infection have been inadvertently set out in the field. When poorly growing plants are lifted, the diagnostic symptoms of club root may be observed (Figure 13-2). Galls are often spindle-shaped and may involve the entire diameter of the diseased root. Branch roots frequently emerge from galls, giving them a hairy appearance.

Etiology

Club root is caused by the apparently primitive fungus *Plasmodiophora brassicae* Woronin, which, like the other primitive fungi, is characterized by holocarpic reproduction and by a naked amoeboid thallus (plasmodium) instead of a mycelial thallus (Figure 13-3).

Structures produced in sequence during the life cycle of *Plasmodiophora brassicae* are the hypnospore, zoospore, amoeboid thallus, zoosporangia, zoospores (zygotes?), plasmodium, and hypnospores. The hypnospore, sometimes simply referred to as the spore or the resting spore, forms within an infected cell and is liberated into the soil upon disintegration of the roots after the growing season. The hypnospore lies dormant throughout the winter in cold climates, and is the means by which the fungus survives between growing seasons. The reason that club root is practically nonexistent in the southern United States may be that hypnospores are unable to survive during the summers between the winter growing seasons. Because spores do not require a period of rest, they probably germinate in warm soils whenever moisture is present. Thus, when succeeding crops are planted, no inoculum has survived to perpetuate the disease. Occasional outbreaks of club root of crucifers in the southeastern United States have been due to the use of infected transplants from more northern regions. There have been no reports of *P. brassicae* surviving in warm soils so infested.

In those regions where club root occurs, the hypnospore germinates to produce a single biflagellate zoospore, which, upon contacting a root hair, loses or withdraws its flagella, becomes amoeboid, and enters the epidermal cell. Presumably, ingress is accomplished through enzymatically catalyzed reactions that result in digestion of the thin wall of the root hair. Once inside the plant, the fungus produces an amoeboid thallus that soon gives rise to zoosporangia. These liberate zoospores that apparently emerge from the diseased root. It has been postulated—though not proved—that these zoospores are of different mating types, and that their copulation gives rise to zygotes that are capable of entering the root. In any event, the cells of infected roots contain naked fungal protoplasts, or plasmodia. The infected cells enlarge and are stimulated to divide. The plasmodium may be capable of moving from one cell to another; it is certain that portions of plasmodia are cut off with suscept-cell cytoplasm during cell division. The extensive hyperplasia and hypertrophy result in the "clubs" or tumefactions diagnostic of the disease. Most cells in galled tissues

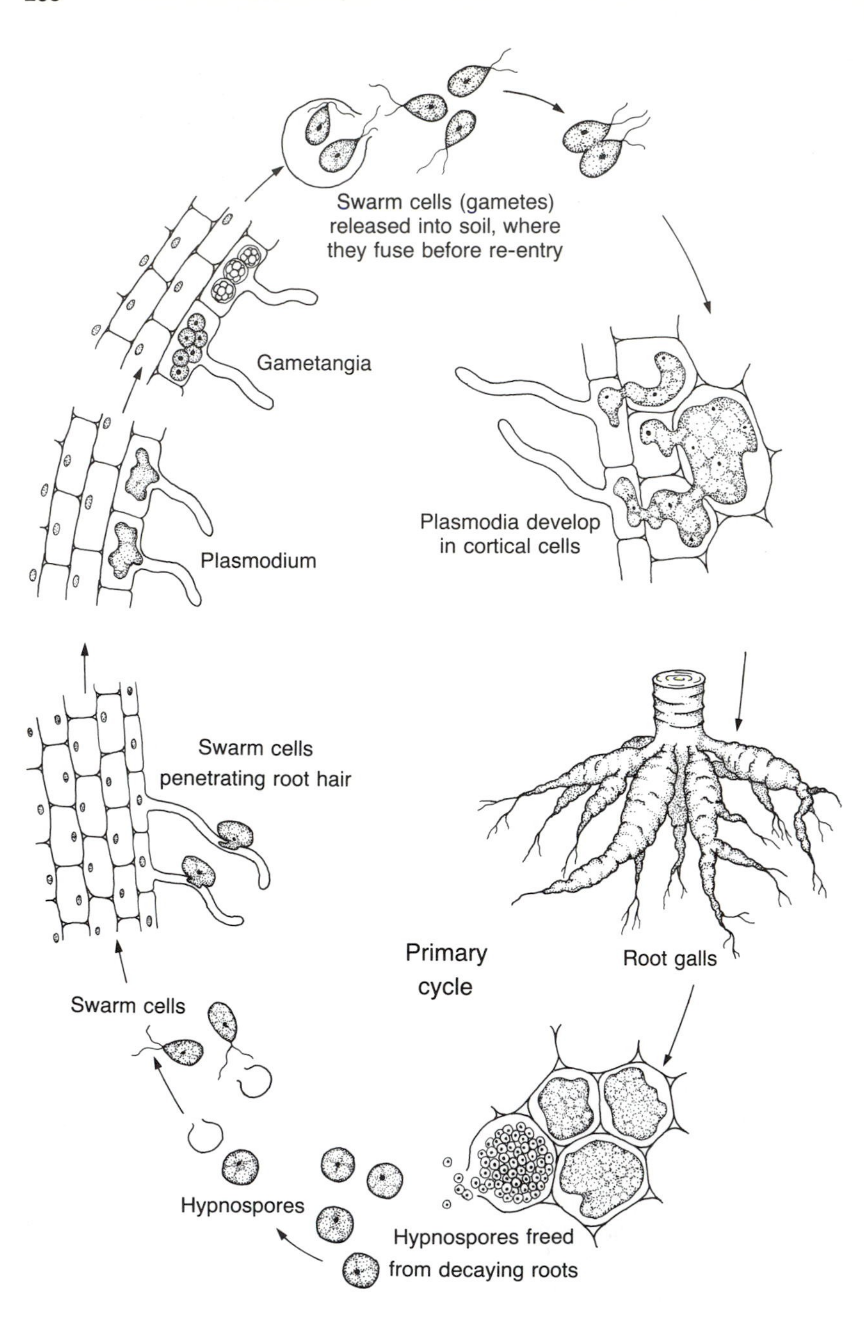

FIGURE 13-3
Club root of crucifers.

fail to differentiate, but remain meristematic throughout the course of the disease.

Toward the end of the fungus cycle, the protoplasm of the plasmodium breaks into units around which walls form, and these units become hypnospores. Presumably, meiosis occurs in the plasmodium prior to spore formation; the hypnospores, therefore, are thought to contain a single haploid nucleus.

Epidemiology

The optimum soil temperature for the development of club root is 27–30°C, but it may occur at any soil temperature between 9°C and 30°C.

Although the development of several diseases caused by root-invading organisms is affected by the *p*H of the soil, club root of crucifers is one of two such diseases so strongly affected by the soil reaction that it can often be controlled simply by adding lime to make the soil alkaline. Hypnospores do not germinate when the *p*H of the soil solution is higher than 7.2. (The other plant pathogen that is strikingly affected by the *p*H of the soil is *Streptomyces scabies,* which causes the scab disease of potatoes. This actinomycete is not active in acid soils, and it is a common practice among growers of potatoes to add acid fertilizers to the soil to lower its *p*H to approximately 5.5.)

Prevention of Disease

The club-root pathogen should be excluded from cabbage fields through the use of disease-free transplants. The fungus can be eradicated from the soil of plant beds by sterilizing the soil with heat or such chemicals as chloropicrin, methyl bromide, Vapam, or Vorlex. It cannot be eradicated profitably from field soils by this method, but good field control is possible through the use of Terraclor (PCNB) in the cabbage transplant water. Eradication through crop rotation is relatively ineffective: in the soils of the cooler regions of the temperate zones, hypnospores can survive for at least ten years. Liming of the soil to make it alkaline (*p*H 7.2) controls the disease in most mineral soils. This protective measure is not recommended when the crop is to be grown in a peat soil of high buffering capacity. Several varieties of cabbage, cauliflower, and broccoli are somewhat resistant to club root, and are recommended for use in infested areas.

PEACH LEAF CURL

The leaf-curl disease affects peaches, apricots, and nectarines, and occurs wherever these suscepts are grown. A severely diseased tree is weakened as a result of losing the leaves of its first flush of growth. Diseased fruit drop prematurely.

The first evidence of disease is the malformation of young leaves. Usually, there is a downward bending of the midrib, accompanied by savoying of laminar tissue, which becomes light green, yellow, or red (Figure 13-4). Diseased leaves become thicker and more brittle than comparable healthy ones. Fruit infection results in malformation, anthocyanescence, and corkiness of affected areas. Because they drop prematurely, however, diseased fruits are seldom observed.

Etiology

Peach leaf curl is caused by the ascomycete *Taphrina deformans* (Berkeley) Tulasne. The fungus produces no fruiting body, the asci

FIGURE 13-4
Leaf curl of peach.

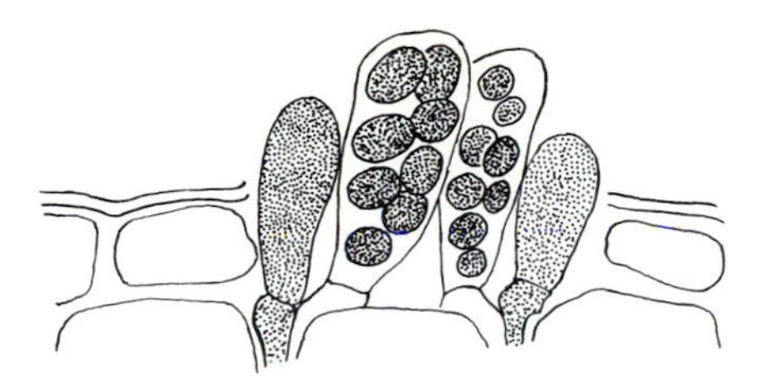

FIGURE 13-5
Asci of *Taphrina deformans* emerging from a diseased leaf.

developing in masses just below the cuticle (Figure 13-5). The asci impart a whitish color to the surfaces of puckered leaves.

When ascospores produced in diseased leaves during the spring are liberated, they may lodge in cracks in the bark of twigs and in the buds of the infected tree. Perhaps some ascospores survive there until the following spring. Others, however, germinate by budding to give rise to secondary unicellular hyaline spores, or conidia. These then produce secondary conidia, some of which survive the winter in bud scales or crevices in the bark. In the spring of the following year, conidia or ascospores that have survived in the tree germinate on the surfaces of immature leaves and send forth germ tubes that penetrate directly through the uncuticularized surfaces of the leaves; the fungus cannot penetrate leaves once cutin has been laid down on the outer walls of their epidermal cells. The fungus develops an intercellular mycelium that ramifies among the cells of the epidermis and palisade parenchyma, and forms a mat under the cuticle. The subcuticular mycelium is dicaryotic, plasmogamy having occurred earlier through anastomosis of the infecting hyphae. Caryogamy occurs in terminal cells of the subcuticular mycelium. These cells enlarge to form asci; ascospores are cut out around the haploid nuclei that result from meiosis in the immature asci.

Epidemiology

Because peach leaf curl is most destructive during the cool, moist weather of early spring, and because its ascospores and conidia germinate best near 20°C, it seems reasonable to assume that the disease requires cool weather for its development. This assumption is invalid, however, in view of the fact that only the immature tissues can be penetrated, and only these tissues are capable of responding pathologically to the pathogen. The disease develops during the cool weather of early spring, not because the weather is cool, but because susceptible tissue is available only at that time.

Prevention of Disease

The fact that *Taphrina deformans* survives the winter only as ascospores or conidia in the tree is the basis for effective control by means of a single eradicative spraying during the dormant season. Although the best results have been obtained with sprays applied in the early spring just before the buds break, spraying in the autumn has also been effective. Bordeaux mixture, ferbam, and ziram are recommended. Lime-sulfur spray has been recommended where scale insects are also to be controlled.

Selected References

CROWN GALL

Banfield, W. M., 1934. Life history of the crown-gall organism in relation to its pathogenesis on the red raspberry. *J. Agr. Res.* 48:761–787.

Braun, A. C., 1954. The physiology of plant tumors. *Annu. Rev. Plant Physiol.* 5:133–162.

Robinson, W., and H. Walkden, 1923. A critical study of crown gall. *Ann. Bot.* (*London*) 37:299–324.

Schroth, M. N., and D. C. Hildebrand, 1968. A chemotherapeutic treatment for selectively eradicating crown gall and olive knot neoplasms. *Phytopathology* 58:848–854.

Smith, E. F., N. A. Brown, and L. McCulloch, 1912. The structure and development of crown gall: a plant cancer. *U. S. Dep. Agr. Bur. Plant Ind. Bull.* (255).

———, ———, and C. O. Townsend, 1911. Crown gall of plants: its cause and remedy. *U. S. Dep. Agr. Bur. Plant Ind. Bull.* (213).

CLUB ROOT OF CRUCIFERS

Ayers, G. W., 1944. Studies on the life history of the club root organism, *Plasmodiophora brassicae. Can. J. Res.* (C)22:143–149.

Chupp, C., 1917. Studies on clubroot of cruciferous plants. *Cornell Univ. Agr. Exp. Sta. Bull.* (387):319–452.

———, 1928. Club root in relation to soil alkalinity. *Phytopathology* 18: 301–306.

Colhoun, J., 1958. Club root disease of crucifers caused by *Plasmodiophora brassicae* Woron. *Commonw. Mycol. Inst. Phytopathol. Pap.* (3).

Gallegly, M. E., and C. F. Bishop, 1955. Pentachloronitrobenzene for control of clubroot of crucifers. *Plant Dis. Rep.* 39:914–917.

Kunkel, L. O., 1918. Tissue invasion by *Plasmodiophora brassicae*. *J. Agr. Res.* 14:543–572.

Walker, J. C., and R. H. Larson, 1960. Development of the first club root resistant cabbage variety. *Wis. Agr. Exp. Sta. Bull.* (547):12–16.

Williams, P. H., 1966. A cytochemical study of hypertrophy in clubroot of cabbage. *Phytopathology* 56:521–524.

Woronin, M., 1878. *Plasmodiophora brassicae,* Urheber der Kohlpflanzen-Hernie. *Jahrb. Wiss. Bot.* 11:548–574. (English transl. by C. Chupp, 1934. The cause of cabbage hernia. *Phytopathol. Classics* 4:1–32).

PEACH LEAF CURL

Fenwick, H. S., 1961. Peach leaf curl. *Idaho Agr. Exp. Sta. Res. Bull.* (366).

Foster, H. H., and D. H. Peterson, 1952. The peach leaf curl epidemic of 1951. *Plant Dis. Rep.* 36:140–141.

Martin, E. M., 1940. The morphology and cytology of *Taphrina deformans*. *Amer. J. Bot.* 27:743–751.

Mix, A. J., 1935. The life history of *Taphrina deformans*. *Phytopathology* 25:41–56.

Wallace, E., and H. H. Whetzel, 1910. Peach leaf curl. *Cornell Univ. Agr. Exp. Sta. Bull.* (276):157–178.

14

Diseases Affecting Meristematic Activity—Smuts

Although all the various smut diseases are diagnosed by signs—black masses of spores in sori—rather than by symptoms, and although they are all similar in that the growth processes of the attacked plant are adversely affected, the smut diseases actually make up a heterogeneous group. For example, onion smut is much like a damping-off disease, in that stem tissues are susceptible for only 10–20 days after emergence; corn smut is a typical hyperplastic gall disease; and others (including loose smut of oats) induce hypoplasia in apical meristems and mild hyperplasia in developing fruit.

On the basis of the nature of infections they cause, the smut fungi may be termed either localized or systemic. Those that are systemic invade the suscept either through seedling stems or through the stigma and style, attacking the ovary or embryo. A number of smut fungi that invade seedling stems survive as smut spores on the seed. Thus, seed dressing with chemicals eradicates those fungi from the seed and excludes them from fields in which treated seed are sown. Chemical seed dressings do not adequately protect seedlings from invasion by most smuts that survive in the soil. Onion smut may be controlled, however, by applying fungicides to the furrow when the seeds are sown: the fungus is eradicated from the soil adjacent to the seed. Until recently, a smut fungus surviving within infected embryos could not be eradicated from the seed except by drastic treatment, such as the use of heat sufficient to kill the fungus but insufficient to kill the embryo. There is now available, however, a seed-treating chemical that acts systemically to eradicate the fungi of loose smut of wheat and loose smut of barley from infected embryos. The chemical, 2,3-dihydro-5-carboxanilido-6-methyl-1,4-oxathiin, or DCMO, is available as a dust, a wettable powder, a liquid, or as a mixture of granules (see von Schmeling and Kulka, 1968).

Localized smut fungi may infect any meristematic tissue. These organisms usually survive in the soil, and control is frequently effected through the eradication of the fungi by means of crop rotation. The smuts are parasitic pathogens having a specialized food relations with their hosts; therefore, it is often possible to obtain disease control through breeding and selection for resistance.

The smuts belong to the subclass Heterobasidiomycetes, and are characterized by distinctive sexual spores, the **smut spores** (chlamydospores). These thick-walled resting spores develop in definite structures, called **sori,** that are at first enclosed by a membrane. The mass of smut spores in a sorus is black, and the presence of a number of sori on suscept tissue gives it a blackened or charred appearance.

Smut spores develop only on binucleate mycelia; each binucleate "cell" contains one nucleus of one mating type (+) and one of the other type (−). At first, the binucleate hyphae branch profusely and form tissues of small cells whose membranes swell; the cells appear to be imbedded in a gelatinous sheath. Single cells become thin-walled, irregularly shaped binucleate spores. Caryogamy occurs during spore development. At maturity, spores consist of thin-walled endospores and thick-walled, dark exospores that are often characteristically sculptured.

The smuts belong to the order Ustilaginales, which contains the families Ustilaginaceae and Tilletiaceae. All of the economically important plant pathogens of the order are members of these two families.

In the Ustilaginaceae, the smut spore forms, upon germination, a **promycelium,** a tubular outgrowth that bears spores termed **sporidia** (Figure 14-1). Promycelia and sporidia are analogous to basidia and basidiospores, respectively. The diploid nucleus moves from the smut spore into the promycelium and undergoes meiosis. Three septa form in the promycelium, separating the four haploid nuclei. Sporidia bud off indefinitely from the cells of the promycelium. Sporidia may

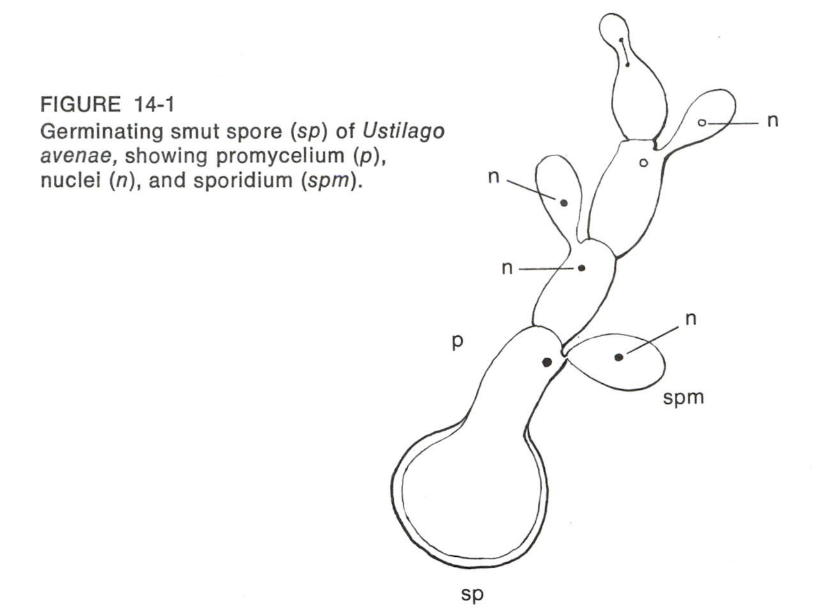

FIGURE 14-1
Germinating smut spore (*sp*) of *Ustilago avenae,* showing promycelium (*p*), nuclei (*n*), and sporidium (*spm*).

copulate directly or by means of copulation tubes, thus effecting plasmogamy. The developing dicaryotic mycelium may then invade susceptible tissues. In the corn-smut fungus, however, plasmogamy can occur within invaded plant tissue through the fusion (**anastomosis**) of haploid hyphae of different mating types.

In the Tilletiaceae, meiosis and one or more mitotic nuclear divisions occur in the smut spore (Figure 14-2). The haploid nuclei migrate into a nonseptate promycelium. In most species, a number of sickle-shaped sporidia develop in a whorl on the top of the promycelium. One haploid nucleus migrates into each sporidium, which is then separated from the promycelium by a septum. Plasmogamy occurs in place when copulation branches from two sporidia fuse. The nucleus from one sporidium moves into the other, which germinates to form slender binucleate hyphae that cut off binucleate, falcate, secondary sporidia (conidia). Secondary sporidia germinate and give rise to binucleate infection hyphae.

CORN SMUT

Corn smut, or boil smut, which occurs only on corn (*Zea mays*) and teosinte (*Euchlaena mexicana*), effects losses in these suscepts wher-

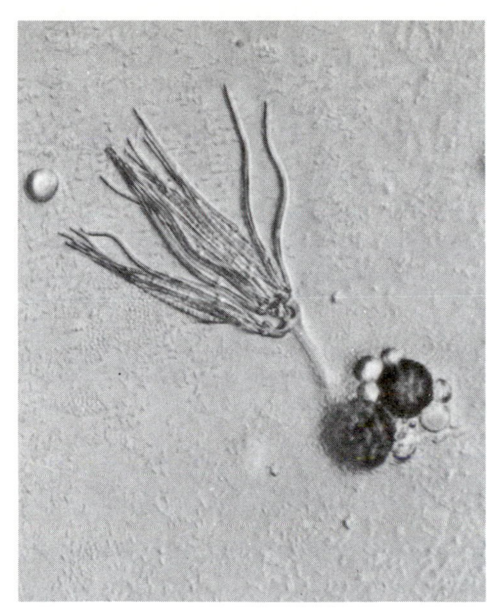

FIGURE 14-2
Germinating smut spore of *Tilletia contraversa*.

ever they are grown, with the possible exception of Australia. Losses sometimes amount to as much as two-thirds of certain crops of highly susceptible sweet corn, but average losses in the United States are only slightly more than 2% of the crop.

Smut galls may occur in any above-ground meristematic tissues, but are most common in tassels, ears, and at the nodes of stems (Figure 14-3). Each gall consists of a whitish membrane of hypertrophied suscept tissue that encloses a black mass of smut spores. The entire structure is termed a sorus. Galls on leaves are small, whereas those on stalks or ears may reach the size of a coconut.

Etiology

Corn smut is caused by *Ustilago maydis* (De Candolle) Corda, a heterobasidiomycete of the order Ustilaginales and the family Ustilaginaceae. The dark, thick-walled smut spores are formed directly on the binucleate mycelium, each cell of which rounds off and forms a spore. Because they are formed directly on the mycelium, smut spores are often called chlamydospores. They function, however, as teliospores, since they are the organs of caryogamy and meiosis. To term such spores chlamydospores is technically incorrect, because

FIGURE 14-3
An ear of corn affected by corn smut.

chlamydospores are thick-walled asexual spores formed directly on mycelium. To call them teliospores, while technically correct, leads to confusion with the rust fungi, for whose resting spores the term is usually reserved. Because these spores are unique to the smut fungi, it is appropriate to call them simply "smut spores."

The spores of *Ustilago maydis* are spherical, 7–12 μ in diameter,

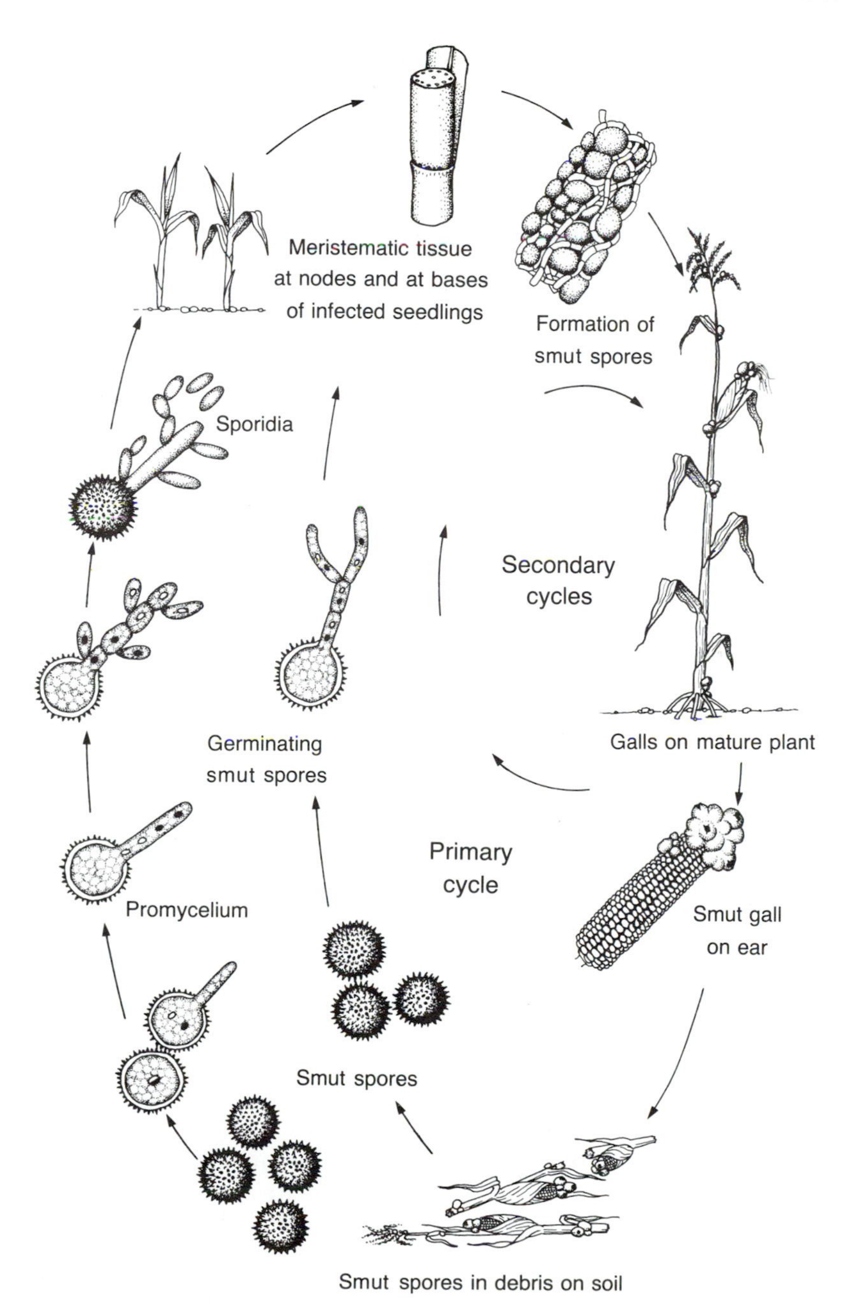

FIGURE 14-4
Corn smut.

with the outer wall marked by spines (echinulations). Galls on diseased plants dry out at the end of the growing season, at which time their spores are released. They survive between seasons in the soil, in crop refuse, and, if ingested by farm animals, in manure. Smut spores do not require a rest period between maturity and germination, but will lie dormant in cold or dry soils (Figure 14-4).

The primary cycle begins when smut spores are splashed by rain onto young plants or when sporidia from germinating smut spores are blown or splashed onto the plants. Spores or sporidia are usually washed downward into the whorl of leaves at the top of a young plant or into the leaf sheaths at the nodes of stems. As described in the preceding section, smut spores of *Ustilago* produce septate promycelia on which successive haploid sporidia are formed. Each sporidium may germinate to produce a germ tube, the tip of which becomes appressed to the surface of the infection court. Uninjured suscept tissue may be penetrated directly by the penetration tube that branches from beneath the tip of the germ tube. Hyphae then ramify between the cells of invaded tissues. Hyperplasia occurs only when hyphae of opposite mating types fuse to produce dicaryotic mycelium. In *Ustilago maydis,* as in other species of the genus, dicaryotization (plasmogamy) may occur in the infection court as a result of copulation by sporidia, by copulation tubes, or by a sporidium and a copulation tube. When plasmogamy occurs in the infection court, the suscept is subsequently penetrated directly by the dicaryotic mycelium that results.

Once within the suscept, the corn-smut fungus produces feeding hyphae. These hyphae, though mostly intercellular, sometimes penetrate suscept cells. The fungus remains localized in the original infection court, whose tissues become hypertrophied and filled with fungal mycelium, most of which becomes a mass of smut spores.

Epidemiology

Corn smut is most severe when growing seasons are warm, possibly because the optimum temperature for spore germination is 26–34°C. The disease is also favored by succulent growth.

A "warm climate," however, is not necessarily favorable to corn smut, because the disease is more severe in the northern plains of the United States than in the deep southeastern states. In Florida, for example, smut spores that are produced in the summer probably germinate and perish before the next growing season. The disease is more serious in Central America, where two successive crops of corn are grown during the rainy season. Spores from the first crop infect the next one; spores from the second crop lie dormant during the mild, dry winter, and germinate at the onset of the rainy season in the spring, when another crop of corn is growing.

Prevention of Disease

Corn smut is controlled chiefly through the use of resistant varieties. Most varieties of field corn are resistant, the dent varieties being more resistant than the flint varieties. Those varieties whose husks are long and completely cover the ear are morphologically resistant.

Most varieties of sweet corn are susceptible, but the disease can be controlled in vegetable farms by three-year rotations with other crops. In those areas where *Helminthosporium* leaf blight is controlled by protective spraying, the same fungicides also give some control of corn smut.

BUNT OF WHEAT

Bunt (stinking smut) occurs in wheat wherever the plant is grown. Although modern control practices keep average losses below 10% of the crop, losses as high as 88% have been reported.

Infected plants are dwarfed and bear flowers with large green pistils and small yellow anthers on short stamens. Diseased heads are usually dwarfed and flattened, with loosely packed kernels. In awned varieties, the awns are usually shed. Diseased kernels may appear normal, but are filled with masses of black, foul-smelling smut spores. When diseased kernels are crushed, the black spores become apparent to the senses of sight and smell. The unbelievable stench, said to resemble that of sour herring, is due to the high concentration of diethanolamine.

Etiology

Bunt is caused by either one of two smut fungi of the family Tilletiaceae. One, *Tilletia caries* (De Candolle) Tulasne, has spherical spores 15–20 μ in diameter with a network of ridges (reticulations) on the outer wall. The other, *T. foetida* (Wallroth) Liro, bears smooth-walled ellipsoidal spores that measure 16–18 $\mu \times$ 19–25 μ.

Smut spores usually survive between crop seasons on the surface of seed, but in cool wet areas, as in the Pacific Northwest of the United

States, they may survive in the soil. When wheat seeds germinate at the beginning of the next growing season, smut spores germinate, and sporidia copulate and give rise to secondary binucleate conidia. These produce germ tubes that may contact the germinating seedling in the region of the mesocotyl. Direct penetration occurs and the intercellular mycelium grows to the apical meristem, as in loose smut of oats. The systemic, almost benign mycelium produces smut spores in sori within maturing pericarps.

Epidemiology

Moisture is particularly favorable to the initiation and development of the disease. Although the range of temperature over which spores germinate coincides with that over which wheat seeds germinate, the optimum temperature for spore germination (16–20°C) is significantly lower than the optimum for seed germination (roughly 25°C). Thus, planting wheat in warm soil might reduce the chances of its being infected by the bunt fungi. Unfortunately, this might increase the chances of its being infected by the seedling blight fungus, *Fusarium moniliforme*.

Prevention of Disease

In all areas where the bunt fungi survive only on contaminated seed, the disease is easily controlled by treating the seed with an organic fungicide, such as carboxin, captan, maneb, or thiram. Where the fungi survive in the soil, the disease is usually controlled through the use of resistant varieties, although crop rotation, when feasible, is also effective. Soil treatment with hexachlorobenzene eradicates soil-borne bunt spores. Many durum wheats are resistant to bunt, whereas the *Triticum compactum* types are usually susceptible. *T. vulgare* types are variable with respect to resistance.

A Note on Dwarf Bunt of Wheat

Dwarf bunt is similar to common bunt, but diseased plants are severely stunted and infected heads are characteristically curved. The causal agent, *Tilletia contraversa*, survives in the soil (as well as on the seed) in all wheat-growing areas of the world, and cannot, therefore, be properly controlled by treating the seed with fungicides. The disease may be controlled by planting seed certified to have been harvested from disease-free fields, by treating the soil with hexachlorobenzene, by rotating crops, and by planting resistant varieties of wheat.

LOOSE SMUT OF OATS

Loose smut, sometimes called naked smut or black heads, occurs in species of *Avena* (oats) wherever they are grown. Entire fields may be devastated when badly infested or infected seeds are sown, and average losses in the United States amount to 3–7% of the crop, despite modern control practices.

Loose smut can be recognized by the black sori that replace the kernels in diseased panicles (Figure 14-5). The membranes of suscept tissues rupture and the black powdery masses of smut spores are exposed. Diseased plants are somewhat dwarfed; they mature earlier than healthy ones, produce fewer culms, and give rise to immature tillers.

Etiology

Loose smut of oats is caused by *Ustilago avenae* (Persoon) Jensen. The smut spores are liberated from sori at about the time that healthy plants are in flower. The spores are blown to healthy plants, where they often lodge between the glumes and kernels; typically, the spores remain dormant in stored grain. Smut spores may germinate in floral parts and give rise to an infection of the glumes, of the immature pericarp, or of the epidermis of the ovary. Ingress is by direct penetration, apparently by chemical action as well as by mechanical pressure. When seeds have been merely infested, the smut spores germinate after the seeds are planted at the beginning of the following season, usually at about the same time that the seeds germinate. The germ tube grows along the mesocotyl of the seedling and a penetration tube enters directly. The intracellular mycelium grows toward the apical meristem and literally "grows up" with the developing plant. Similarly, mycelium from infected pericarps (and, possibly, mycelium from infected glumes) may penetrate the seedling and invade meristematic tissues. When the epidermis of the embryo has been infected during the preceding season, mycelium apparently begins to grow from this region to the apical meristem of the young seedling. In any event, the binucleate mycelium, which arises directly from the germinating smut spore, ultimately becomes associated with the apical meristem. It is also found in nodal regions of maturing dis-

FIGURE 14-5
Loose smut of oats: healthy panicle (*left*) and four diseased panicles in different stages of disease development (*right*).

eased plants. The mycelium replaces the floral parts, and the black sori of smut spores constitute the entire panicle of a diseased plant.

Prevention of Disease

The pathogen can be excluded by sowing pathogen-free seed. The fungus can be eradicated from the seed by treating it with organic fungicides (captan or maneb); such treatment is effective with infected as well as infested seed.

The disease may also be controlled by planting resistant varieties. In some such varieties, there is a suppression of mycelial development; in others, there is hypersensitivity, the rapid death of invaded cells resulting in the "starvation" of the parasitic pathogen. In still other resistant varieties, the cytoplasm of invaded cells may surround and ingest the hyphal tips that enter, or the walls of attacked cells may thicken to form pads that prevent their invasion by the hyphae.

Selected References

von Schmeling, B., and M. Kulka, 1968. The systemic fungicide 2,3-dihydro-5-carboxanilido-6-methyl-1,4-oxathiin. *In* J. G. Moseman, ed., *Fungicidal Control of Smut Diseases of Cereals,* pp. 1–2. Beltsville, Md.: USDA Crops Research Division.

CORN SMUT

Christensen, J. J., 1963. Corn smut caused by *Ustilago maydis. Amer. Phytopathol. Soc. Monogr.* (2).

Hanna, W. F., 1929. Studies in the physiology and cytology of *Ustilago zeae* and *Sorosporium reilianum. Phytopathology* 19:415–442.

Potter, A. A., and I. E. Melchers, 1925. Study of the life history and ecologic relations of the smut of maize. *J. Agr. Res.* 30:161–173.

Stakman, E. C., J. J. Christensen, C. J. Eide, and B. Peturson, 1929. Mutation and hybridization in *Ustilago zeae. Minn. Agr. Exp. Sta. Tech. Bull.* (65).

Walter, J. M., 1934. The mode of entrance of *Ustilago zeae* into corn. *Phytopathology* 24:1012–1020.

BUNT OF WHEAT

Churchward, J. G., 1940. The initiation of infection by bunt of wheat (*Tilletia caries*). *Ann. Appl. Biol.* 27:58–64.

Dastur, J. F., 1921. Cytology of *Tilletia tritici* (Bjerk.) Wint. *Ann. Bot.* (*London*) 35:399–407.

Hansen, F., 1958. Anatomical studies on the penetration and the spread of *Tilletia* spp. in cereal plants in relation to the state of development of the host plant. *Phytopathol. Z.* 34:169–208.

Heald, F. D., and H. M. Woolman, 1915. Bunt or stinking smut of wheat. *Wash. Agr. Exp. Sta. Bull.* (126).

Purdy, L. H., E. L. Kendrick, J. A. Hoffman, and C. S. Holton, 1963. Dwarf bunt of wheat. *Annu. Rev. Microbiol.* 17:199–222.

Sartoris, G. B., 1924. Studies in the life history and physiology of certain smuts. *Amer. J. Bot.* 11:617–647.

LOOSE SMUT OF OATS

Gage, G. R., 1927. Studies of the life history of *Ustilago avenae* (Pers.) Jensen and of *Ustilago levis* (Kell. & Swing.) Magn. *Cornell Univ. Agr. Exp. Sta. Mem.* (109).

Holton, C. S., 1932. Studies in the genetics and the cytology of *Ustilago avenae* and *Ustilago levis*. *Minn. Agr. Exp. Sta. Tech. Bull.* (87).

Kolk, L. A., 1930. Relation of host and pathogen in the oat smut, *Ustilago avenae*. *Bull. Torrey Bot. Club* 57:443–506.

Tapke, V. F., 1948. Environment and the cereal smuts. *Bot. Rev.* 14:359–412.

Zade, A., and A. Arland, 1933. The relation of host and pathogen in *Ustilago avenae:* a reply. *Bull. Torrey Bot. Club* 60:77–87.

15

Diseases Affecting Meristematic Activity—Nematode-Induced Diseases

Nematodes are ubiquitous animals; most are free-living saprophytes, but several hundred species are known to parasitize green plants, in which they cause some of the most important plant diseases. The three nematode-induced diseases discussed in this chapter illustrate the several kinds of damage that nematodes cause in plants.

ROOT KNOT

Root knot is distributed throughout the world and occurs in a wide variety of crop plants. It is most severe in areas of warm climate, and causes extensive economic loss in nursery plants as well as in row crops.

The diagnostic symptom of root knot is tumefaction: irregularly shaped galls, which usually involve the entire cross section of the affected root, develop immediately behind the growing tip (Figure 15-1). Apical root growth ceases once the galls begin to form, but root branches often develop from the galls, especially in plants infected by the species of root-knot nematode that is prevalent in the

FIGURE 15-1
Roots of tomato affected by root-knot nematodes.

northern United States. Necrosis occurs in roots of unsuitable hosts at the points of penetration.

Etiology

It has been known for more than a century that root knot is caused by a nematode, and for most of this time it was thought that the causal agent was a single species, known after 1932 as *Heterodera marioni* Goodey. It has long been apparent that different populations of the root-knot nematode varied in their pathogenicity to different species of crop plants, but it was not until 1949 that B. G. Chitwood described five separate species of the root-knot nematode, at the same time separating them from the genus *Heterodera* and placing them in the genus *Meloidogyne*. Since then, several other root-knot nematodes have been described: the genus *Meloidogyne* is now known to contain more than a dozen species. *M. hapla* is common in the northern United States, whereas *M. incognita* and its subspecies are prevalent in the southern states.

The root-knot nematode survives between crop seasons as eggs in the soil. In a warm, moist environment, the eggs hatch, liberating vermiform immature nematodes that are in the second larval stage. The chance encounter of these larvae with the succulent tissues of the root tips of susceptible plants results in inoculation. The short prepenetration stage of pathogenesis consists of the attack of the root tissue by the larvae and is completed when the larvae gain ingress as a result of having pierced the epidermis with their stylets.

Male nematodes live parasitically for several weeks and then undergo three molts in rapid succession before emerging from the root. Presumably, the adult males live free in the soil. Females that have entered roots remain there after molting and increase in size, particularly in girth (see Figure 3-12). They become pear-shaped about three weeks after penetration, with their heads protruding into what would have become vascular tissue. Feeding by both male and female nematodes stimulates hyperplasia and hypertrophy in the invaded tissues, and the galls or knots develop rapidly. Cell walls in the vicinity of the head of sedentary females are digested and the several adjoining cells coalesce to form "giant cells." Food materials essential to the continued nourishment of the nematode collect in these cavities. The nematodes fail to mature if giant cells do not form or if they develop too slowly. At maturity, the posterior end of the female either protrudes through the surface of the gall tissue or lies very near the surface. Eggs are laid in a gelatinous matrix extruded from the vulva (Figure 15-2).

Epidemiology

The nematode requires a moist environment for all of its vital activities, and the rate of its development in a moist environment is determined by the temperature. At temperatures below 15.5°C or above 33.5°C, little growth occurs and females fail to reach maturity. Development is most rapid between 27°C and 30°C, less than three

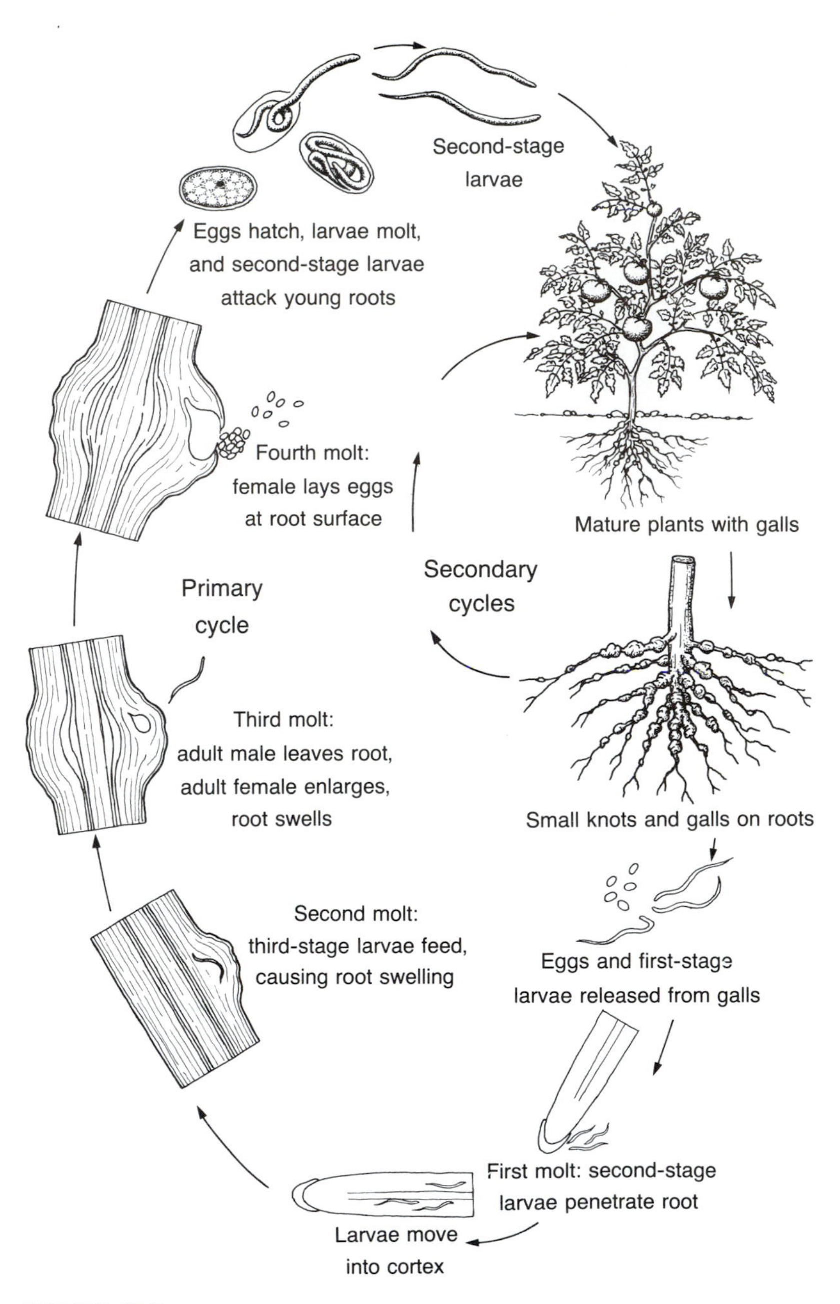

FIGURE 15-2
Root knot of tomato.

weeks being required for development from infective larvae to egg-laying females. Thus, in warm climates where the growing season is long, nematode populations may be very large, and there may be many successive generations in a single season.

Prevention of Disease

Most measures used to control root knot are eradicative, but the pathogen can often be excluded from uninfested areas by planting healthy stock. Nursery stock may be inspected and certified to be free of root knot, thereby excluding the pathogen from orchards and plantings of ornamentals.

In some instances, heat treatments have been used to rid transplants or propagative parts of the root-knot nematode. However, because the different species of the root-knot nematode and their different suscepts vary widely with respect to their tolerance of heat, the effectiveness of hot-water treatments for the control of root knot cannot be predicted. In general, plant parts to be treated are immersed in water at 50°C. Living roots may be kept there for approximately ten minutes, whereas bulbs, corms, tubers, and storage roots may be kept at 50°C for as much as an hour. Unless extension service personnel make detailed recommendations, based upon experimental results, for the hot-water treatment of specific plant materials, the grower contemplating using such a control measure should test it on a few plants before jeopardizing large numbers of transplants or rootstocks.

The root-knot nematode may be eradicated by rotating susceptible crops with such insusceptible plants as small grains and certain grasses. The most satisfactory method of eradication, however, is the fumigation of the soil with nematicidal chemicals. Among the more effective chemicals are such halogenated hydrocarbons as DD (1,2-dichloropropene-1,2-dichloropropane) and such dithiocarbamates as Vapam (sodium methyldithiocarbamate hydrate).

Fallow and dry tillage may be effective in warm climates, and "flooding" the land has been effective in Florida. If soil is kept wet to the point of saturation for several months, nematodes are greatly reduced in number. This control measure is most effective during warm weather, because nematode eggs are stimulated to hatch, and the young are much more sensitive to "flooding" than are the eggs.

A curious method of eradicating nematodes is the planting of trap crops—the idea being to permit nematodes to enter the root systems of susceptible plants, which are then plowed under before the nematodes have had an opportunity to reproduce. Careful timing is critical, because the nematodes will increase in number if the crop is allowed to grow too long. Moreover, females that are almost mature will continue to develop and lay eggs in excised roots. The modern concept of trap crops includes the use of certain plant species that larvae may enter but in which they fail to survive or reproduce. Such plants as *Crotalaria spectabilis* or the common marigold are automatic trap crops for root-knot nematodes. Both species are readily invaded by larvae, but neither permits continued growth or reproduction.

Varietal resistance to the various species and subspecies of the root-knot nematode occurs in a number of plants, including tomato, pepper, tobacco, bean, soybean, lima bean, sweet potato, lespedeza, peach, and rose. In several herbaceous species, resistance appears to be determined by a single dominant gene. Increasing attention is being given to the breeding and selection of plant varieties resistant to root-knot nematodes, and this method of disease control offers much promise.

FOLIAR NEMATODE-INDUCED DISEASE OF CHRYSANTHEMUM

That a necrosis of chrysanthemum foliage is caused by a leaf-invading nematode was discovered in England and the United States late in the nineteenth century. Since that time, the foliar nematode has become known as a dangerous pathogen of chrysanthemums throughout Europe and the United States. In addition to the primary host, the nematode parasitizes some thirty other species of plants, including aster, dahlia, delphinium, tobacco, strawberry, and zinnia.

The nematode survives the winter in the dormant buds near the base of the infected plant. Consequently, the earliest symptoms in the spring are the dwarfing of shoots and savoying of leaves. As the growing season progresses, necrotic symptoms appear in the laminar tissues of mature leaves. Chloranemic spots between veins later enlarge and become dark brown. Necroses are often angular, being limited by large veins (Figure 15-3).

Etiology

Foliar nematode-induced disease of chrysanthemum is caused by *Aphelenchoides ritzemabosi* (Schwartz) Steiner. Adult nematodes or

FIGURE 15-3
Foliar nematode-induced disease of chrysanthemum: healthy leaf (*left*) and diseased leaves (*center, right*).

eggs survive between growing seasons in the dormant buds near the base of the plant. As a result of the feeding activity of the adults and the young that hatch in the spring, the early growth is stunted and distorted. Later, the nematodes migrate up the stem of the plant when it is covered with a film of water in rainy or humid weather. When the nematodes reach mature leaves, they enter through stomata and initiate infection in the mesophyll, which responds necrotically. The epidermis is not parasitized. The nematode completes its life cycle, from egg to egg, in approximately two weeks.

Although the nematode is not soil-inhabiting, it reaches the soil in diseased leaf tissue. It is not known how long the nematode will remain viable in foliage that has fallen to the ground. It will survive for several years in dry leaves kept in a laboratory or herbarium.

Prevention of Disease

The disease can be controlled during the growing season by weekly applications of a spray containing malathion or parathion in water, mixed at the rate of 0.25 lb/gal.

Propagative material should be taken from disease-free plants, but cuttings can be freed of nematodes by immersing them for five minutes in water at 50°C.

WHEAT SEED-GALL DISEASE

Certain nematodes invade the flower primordia of plants—usually grasses—and induce the formation of galls instead of seed. Seed galls of wheat, rye, emmer, and spelt occur throughout Europe and Asia, and sporadically in North America. Losses result from damage to growing plants as well as from reduced yields of healthy grain.

The wheat seed-gall nematode was first observed in 1743 by Needham, who described longitudinal fibers that had been removed from "the small black grains of smutty wheat." He observed that these "fibers" were actually living creatures, even though the galls from which they come were twenty-five years old. The causal relationship between these creatures and the seed-gall disease of wheat was demonstrated in 1775 by Rofferdi, but it was not until 1857 that Davaine published his classical paper on the life history of the nematode.

The symptoms of the disease are similar to those induced by other bud-invading nematodes. Diseased plants are dwarfed and their leaves are variously distorted, being wrinkled, curled, and twisted. Foliar symptoms occur on one or more culms, and are most severe in young plants. Galls replace kernels in diseased heads. They are usually shorter and darker than healthy kernels, which they resemble superficially.

Etiology

The wheat seed-gall disease is caused by the nematode *Anguina tritici* (Steinbuch) Filipjev. Galls that fall to earth release second-stage larvae into the soil as they disintegrate. This usually occurs in the autumn, and the larvae survive the winter either in the soil or in young plants of winter wheat. Inoculation occurs as larvae come into contact with the leaves that enclose the growing point of the wheat plant. The larvae are protected by the leaves and leaf sheaths, and usually parasitize the plant ectoparasitically. Occasionally, however, the larvae enter leaves and complete their life cycles within. When this happens, leaf galls are produced. In most instances, the nematodes do not enter the plant until flower primordia have been formed. Feeding activity there stimulates the production of a gall instead of a seed. The nematodes mature rapidly in these galls and the females lay their eggs there.

Wheat plants become susceptible to attack by the seed-gall nematode when the coleoptile becomes loose on the stem; the tissue that is invaded remains vulnerable only so long as the growing point is near the base of the plant. Cool, wet weather retards the rate of growth of the plant and thereby favors initiation of the disease.

Prevention of Disease

The nematode can be excluded by planting only healthy kernels. Galls can be separated from healthy seed by flotation in brine (40 pounds of salt in 25 gallons of water) or by means of mechanical separators. The nematode can be eradicated from the soil by a short crop rotation. Although the larvae will remain viable for more than twenty-five years in dry galls, they cannot survive more than a year in soil devoid of host plants.

Selected References

ROOT KNOT

Christie, J. R., 1936. The development of root-knot nematode galls. *Phytopathology* 26:1–22.

———, 1946. Host-parasite relationships of the root-knot nematode, *Heterodera marioni,* II. Some effects of the host on the parasite. *Phytopathology* 36:340–352.

Dropkin, V. H., and P. E. Nelson. 1960. The histopathology of root knot nematode infections in soybeans. *Phytopathology* 50:442–447.

Godfrey, G. H., and J. M. Oliviera, 1932. The development of the root-knot nematode in relation to root tissues of pineapple and cowpea. *Phytopathology* 22:325–348.

Melendez, P. L., and N. T. Powell, 1967. Histological aspects of the *Fusarium* wilt–root knot complex in the flue-cured tobacco. *Phytopathology* 57:286–292.

FOLIAR NEMATODE-INDUCED DISEASE OF CHRYSANTHEMUM

Allen, M. W., 1952. Taxonomic status of the bud and leaf nematodes related to *Aphelenchoides fragariae* (Ritzema Bos, 1891). *Proc. Helminthol. Soc. Wash.* 19:108–120.

Bryden, J. W., and W. E. H. Hodson, 1957. Control of chrysanthemum eelworm by parathion. *Plant Pathol.* 6:20–24.

Dimock, A. W., and C. H. Ford, 1950. Control of foliar nematode of chrysanthemums with parathion sprays. *Phytopathology* 40:7. (Abstr.)

Franklin, M. T., 1950. Two species of *Aphelenchoides* associated with strawberry bud diseases in Britain. *Ann. Appl. Biol.* 37:1–10.

Hesling, J. J., and H. R. Wallace, 1961. Observations on the biology of chrysanthemum eelworm, *Aphelenchoides ritzema-bosi* (Schwartz) Steiner in florists' chrysanthemum. I. Spread of eelworm infestation. II. Symptoms of eelworm infestation. *Ann. Appl. Biol.* 49:195–209.

WHEAT SEED-GALL DISEASE

Bloom, J. R., 1963. Effect of temperature extremes on the wheat gall nematode, *Anguina tritici. Plant Dis. Rep.* 47:938–940.

Byars, L. P., 1920. The nematode disease of wheat caused by *Tylenchus tritici. U. S. Dep. Agr. Bull.* (842).

Leukel, R. W., 1924. Investigations on the nematode disease of cereals caused by *Tylenchus tritici. J. Agr. Res.* 27:925–955.

16

Diseases Affecting Water Conduction—Vascular-Wilt Diseases

Vascular wilts ensue when certain bacteria or fungi develop in the xylem and xylem parenchyma of susceptible plants. The xylem becomes plugged, presumably with enzymatically hydrolyzed constituents of affected cell walls or with by-products of the metabolism of the pathogen. Toxic metabolic products from the pathogen often contribute to the syndrome. Vascular tissues turn brown, providing a diagnostic symptom for the vascular-wilt diseases.

Herbaceous plants are invaded by soil-borne wilt-inducing bacteria through their roots, whereas woody plants are usually invaded through wounds in their stems. Insects are usually the agents of inoculation for those pathogens that cause vascular wilts of trees. These facts are the bases for several control measures, notably crop rotation of herbaceous plants, and insect control in fields, orchards, and forests. A number of wilt diseases of herbaceous plants are controlled through the use of resistant varieties.

SOUTHERN BACTERIAL WILT OF SOLANACEOUS PLANTS

Bacterial wilt occurs throughout the world in the torrid zone and the warm parts of the temperate zones. It is most severe in solanaceous plants, but it is also responsible for the moko disease in banana plants in Central America. In the United States, the disease in tomato is commonly called southern bacterial wilt; the disease in potato is frequently called brown rot; and the disease in tobacco is often referred to as Granville wilt, because it first reached epiphytotic proportions in Granville County, North Carolina. The disease has also been called the bacterial-ring disease and the slime disease.

Foliage of infected plants become flaccid and leaves and shoots droop in the hot part of the day. At first, wilting plants may recover turgor at night, but, as the disease progresses, affected shoots are permanently wilted. The symptoms progress downward from the youngest leaves, which are the first to show symptoms. The brown discoloration that develops in the vascular system can be readily detected by splitting the stem longitudinally. This vascular browning is diagnostic for wilt diseases, but does not distinguish Southern bacterial wilt from wilt diseases caused by other bacteria or fungi. Signs of the pathogen—bacterial ooze—can sometimes be seen on the cut surfaces of excised stems of severely diseased plants.

Etiology

Southern bacterial wilt is caused by *Pseudomonas solanacearum* E. F. Smith, a motile rod-shaped bacterium that grows in cream-colored or white colonies in culture. The bacteria survive in the soil between crop seasons, and enter roots through wounds caused by insects or mechanical injuries, or through the natural ruptures that occur when branch roots emerge from their origins in the pericyclic regions of primary roots. Inoculation occurs through the chance contact between bacteria in the soil and the wounded tissue of the root. Incubation is of short duration, because plant cells at the point of wounding react almost immediately to the pathogen.

Once inside the susceptible root, the bacteria grow inward and soon reach the young cells of the xylem. There they remain, obtaining nourishment from the xylem sap and from the adjacent xylem parenchyma. The vessels of diseased plants become plugged by the polysaccharides produced by the multiplying bacteria and by the bacteria themselves.

Although the results of experiments with wilt-inducing fungi support the toxin theory of pathologic wilting—that pathogens induce wilting at least partly through the action of toxins that upset the osmotic relations of leaf parenchyma, causing the tissue to lose water rapidly—there is no evidence that bacteria induce wilting by the same means.

Epidemiology

The disease is serious only in the tropics and the warmer regions of the temperate zones. The disease develops most rapidly in soils at 25–35°C, and the bacteria seem unable to survive in soils subjected to freezing and thawing. The disease may be more severe in soils infested by plant-parasitic nematodes than in nematode-free soils. Under controlled conditions, the nematode *Meloidogyne incognita* predisposes tomato plants to bacterial wilt. In some way, as yet not understood, feeding by the nematode so alters the metabolism of the tissues of resistant plants that they become highly susceptible to the disease. Nematode-induced predisposition to wilt is also documented in connection with the vascular disease caused by *Fusarium oxysporum,* which is discussed in the next section.

Prevention of Disease

The bacterium cannot be readily eradicated from the soil through short-term crop rotation: once infested, a field soil might harbor viable bacteria for years. Perhaps the organism grows saprophytically in the soil; it may also maintain itself in a given area through continued invasion of solanaceous weeds. Only long rotations of five to seven years are effective, and this is currently the principal control measure.

Eddins (1936) found that the bacterium could be eradicated from potato soils near Hastings, Florida, by alternately lowering and raising the *p*H of the soil. In June, the soil reaction was lowered to a *p*H of 4 or 5 by the addition of sulfur at the rate of 800 lb/acre; then, in November, dolomitic limestone was added to the soil at the rate of 3000 lb/acre to raise the *p*H to near 6.0. It has not been feasible, however, to use this procedure on a commercial scale; costs of materials and labor have been too high to permit growers using this procedure to compete with those whose soils are not contaminated with the pathogen.

Resistant varieties offer the most satisfactory means of control. Although the resistant varieties that are currently available are not commercially acceptable, there is extensive research now underway aimed at developing resistant varieties well suited to commercial production.

FUSARIUM AND *VERTICILLIUM* WILTS

Vascular-wilt diseases induced by either *Fusarium oxysporum* Schlechtendahl or *Verticillium alboatrum* Reinke and G. D. W. Berthold occur

FIGURE 16-1
Fusarium-induced wilt of tomato. (Courtesy of H. D. Thurston, Cornell University.)

in many crop plants throughout the world. *Fusarium* usually predominates in the relatively warm soils of the temperate zones and those of the torrid zone, whereas *Verticillium* is the pathogen most likely to be found in the cool soils of the temperate zones. One form or another of either fungus might attack almost any crop plant, including woody shrubs and trees. *Fusarium,* for example, causes important wilt diseases in such diverse plants as tomato, banana, mimosa, and cotton.

As with all vascular-wilt diseases, symptoms of *Fusarium* and *Verticillium* infections include drooping of above-ground parts (Figure 16-1). In the early stages of disease, plants wilt during the day and recover their turgor at night. As the disease progresses, the permanent wilting point is reached and turgor is never regained. In some woody plants—as in mimosa infected by *Fusarium* or stone-fruit trees infected by *Verticillium*—the foliage on affected branches becomes yellow and drops off before extensive wilting has occurred. The diagnostic symptom of any pathological-wilt disease is the brown discoloration of the vascular region, visible in cross-sections of infected stems or roots and in sections tangential to the xylem. Diagnosis of the particular wilt disease depends upon recognition of the causal agent.

Etiology

Fusarium wilt is induced by any of the several formae speciales (*sing.* forma specialis, *abbr.* f. sp.) of *Fusarium oxysporum.* These forms, structurally indistinguishable from one another, vary with respect to their pathogenicity to different species of suscept plants. Thus, for example, *F. oxysporum* f. sp. *lycopersici* attacks only tomato (*Lycopersicon*), *F. oxysporum* f. sp. *niveum* attacks only watermelon, and *F. oxysporum* f. sp. *conglutinans* attacks only cabbage. Any one of these forms may gain ingress into many different plants; it can establish and maintain itself, however, only within the host plant for which it is specific, and it is only in that plant, therefore, that the form can incite disease.

Verticillium alboatrum induces vascular wilt in a wide variety of suscepts. The wilt-inducing *V. dahliae* is considered synonymous with *V. alboatrum* by some authors, whereas others treat it as a separate species.

Fusarium and *Verticillium* are easily identified. Both fungi can be isolated and grown in pure culture, where they usually sporulate profusely. Also, the sporulating stage of each fungus will develop within one or two days on the split stems of diseased plants, provided the stems are kept moist. *Fusarium* produces sickle-shaped, multiseptate conidia on sporodochia (Figure 16-2), whereas *Verticillium* produces smaller, single-celled conidia on conidiophores that develop in whorls on strands of mycelium. Both fungi produce chlamydospores, in which form they may survive in the soil; *Verticillium* may also survive in the soil in the form of microsclerotia.

Very little saprophytic growth of *Verticillium* occurs in the soil, but *Fusarium* is said to be capable of considerable saprophytic growth. It is not clear whether *Fusarium* spp. that infect vascular systems of plants nourish themselves, during periods of survival, largely from food materials stored in the chlamydospore or from materials in the soil solution. Conceivably, *Fusarium* survives for five to ten years in the soil as a "saprophyte," because a percentage of the chlamydospores fail to germinate for years and because the fungus quickly produces new chlamydospores on hyphae shortly after germination of the "parent" spores. Food reserves in the original spore might be sufficient for several successive generations of spores, and, if these

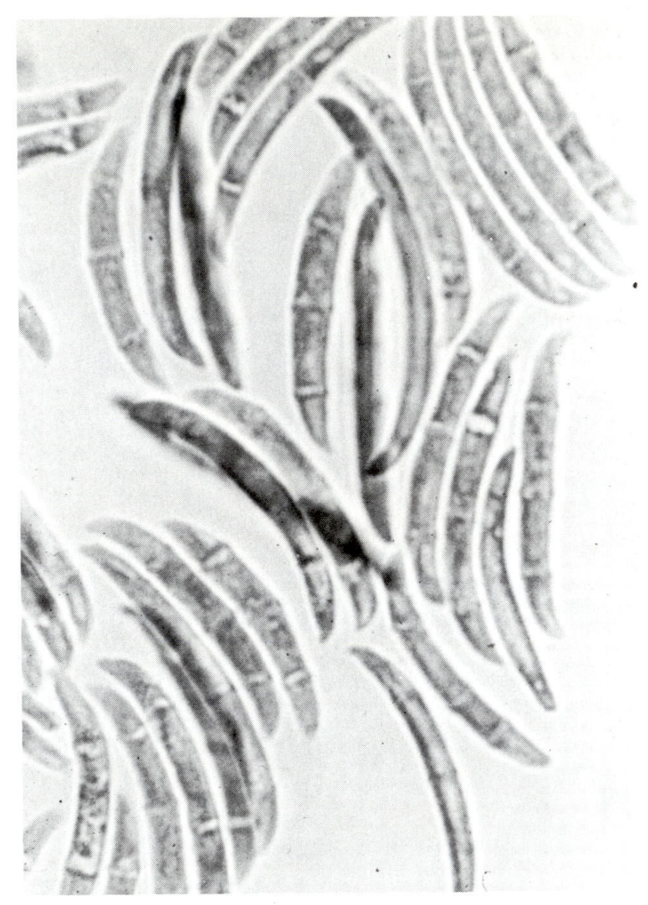

FIGURE 16-2
Conidia of *Fusarium oxysporum.*

generations occurred as infrequently as one or twice a year, the fungus could survive for some years without obtaining anything more than water from the soil solution. In any event, soil, once infested with *Fusarium oxysporum,* is likely to remain infested for five to ten years. *Verticillium alboatrum,* however, survives in the soil for only six months or so in the form of mycelium, or for one or two years in the form of microsclerotia.

The primary cycle begins when the roots of susceptible plants contact hyphal strands developing from germinating chlamydospores or microsclerotia. Ingress by either fungus may occur through the openings made by the emergence of lateral roots, through mechanically injured areas, or by direct penetration of hyphae through the tender root tissues in the regions of cell elongation or meristematic activity.

Once inside the root, without causing any apparent damage to the cells between which it grows, the wilt-inducing fungus grows to the region of the xylem. Once in the xylem, the fungus develops in the tracheids and vessels, and also invades the xylem parenchyma. *Fusarium* and *Verticillium* secrete pectolytic enzymes that catalyze the hydrolytic reactions that result in partial destruction of the middle lamellae of the xylem parenchyma and in the degradation of pectic compounds in the walls of vessels and tracheids. Parenchymatous cells are killed and turn brown, and a brown-staining reaction occurs in the walls of affected vessels and tracheids. This brown discoloration, which is the diagnostic symptom of pathological wilting, apparently results from the oxidation of polyphenols, giving rise to the dark melanin pigments.

Although considerable mycelium may form in the vessels, and although spores may be produced there, the amount of fungus produced is insufficient to account for wilting as a result of mechanical plugging of the tracheary elements. Wilting probably results from the blocking of tracheary elements by the gel formed by the mixture of pectolytic breakdown products and the cell contents. Also, *Fusarium* produces extracellular toxins that move to the leaves, where they induce loss of turgor in the ground parenchyma. The combined effects of toxins, plugging by the fungus, and plugging by the gel of pectic compounds probably account for the wilting of plants infected by *Fusarium oxysporum.* It is becoming increasingly evident that similar effects result in pathological wilting, irrespective of the causal fungus.

Epidemiology

The initiation of the vascular-wilt diseases caused by *Fusarium* and *Verticillium* is favored by high soil moisture, which is necessary for spore germination and subsequent growth of mycelium. The temperature of the soil, however, affects the two fungi differently: whereas *Fusarium* is most active in warm soils, the growth of *Verticillium* is retarded by temperatures above 30°C.

Prevention of Disease

In some instances, the wilt-inducing *Fusarium* and *Verticillium* species can be excluded from uninfested soils by the planting of disease-free plants. The eradicative measure, crop rotation, is moderately effective against *Verticillium;* it is usually not effective against *Fusarium,* however, unless insusceptible species are grown in the infested soil for five to eight years. Chemical treatments of the soil have not been generally practical, except on a small scale, as in plant beds or greenhouses, where fumigation is effective.

The only protective control measures used are cultural practices that permit the plant to grow and develop despite infection. These include planting in well-drained soils and the use of adequate amounts of commercial fertilizers.

The most effective method of controlling these diseases in field-grown row crops is the use of resistant varieties. Programs of breeding and selection for *Fusarium*-resistant tomatoes, for example, have been eminently successful. Resistance is conferred by several genes in some varieties, by one gene in others. Two different races of *F. oxysporum* f. sp. *lycopersici* have been discovered, and any one of these can infect varieties that are resistant to the other race. As new races occur, the breeder-pathologist must adapt his program to include selection for resistance to the different race and thus keep one step ahead of the pathogen.

Wilt-resistant varieties of such crop plants as cotton and tomato have, upon occasion, appeared to have lost their resistance even to the common race of *Fusarium oxysporum.* In some instances, this "loss of resistance" has resulted from subtle changes in the metabolism of the plant that were induced by the parasitic activities of root-knot nematodes, *Meloidogyne* spp. The nematode predisposes the "resistant" plants to infection by the wilt fungi. Nematode control, then, may be an important concomitant to the use of wilt-resistant varieties of crop plants.

DUTCH ELM DISEASE

Because the vascular wilt of American and European elms decimated the elms in the Netherlands, the disease was the subject of much research by Dutch plant pathologists. Consequently, this vascular wilt came to be known as the Dutch elm disease. The disease, which is common in all regions of Europe and America where native elms are grown, poses a serious threat, because susceptible elms would be eradicated by the disease, were it not for the control measures that are employed against it. Those elms that are native to Asia are resistant to the disease.

The foliage of infected branches loses its turgor, wilts, and later becomes chloranemic. When it occurs on individually infected branches among healthy ones, this symptom syndrome is called flagging. Once infection has taken place, death of the entire tree can be expected within several years (see Figure 5-2). The elm disease is typical of the vascular-wilt diseases in that the diagnostic symptom is the brown discoloration of tissues in the vascular region. The outer layer of wood may appear streaked, mottled, or uniformly brown to dark gray. The Dutch elm disease and *Verticillium* wilt of elm may be distinguished only by identifying the causal agent.

Etiology

Dutch elm disease is caused by the ascomycete *Ceratocystis ulmi.* The imperfect stage, which is usually referred to by the name *Graphium ulmi,* develops in the infected plant and sporulates beneath the bark of dead elm wood. Conidia are borne on coremia and are held together in heads by a sticky secretion. The perfect stage of the fungus occurs less often than the imperfect one and is rare in the United States. When ascospores are produced, they form in asci in globose perithecia that have long beaks. Perithecia are imbedded in the bark of dead wood with the upper portion of the beak protruding. Ascospores are not important inocula because they are not disseminated to infection courts, which are open wounds in the bark of living trees.

The coremia produced under the bark of dead elm wood form in the brood galleries of the elm bark beetle (see Figure 5-3). The beetles deposit eggs deep in the bark of recently killed elm branches. The larvae that hatch from the eggs eat the outer layer of wood and the inner layer of bark, forming characteristic galleries. Later, these larvae form pupae at the extremities of the feeding tunnels. Meanwhile, if such wood was infected by *Ceratocystis ulmi,* the fungus sporulates profusely in the grooves of the galleries. Adult beetles develop from the pupae; they bore holes through the bark and emerge, with the sticky conidia of the fungus clinging to them. The beetles then fly to healthy elm trees and feed upon the bark, making the wounds through which the trees are inoculated with conidia (Figure 16-3). Beetles that are agents of inoculation in the United States are *Scolytus multistriatus, S. scolytus,* and *Hylurgopinus rufipes.*

Once deposited in the open wounds, the conidia germinate, provided the infection court remains moist. The resulting mycelium grows in the xylem, affecting xylem parenchyma, tracheids, and vessels. A second conidial form (the *Cephalosporium* stage) develops; its spores may be transported for some distance in the vascular sys-

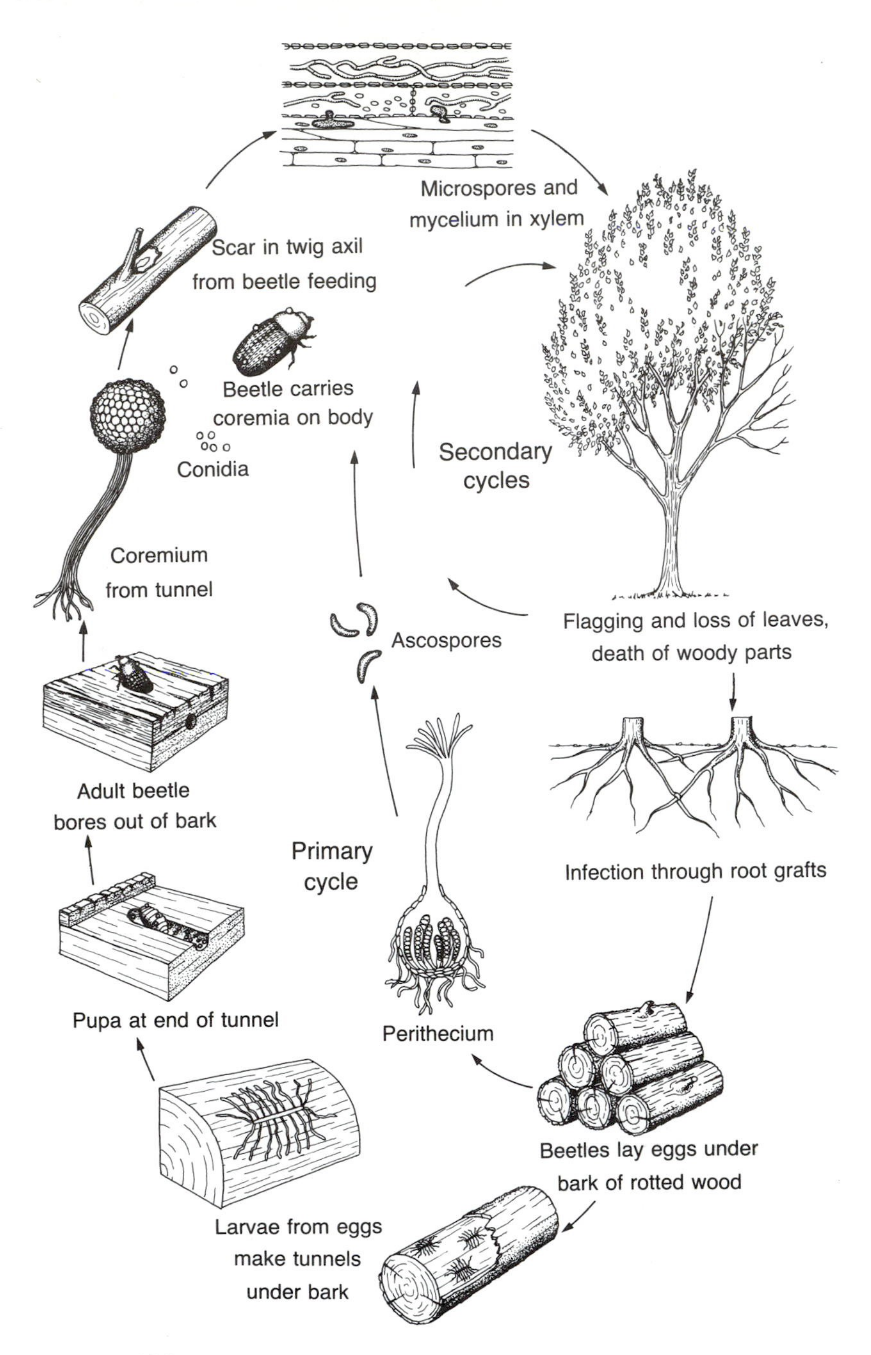

FIGURE 16-3
Dutch elm disease.

tem. The mycelium is frequently transmitted from tree to tree by naturally occurring root grafts. The fungus induces wilting, possibly much in the same ways that *Fusarium* and *Verticillium* affect their hosts. Moreover, numerous tyloses form in the vessels of affected elm trees, and these contribute to the adverse effect of the disease on water conduction.

Initiation and development of infection are favored by high relative humidity in the infection courts. Because this condition is most likely to occur in the early spring, the early-season feedings of bark beetles are the most infective.

Prevention of Disease

If beetle-infested wood from diseased elms is kept as far as 500–600 feet away from healthy trees, there is a good chance that the pathogen can be excluded from an area. Every effort should be made, therefore, to eradicate recently killed elm wood (and the fungus within it) in the vicinity of healthy elms. The roots of trees should be pruned to prevent natural grafts with adjacent trees.

Prized plantings can be protected by applying insecticidal sprays (methoxychlor, for example) early in the spring, during which time the newly formed adult beetles are emerging from dead wood.

As mentioned earlier, elms from Asia are resistant to Dutch elm disease. The planting of Asian elms is not an entirely satisfactory solution, however: as useful as they may be, the Asian elms are not aesthetically satisfactory substitutes for the graceful American elm, which is one of our favorite shade trees.

Selected References

SOUTHERN BACTERIAL WILT OF SOLANACEOUS PLANTS

Eddins, A. H., 1936. Brown rot of Irish potatoes and its control. *Fla. Agr. Exp. Sta. Tech. Bull.* (299).

Gallegly, M. E., and J. C. Walker, 1949. Relation of environmental factors to bacterial wilt of tomato. *Phytopathology* 39:936–946.

Kelman, A., 1953. The bacterial wilt caused by *Pseudomonas solanacearum*. *N. C. Agr. Exp. Sta. Tech. Bull.* (99).

Lucas, G. B., J. N. Sasser, and A. Kelman, 1955. The relationship of root-knot nematodes to Granville wilt resistance in tobacco. *Phytopathology* 45:537–540.

Winstead, N. N., and J. C. Walker, 1954. Production of vascular browning by metabolites from several pathogens. *Phytopathology* 44:153–158.

FUSARIUM AND *VERTICILLIUM* WILTS

Bradfoot, W. C., 1926. Studies on the parasitism of *Fusarium lini* Bolley. *Phytopathology* 16:951–978.

Brandes, E. W., 1919. Banana wilt. *Phytopathology* 9:339–389.

Edgerton, C. W., and C. C. Moreland, 1920. Tomato wilt. *La. Agr. Exp. Sta. Bull.* (174).

Gäumann, E., and O. Jaag, 1947. Die physiologischen Grundlagen des parasitogenen Welkens. *Ber. Schweiz. Bot. Ges.* 57:3–34, 132–148, 227–241.

Gilman, J. C., 1916. Cabbage yellows and the relation of temperature to its occurrence. *Ann. Mo. Bot. Gard.* 3:25–84.

Holdeman, Q. L., and T. W. Graham, 1954. Effect of the sting nematode on expression of *Fusarium* wilt in cotton. *Phytopathology* 44:683–685.

Ludwig, R. A., 1952. Studies on the physiology of hydromycotic wilting in the tomato plant. *McGill Univ. Macdonald Coll. Agr. Tech. Bull.* (20).

Parker, K. G., 1959. *Verticillium* hadromycosis of deciduous tree fruits. *Plant Dis. Rep. Suppl.* (255):38–61.

Rudolph, B. A., 1931. *Verticillium* hadromycosis. *Hilgardia* 5:197–353.

Scheffer, R. P., S. S. Gothoskar, C. F. Pierson, and R. P. Collins, 1956. Physiological aspects of *Verticillium* wilt. *Phytopathology* 46:83–87.

———, and J. C. Walker, 1953. The physiology of *Fusarium* wilt to tomato. *Phytopathology* 43:116–125.

Waggoner, P. E., and A. E. Dimond, 1955. Production and role of extracellular pectic enzymes of *Fusarium oxysporum* f. *lycopersici*. *Phytopathology* 45:79–87.

Wardlaw, C. W., 1930. The biology of banana wilt (Panama disease), I. Root inoculation experiments. *Ann. Bot.* (*London*) 44:741–766.

———, 1931. The biology of banana wilt (Panama disease), III. An examination of sucker infection through root-bases. *Ann. Bot.* (*London*) 45:382–399.

Winstead, N. N., and J. C. Walker, 1954a. Production of vascular browning by metabolites from several pathogens. *Phytopathology* 44:153–158.

———, and ———, 1954b. Toxic metabolites of the pathogen in relation to *Fusarium* resistance. *Phytopathology* 44:159–166.

DUTCH ELM DISEASE

Banfield, W. M., 1968. Dutch elm disease recurrence and recovery in American elm. *Mass. Agr. Exp. Sta. Bull.* (568).

Clinton, G. P., and F. A. McCormick, 1936. Dutch elm disease, *Graphium ulmi. Conn. Agr. Exp. Sta. Bull.* (389):701–752.

Collins, D. L., K. G. Parker, and H. Dietrich, 1940. Uninfected elm wood as a source of the bark beetle (*Scolytus multistriatus* Marsham) carrying the Dutch elm disease pathogen. *Cornell Univ. Agr. Exp. Sta. Bull.* (740).

Dimond, A. E. 1962. Dutch elm disease control by chemotherapy. *Frontiers Plant Sci.* 14(2):4–5.

Sinclair, W. A., and J. A. Weidhass, Jr., 1967. Dutch elm disease and its control in New York State. *N. Y. Agr. Exp. Sta. Geneva. Bull.* (1185).

17

Diseases Affecting Photosynthesis—Bacterial Leafspots and Blights

Species of *Erwinia, Pseudomonas,* and *Xanthomonas* may induce leafspots and blights of plants. Leaf-infecting bacteria are often seed-borne, and the primary cycle of each such pathogen occurs in the cotyledons. Typically, the bacteria ooze out of infected spots, and are splashed by rain or carried by insects to infection courts on true leaves, in which disease cycles continue after the bacteria have entered through stomata, hydathodes, or wounds. Many blight-inducing bacteria infect the twigs of woody plants and survive in twig cankers between growing seasons. Inoculum for primary cycles develops in the form of bacterial ooze when the organisms resume growth after dormancy.

The blight- and leafspot-inducing bacteria must be transported to the surfaces of leaves or blossoms before prepenetration and infection phenomena can proceed, and it is often possible, therefore, to achieve disease control by protective spraying or dusting. Moreover, various methods of obtaining pathogen-free seed ensure control of those bacterial diseases in which the pathogens are seed-borne. Removal of diseased plant tissue in which bacteria might survive between growing seasons is also effective against some bacterial leafspots. Finally, crop-rotation programs and the planting of resistant varieties may be helpful in controlling certain bacterial infections of leaves, blossoms, fruit, and stems.

BACTERIAL BLIGHTS OF BEAN

Of the several bacterial spots and blights of the common bean (*Phaseolus vulgaris*), two, the common bacterial blight and halo blight, are economically important. Both diseases occur in bean-growing areas throughout the world, but are epiphytotic only in the more humid regions. Both diseases affect the sieva bean and the scarlet runner bean as well as in the common, or French, bean.

When leaves are invaded from the outside, both diseases appear as small hydrotic spots that later become necrotic. In common blight (Figure 17-1), the necroses enlarge to irregularly shaped lesions up to one-fourth of an inch across, which, in severe infections, may coalesce. The lesions are surrounded by narrow chloranemic margins enclosed by narrow, pale green bands. In halo blight (Figure 17-2), the angular

FIGURE 17-1
Common bacterial blight of bean.

FIGURE 17-2
Halo blight of bean.

necrotic lesions remain small, and conspicuous yellow halos, often one-half inch across, develop early around the brown necroses.

Dry, sunken, red necroses develop on infected pods. Seeds are also invaded, either by bacteria from deep lesions on the pod or by bacteria that have invaded the plant systemically. Yellowish spots can sometimes be seen on pods of white-seeded varieties. Severely affected seeds are small and shriveled.

Seed-borne bacteria may give rise to systemic infections in seedlings. The bacteria are in the cotyledons, usually just beneath the pericarp, irrespective of whether the seed had been infected locally or systemically. Upon germination of the seed, the bacteria may invade the hypocotyl and reach the water-conducting elements, causing wilting and death of the young plant. The symptom "snake head" occurs in those seedlings whose apical meristems are destroyed by the bacteria. Such plants fail to develop.

In other plants from infected seed, only the cotyledons of the seedlings will show symptoms. Bacteria that ooze to the surface of necrotic lesions on the cotyledons are splashed to leaves and stems, in which they then induce the typical symptoms. Both diseases may cause reddish necrotic lesions on stems, and both may reach the xylem beneath stem lesions. Once in the xylem, either bacterium may be carried to the foliage, where leaf-spotting occurs.

During humid weather, the bacteria ooze out on the surface of diseased tissue. If the disease is the common blight, the ooze is yellowish; if it is the halo blight, the ooze is creamy.

Etiology

Common blight of bean is caused by *Xanthomonas phaseoli* (E. F. Smith) Dowson; halo blight is caused by *Pseudomonas phaseolicola* (Burkholder) Dowson. Both are rod-shaped bacteria with single polar flagella; both measure approximately $1 \times 2\ \mu$. On nutrient culture media, *X. phaseoli* produces yellow colonies, whereas *P. phaseolicola* produces creamy colonies.

The bacteria may survive between crop seasons either in diseased seed or in the soil in association with crop residue from the preceding season. The primary cycle, therefore, may be initiated either within infected cotyledons as soon as dormancy is broken, or in young plants upon which the bacteria are splashed from crop residues. Bacteria in diseased cotyledons become active when the seeds are placed in the moist, warm environment conducive to their germination. The bacteria are saprophytic pathogens; they produce pectolytic enzymes that digest middle lamellae and thus macerate the infected tissues. The bacteria proliferate rapidly, nourishing themselves on the killed suscept cells. Large cavities form and become filled with the bacteria. During the course of cotyledonary infection, the bacteria reach the surface of the cotyledon or, as indicated earlier, reach the hypocotyl and penetrate to the xylem. Bacteria on the surfaces of cotyledons may be splashed to the foliage and stems, where secondary cycles are initiated in exactly the same way that primary cycles are begun by bacteria that have survived between seasons in crop residues. In these infections from the outside, the suscept is entered through stomata. The bacteria multiply in the substomatal cavities and produce enzymes that attack the surrounding parenchyma. The infection may become systemic if the bacteria invade the xylem from the leaves or stems.

The common blight is favored by temperatures above 25°C, whereas halo blight is most severe during cooler weather. High humidity is essential during the inoculation and infection stages of both diseases, and the rate of their spread in the field is increased by wind-blown rain.

Prevention of Disease

Bacterial-blight pathogens can be excluded by planting healthy seed. In the United States, seeds grown in the irrigated lands of the Far West are likely to be disease-free, since the climate of that region is unfavorable to the development of the bacterial blights of bean. The pathogens can be eradicated from crop residues by means of short crop rotations of two or three years duration. Varieties resistant to one or both of these diseases are available.

ANGULAR LEAFSPOT OF COTTON

Angular leafspot—which is also known as bacterial blight, bacterial boll rot, vein blackening, and black arm—occurs in cotton throughout the world.

In the early stages of infection, hydrotic spots appear on the adaxial surfaces of cotyledons and leaves and on the stems. The hydrosis gives way to necrosis, which, in the leaves, occurs throughout the cross-section and takes the form of an angular, vein-delineated spot. Lesions on stems are dark reddish brown with anthocyanescent margins. Necrotic streaks occur in diseased stems of certain varieties of the long-staple Sea Island and Egyptian cottons. Lesions may girdle the stems, causing blackening and death of all tissues above the girdled region. The infection of a young plant usually results in the death of the entire plant. Young bolls blast upon infection; circular, black, necrotic spots develop on older bolls. The bacterial exudate that oozes out of infected boll tissue often acts as a glue and prevents their opening. In any event, the lint of infected bolls is stained brown or black.

Etiology

Angular leafspot is caused by *Xanthomonas malvacearum* (E. F. Smith) Dowson, a rod-shaped bacterium, with one to four flagella at each end, that produces yellow colonies in culture.

The bacterium survives in infected seed between crop seasons. There is no good evidence for its survival in diseased crop residue in the soil. As a diseased seed germinates, the cotyledons become infected by bacteria in the basal cap, which is carried up out of the soil with the cotyledons. In moist weather, the bacteria ooze out of necrotic areas of the cotyledons and are splashed by rain to infection courts. Ingress is through stomata of stems and leaves. The bacterium macerates parenchymatous tissue and acts as a saprophytic pathogen. Mature seeds are invaded through the chalaza.

Epidemiology

The disease is favored by wet weather, and dry weather effectively checks its development and spread in the field: a relative humidity of 85% or more is essential to the initiation of infection. Although cool weather has been reported favorable to the development of the disease, the optimum temperature for secondary cycles may be as high as 35°C. Under field conditions, cool weather may indirectly favor the development of disease through the influence of temperature on relative humidity.

Prevention of Disease

The pathogen can be excluded through the use of bacteria-free seed. The delinting of seed with sulfuric acid is effective in removing bacteria, but the germination of acid-treated seed is poor. Treatment of delinted seed with an organic-mercury dust aids in the control. Volunteer plants should be destroyed because they are the most important source of inoculum for secondary cycles of the pathogen in the field.

Resistant varieties should be used whenever possible. Although the bacterium exists in several pathogenic forms, the use of resistant varieties offers a promising means for the control of angular leafspot. Resistant lines of the usually susceptible long-staple types of cotton have been developed in breeding programs. Resistance is polygenic, and it seems likely that, once developed, resistant varieties might remain so for many years.

FIRE BLIGHT OF POME FRUIT

Fire blight of pome fruit is apparently indigenous to North America. The disease is economically important in pear and some varieties of apple, particularly east of the Rocky Mountains. It also occurs in Japan, New Zealand, and Italy, and has only recently been observed in Great Britain. The disease is most severe in warm, humid regions. In addition to pear and apple, the fire-blight bacterium attacks a

FIGURE 17-3
Fire blight of pear blossoms.

number of other rosaceous plants, including quince, loquat, service berry, firethorn, mountain ash, strawberry, blackberry, raspberry, hawthorn, spiraea, and rose.

The bacterium attacks the blossoms, leaves, fruit, limbs, and stems of susceptible plants. Affected blossoms turn brown, wilt, and die, and the necrosis extends down the pedicel to involve the foliage, which turns dark brown to black, wilts, and clings around the affected blossoms (Figure 17-3). The entire fruit spur is often blighted, and, in highly susceptible plants, necrotic lesions develop in twigs at the bases of blighted spurs. When young leaves are invaded, angular necroses develop as the tissues are destroyed between the larger veins. Blight of succulent shoots results from the invasion of young stems directly or indirectly from blighted leaves. The bacteria often progress through young shoots to infect larger branches. When large branches or the trunks of trees are invaded, the progress of the bacteria is slowed, and blight cankers occur (Figure 17-3). Complete girdling of the trunk results in the collar-rot phase of the disease. Young fruit are blasted as a result of the infection, but older infected fruit may cling to the tree. Once a pear fruit has been infected, the bacteria usually invade the entire fruit and cause a black rot. The bacteria are checked in diseased apple fruit, however, and cause sunken, black, necrotic lesions in limited areas.

Bacterial exudate oozes out of infected tissues in moist, warm weather, and this exudate constitutes the only sign of the pathogen itself in diseased tissue.

Etiology

Fire blight (Figure 17-4) is caused by *Erwinia amylovora* (Burrill) Winslow et al., a rod-shaped bacterium with peritrichous flagella. It is of historical interest that *E. amylovora* was the first bacterium proved to be pathogenic to plants. T. J. Burrill, in a paper published in 1880, provided the experimental evidence that proved the point.

The bacteria survive the winter in limb cankers. During the wet weather of early spring, they become active, particularly at the edges of the cankers. They ooze out of diseased limbs and accumulate on the surface in drops of amber-colored exudate. This inoculum for the primary cycles of the pathogen is splashed and washed from the cankers to the surfaces of infection courts on blossoms, leaves, and young shoots by rain. Although the inoculum may be carried by crawling insects, splashing rain is undoubtedly the most important agent of inoculation for the primary cycle. The bacteria enter the nectaries of blossoms through stomatalike openings, and enter leaves and young shoots through the stomata. Once inside the nectary or the substomatal cavities, the bacteria multiply and secrete pectolytic enzymes that macerate the adjacent tissues. Plasmolysis of affected cells is soon followed by necrosis and a blackening of diseased tissue. The bacteria then migrate intercellularly through the infected tissue. When twigs are reached, the bacteria are restricted to the cortical tissues, which are killed during the summer as the cankers form.

The secondary cycles occur throughout the flowering season as the bacteria are carried from the nectaries of diseased blossoms to healthy blossoms by pollinating insects. M. B. Waite's discovery that bees are the agents of inoculation for the secondary cycles of the fire-blight bacteria was the first direct evidence of an insect acting as the agent of inoculation for a plant disease.

Fire blight is favored by wet weather and by any condition that results in succulent growth. Rain, especially when it is followed by relatively warm (25°C), cloudy days, is particularly favorable to the production of bacterial exudate. High-nitrogen fertilization, which increases the succulence of plants, is likely to increase their susceptibility to fire blight.

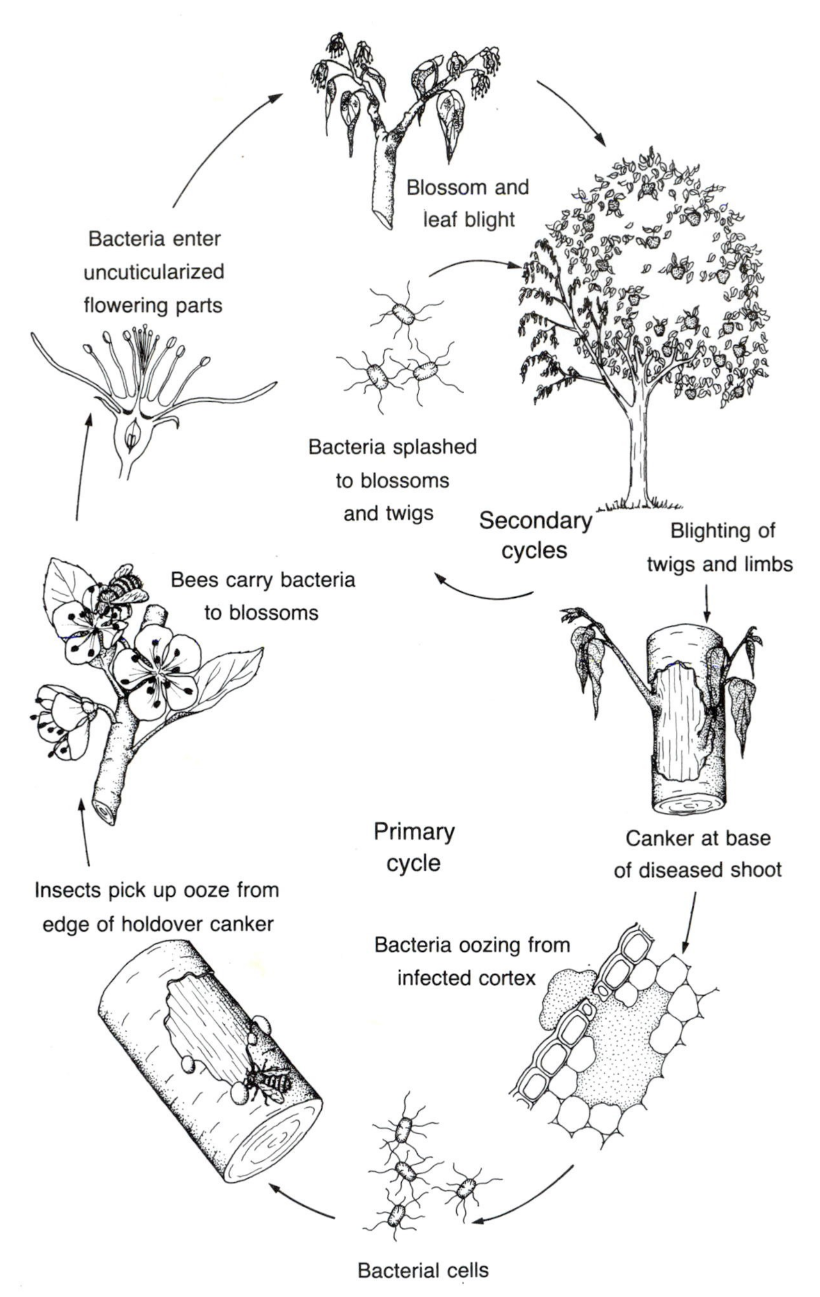

FIGURE 17-4
Fire blight of pome fruit.

Prevention of Disease

The pruning of diseased limbs is an eradicative control measure that will reduce the amount of inoculum and thereby lessen the severity of primary cycles. Blossoms must be protected, however, from secondary cycles in which bees are the agents of inoculation. Leaves in early silver-tip to green-tip stages should be sprayed with 8–8–100 Bordeaux mixture. During the bloom period, satisfactory control has been obtained with sprays of antibiotics, usually streptomycin, or a mixture of streptomycin and terramycin. Bordeaux may also be used during bloom, but it should be diluted (2–6–100) to avoid injury to the foliage.

No satisfactory variety of pear has been developed that is truly resistant to fire blight. Although the cultivar 'Kieffer' is somewhat resistant, it may be severely blighted in regions where the environment is most favorable to the disease. In regions where the collar-rot phase of the disease is more severe than the blossom- and twig-blight phases, varieties and species that resist collar rot can be used as rootstocks.

Selected References

BACTERIAL BLIGHTS OF BEAN

Burkholder, W. H., 1930. The bacterial diseases of the bean, a comparative study. *Cornell Univ. Agr. Exp. Sta. Mem.* (127).

Patel, P. N., and J. C. Walker, 1963. Relation of air temperature and age and nutrition of the host to the development of halo and common blights of bean. *Phytopathology* 53:407–411.

Zaumeyer, W. J., 1930. The bacterial blight of beans caused by *Bacterium phaseoli*. *U. S. Dep. Agr. Tech. Bull.* (186).

———, 1932. Comparative pathological histology of three bacterial diseases of bean. *J. Agr. Res.* 44:605–632.

ANGULAR LEAFSPOT OF COTTON

Brinkerhoff, L. A., and R. E. Hunter, 1963. Internally infected seed as a source of inoculum for the primary cycle of bacterial blight of cotton. *Phytopathology* 53:1397–1401.

Faulwetter, R. C., 1919. The angular leaf spot of cotton. *S. C. Agr. Exp. Sta. Bull.* (198).

Stoughton, R. H., 1932. The influence of environmental conditions on the development of the angular leaf spot disease of cotton. The influence of atmospheric humidity on the infection. *Ann. Appl. Biol.* 19:370–377.

Thiers, H. D., and L. M. Blank, 1951. A histological study of bacterial blight of cotton. *Phytopathology* 41:499–510.

FIRE BLIGHT OF POME FRUIT

Burrill, T. J., 1880. Anthrax of fruit trees, or the so-called fireblight of pear and twig blight of apple trees. *Proc. Amer. Ass. Adv. Sci.* 29:583–597.

Dunegin, J. C., J. R. Kienholz, R. A. Wilson, and W. T. Morris, 1954. Control of pear blight by a streptomycin-terramycin mixture. *Plant Dis. Rep.* 38:666–669.

Goodman, R. N., 1955. Antibiotics for control of fire blight. *Proc. Amer. Soc. Hort. Sci.* 64:186–190.

Hildebrand, E. M., 1939. Fire blight and its control. *Cornell Univ. Ext. Bull.* (405).

Luepschen, N. S., 1960. Fire blight control with streptomycin, as influenced by temperature and other environmental factors and by adjuvants added to sprays. *Cornell Univ. Agr. Exp. Sta. Mem.* (375).

Rosen, H. R., 1929. The life history of the fire-blight pathogen, *Bacillus amylovorus,* as related to means of overwintering and dissemination. *Arkansas Agr. Exp. Sta. Bull.* (244).

Whetzel, H. H., 1906. The blight canker of apple trees. *Cornell Univ. Agr. Exp. Sta. Bull.* (236):104–138.

18

Diseases Affecting Photosynthesis —Fungal Leafspots and Blights

Fungal leafspots and blights differ from similar bacterial diseases only in the matter of their etiology. Thus, the generalizations applied to the bacterial blights in Chapter 17 are also applicable here.

Leaf infections are of two types, depending on the activity of the pathogen: some leaf-infecting pathogens are actually saprophytic, killing suscept cells by enzymatic and toxic action well in advance of themselves, while others are truly parasitic, obtaining nourishment from living suscept cells, which die later as a consequence of the attack. In addition to the exclusionary, eradicative, and protective measures that may be used successfully against any kind of leafspot or blight, those that are caused by parasitic pathogens are frequently controlled through the use of resistant varieties of crop plants.

BEAN ANTHRACNOSE

Anthracnose, also known as pod spot, leafspot, or canker, occurs throughout the world in the garden, or French, bean (*Phaseolus vulgaris*). Although commercially important only in the garden bean,

FIGURE 18-1
Bean leaves infected by the anthracnose fungus.

the disease also occurs in lima bean, tepary bean, mung bean, runner bean, cowpea, garden pea, broad bean, and jack bean. The disease may destroy entire crops, but average losses today amount to less than 5% of the bean crop.

Necrotic lesions appear on the cotyledons of seedlings grown from infected seed. In moist weather, flesh-colored to pink masses of spores cover the interior surface of the necrotic spots. Pits and cankers often develop on the seedling stems below and above the cotyledons and on the petioles of the leaves. Stem lesions may result in death of the young plant; lesions on the petioles cause premature abscission. On the leaves, the most striking symptom is streak, or necroses of the veins (Figure 18-1). Laminar tissue, however, may become flaccid and necrotic, and angular necrotic areas may tear out. Roots are rarely affected. On the pods, brown spots enlarge to become necroses more than a centimeter in diameter; these lesions are dark with lighter-colored borders (Figure 18-2). Infection often extends through the endocarp to the seed, on which there develops a buff-colored spot with a raised, dark border; the border is surrounded by a ring of reddish-brown tissue.

The signs of the pathogen are the conidia that are borne in acervuli. The masses of flesh-colored conidia are held together by a mucilagin-

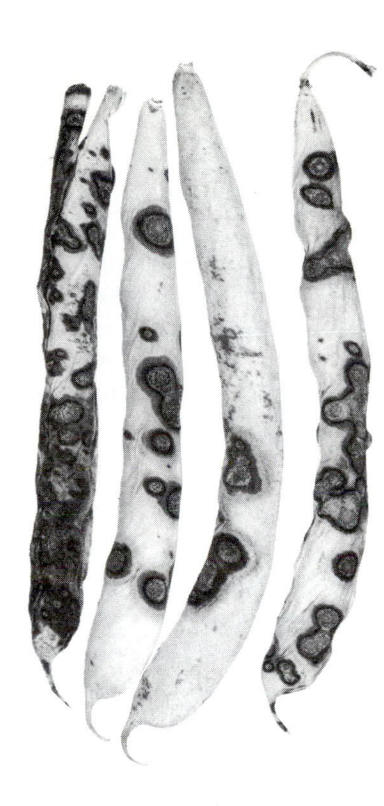

FIGURE 18-2
Bean pods infected by the anthracnose fungus.

ous substance and are readily visible on the surfaces of all lesions during wet weather.

Etiology

Bean anthracnose is caused by the imperfect fungus *Colletotrichum lindemuthianum* (Saccardo and Magnus) Briosi and Cavara. Stroma-like masses of mycelium give rise to acervuli in the epidermal tissues. Dark, septate setae, 30–90 μ in length, usually develop in the periphery of each acervulus. The individual conidia are hyaline, single-celled, and measure approximately $5\mu \times 15\mu$.

The fungus survives between growing seasons both as mycelium in diseased seed and as mycelium (and, possibly, spores) in crop debris in the soil. The primary cycle of the pathogen is initiated when mycelium in the cotyledons becomes active or when spores from plant debris are splashed to infection courts on young plants. Secondary cycles are initiated by conidia that are splashed from plant parts infected earlier in the season.

Penetration and infection phenomena proceed in the same way in the primary and secondary cycles. The spore germinates in a film of water in the infection court, producing a germ tube. An appressorium, which forms at the tip of the tube, is attached firmly to the substrate by a mucilaginous secretion. A fine penetration peg branches from the center of the appressorium and grows through the uninjured cuticle and into the epidermal cell. Apparently, the cuticle and cell wall of the epidermal cell are penetrated by mechanical pressure. During infection proper, the fungus acts as a parasitic pathogen, penetrating living cells and causing disintegration of the protoplasts. After infecting an area of tissue sufficient to produce a visible lesion, the fungus produces acervuli that bear the conidia.

Epidemiology

Vast numbers of conidia are produced during wet weather. Millions of spores are produced on a diseased leaf; up to a billion spores have been formed on some infected pods. Because the production and dissemination of inoculum and the infection of previously healthy plants both depend upon rain, abundant rainfall during the growing season may result in an epiphytotic. The disease develops rapidly over a wide range of relatively cool temperatures. (15–25°C).

Prevention of Disease

The pathogen can be excluded by planting healthy seed. Most seed grown in the Western United States is free from disease, because the dry weather of the growing season there is unfavorable to disease development.

The fungus can be eradicated from the crop residue in the soil by short crop rotations of only two years. Although secondary cycles can be controlled through protective spraying, this control measure is not required if disease-free seed is sown in uninfested soil. If the disease is present, cultural operations should be carried out only when the plants are dry.

Varieties of bean have been developed that are resistant to anthracnose. Resistance is multigenic, and the further development of a successful breeding program is complicated by the fact that the pathogen exists in a number of pathological strains. It is of historical interest that one of the first proofs of the existence of strains among plant pathogens was obtained by M. F. Barrus, who, in 1911, showed that *Colletotrichum lindemuthianum* occurred in two distinct physiological, or pathological, strains. Since that time, at least six pathological strains of *C. lindemuthianum* have been identified.

HELMINTHOSPORIUM LEAFSPOTS OF CORN

Three species of *Helminthosporium* cause leaf diseases of corn in the United States. Two of these, *H. turcicum* Passerini and *H. maydis* Nishikado and Miyake are worldwide in distribution, the former predominating in cool climates, the latter in warmer ones. The third species, *H. carbonum* Ullstrup, appears to be common only in the United States. The diseases presently cause little loss in field corn because resistant hybrids are planted almost exclusively. Most commercial varieties of sweet corn, however, are susceptible to at least one of the diseases; in Florida in the 1950s, losses of winter-grown sweet corn to the disease caused by *H. turcicum* amounted to 20–90% per field.

The "northern leaf blight," which is the disease caused by *H. turcicum*, occurs in the cooler corn-producing areas and in Southern Florida, where sweet corn is grown during the cool winter season. Large, elliptical, necrotic lesions form on leaves and sheaths (Figure 18-3). They appear hydrotic at first, then become gray to olive-brown as necrosis develops. Later, as they dry out, the lesions appear straw-colored. Individual lesions may be as much as 2 cm across and 10–15 cm long.

"Southern leaf spot," the disease caused by *H. maydis*, is characterized by relatively small, reddish-brown, necrotic lesions that are limited by the veins. Similar lesions are induced by *H. carbonum*. *H. maydis* and *H. carbonum* produce a black, moldy growth over the kernels of infected ears. Ear infections by *H. turcicum* are rare.

All three pathogens produce conidia that appear as dark masses on the surfaces of lesions. Perithecial stages of all three species have been described: *Trichometasphaeria turcica* Luttrell for *H. turcicum* Passerini, *Cochliobolus heterostrophus* Dreschler for *H. maydis* Nishikado and Miyake, and *Cochliobolus carbonum* (Ullstrup) R. R. Nelson for *H. carbonum* Ullstrup.

Etiology

The conidiophores of *T. turcica* emerge through stomata bearing conidia that are 15–25 $\mu \times$ 45–132 μ; the conidia are 3- to 10-septate. The conidia of *C. heterostrophus* are somewhat smaller, measuring

FIGURE 18-3
Leaves of corn showing symptoms of northern leaf blight.

10–17 μ × 30–115 μ. The sexual (or perithecial) stage of *H. maydis* (*C. heterostrophus*) is characterized by beaked perithecia that bear numerous asci, each of which contains filamentous ascospores arranged within the ascus in coils. *C. carbonum* produces conidia that measure 7–18 μ × 25–100 μ, and the perithecial stage is similar to that of *C. heterostrophus.*

All three fungi survive between growing seasons in crop refuse in the soil. In wet weather, sporulation is abundant on old stalks, and the conidia are splashed or blown thence to young corn plants. Spores germinate by producing germ tubes from the polar cells. Ingress may be either through stomata or by direct penetration of the cuticle. Once inside the leaf, the fungi grow intercellularly and kill the parasitized cells. Spores are produced about two weeks after the onset of infection. Secondary cycles proceed throughout the growing season.

All of the *Helminthosporium* leafspots of corn are favored by wet weather. Water, in the form of splashing rain, is the agent of dispersal, and free water is required in the infection court for spore germination. Northern leaf blight develops most rapidly at 20°C and below, whereas southern leaf spot develops most rapidly above 20°C.

Prevention of Disease

As indicated earlier, the use of resistant hybrids effectively controls these diseases in field corn. In sweet corn, however, growers must rely both upon sanitation and crop rotation as eradicative measures and upon spraying or dusting with fungicides as protective measures. Maneb or zineb, applied at five- to seven-day intervals throughout the growing season, has proved to be an effective and economical control of the diseases in winter-grown sweet corn in southern Florida. Spraying is begun when the plants are 4–6 inches high and is continued until 10–14 days before harvest.

APPLE SCAB

Scab is the most important disease of the above-ground parts of apple trees. Fruit are blemished, foliage may be so severely affected that defoliation occurs, and young twigs may be adversely affected. The disease occurs throughout the world wherever apples are grown, but is most severe in the cooler apple-growing regions.

Symptoms appear first on the lower sides of young leaves that were infected before the opening of the buds. Later, the necrotic lesions develop on the upper surfaces of young leaves and on sepals and petals. Necroses in leaves (Figure 18-4) are more or less circular and are a dark, olivaceous brown. Margins of mature lesions are dendritic, owing to the dark-colored hyphae that can be seen beneath the cuticle. In moist weather, the lesions become felty as the mycelium, conidiophores, and conidia develop on their surfaces. This combination of symptom (spot) and sign (the fungus itself) is termed blotch. Symptoms on the sepals are much the same as those on the leaves. Dark, circular necroses develop on infected fruit. Lesions become blotchy and then corky; in old lesions, the surfaces crack. Fruit infected early in the season are small and deformed.

Infection of twigs is common in Great Britain but rare in the United States. The bark of infected twigs is ruptured by blisters that form just beneath the surface.

FIGURE 18-4
Scab lesions in leaves and fruit of apple.

Etiology

Apple scab is caused by the ascomycete *Venturia inaequalis* (Cooke) Winter. The fungus produces conidia on short conidiophores that break through the surfaces of lesions on leaves, fruit, and twigs. The conidiophores arise from dark, stromatic fungal tissue between the epidermis and the cuticle. The conidia are dark at maturity and measure approximately $7\ \mu \times 17\ \mu$; they may be either one- or two-celled. Ascospores develop within asci borne in perithecia that form in fallen diseased leaves. The dark, two-celled ascospores are somewhat smaller than the conidia, measuring approximately $5\ \mu \times 13\ \mu$. One cell of the ascospore is noticeably larger than the other.

In most parts of the United States, the pathogen survives the winter in fallen leaves, which bear mature perithecia the following spring. In other regions, particularly in Great Britain, the pathogen also survives as mycelium on diseased twigs.

In the spring, ascospores are forcibly discharged from perithecia in leaves on the ground. As hydrostatic pressure builds up in the peri-

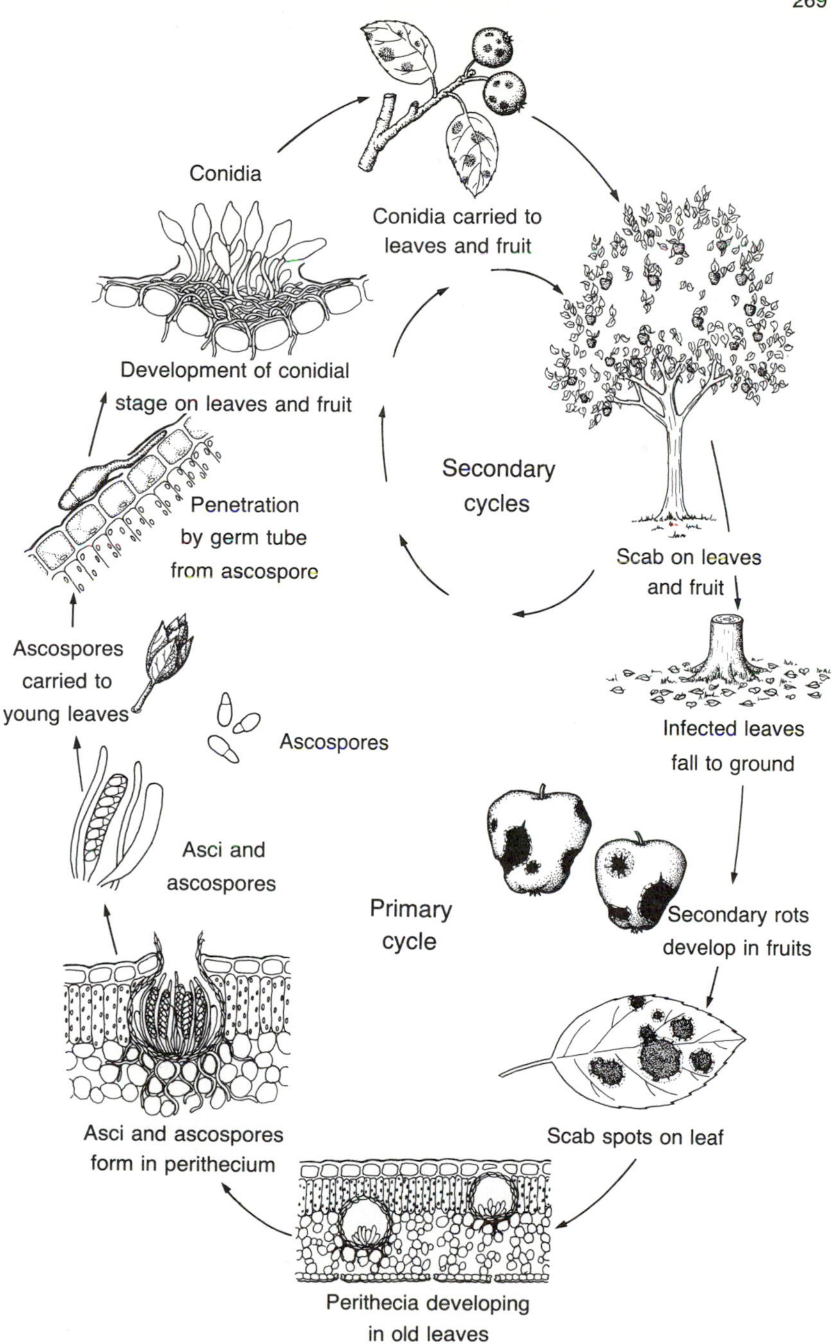

FIGURE 18-5
Apple scab.

thecium, a mature ascus is pushed into the ostiole so that its tip protrudes. The apical plug in the wall of the ascus is digested and the spores are shot out into the air. They are carried by wind currents to the infection courts. When the fungus overwinters in diseased twigs, conidia are produced in the early spring and are splashed by rain to the infection courts.

Either conidia or ascospores germinate in free water in the infection court. A flangelike appressorium, enclosed by a mucilaginous sheath, develops at the tip of a short (15–25 μ) germ tube. A tiny penetration peg grows into the cuticle, but does not pierce the wall of the underlying epidermal cell.

Once through the cuticle, the fungus develops dark, thick-walled hyphae that extend beyond the margins of the dark lesions. This mycelium soon becomes stromatic and gives rise to conidiophores and conidia. The latter are splashed to fruit and young leaves, where secondary cycles are initiated. Fruit remain susceptible throughout the growing season, but leaves become resistant as they mature.

The fungus obtains nourishment from underlying cells of the epidermis and the palisade parenchyma. Later, when depleted of nutrients, these cells die. As long as the leaf remains alive, the pathogen is restricted to the subcuticular areas that were initially invaded. Indeed, mesophyll cells are often stimulated to divide, producing daughter cells with suberized walls that constitute a histogenic demarcation that defends the living leaf against further invasion. When the leaf dies and falls, however, the fungus ramifies throughout the tissue, living as a saprophyte. Perithecia develop in these leaves in the autumn, and ascospores mature the following spring (Figure 18-5).

Epidemiology

The development of apple scab is favored by cool, wet weather. Rain in the early spring is necessary to provide free water in the infection courts and to provide the moisture necessary to the maturation of ascospores. The optimum temperature for infection by ascospores is 15–20°C, whereas that for infection by conidia is somewhat less. When ascospores from perithecia in fallen leaves are the inocula for primary cycles, the severity of outbreaks of apple scab can be reliably forecast on the basis of the length of the wetting period and the temperature.

Prevention of Disease

The pathogen cannot be excluded, but eradicative control measures are often applied. When infection during the preceding season has been severe, the fungus can be eradicated from fallen leaves by spraying the floor of the orchard with an eradicative fungicide, such as Elgetol (sodium dinitrocresylate). The fallen leaves must be saturated with the fungicide just before buds begin to break on the trees. When the fungus survives the winter as mycelium on diseased twigs, as it may in Great Britain, pruning of the infected twigs serves as an eradicative control measure.

The chief control measure for apple scab is protective spraying throughout the growing season. Spraying is begun during the "green-bud" stage, when the buds are not yet open. A second application is made in the "pink" stage, when the petals are showing but have not yet opened. The third spray is made at the "petal-fall" stage, and sprays are continued after blossoming at ten-day intervals. Such fungicides as captan, dodine, ferbam, glyodin, maneb, and wettable sulfur are among those presently recommended. It is often recommended that an eradicative fungicide (dichlone, for example) be used when infection is presumed to have taken place.

Some species of *Malus* are resistant to scab, and these have been used in breeding programs for the development of resistance in commercially acceptable varieties of apple. Resistance is governed by multiple genes. As might be expected, the pathogen exists in several different pathological races, three of which have been characterized.

BLACK SPOT OF ROSE

Black spot, the most important disease of rose, occurs wherever roses are grown. Losses result from premature defoliation and a consequent loss of vigor. Badly affected plants are predisposed to winter killing. The disease has also been called black blotch, leaf blotch, rose actinonema, and star-shaped leafspot.

Roughly circular necrotic spots, up to 1 cm in diameter, develop in diseased leaves (Figure 18-6). The margins of the lesions are fibrillose, owing to the strands of dark mycelium that are visible beneath the cuticle in the areas surrounding the necroses. Yellow bands usually surround the lesions. Black lesions develop also on the stem, receptacles, and sepals, and inconspicuous spots often form on the petioles. The petals are often distorted. Mounds of acervuli develop, mostly near the margins, in old lesions. Apothecia sometimes form in fallen leaves during the winter and early spring, especially in the colder regions of the temperate zones.

FIGURE 18-6
Leaves of rose infected by the black-spot fungus.

Etiology

Black spot is caused by *Diplocarpon rosae* Wolf; before its perfect stage was known, its imperfect stage was referred to by the name *Actinonema rosae*. The imperfect stage is by far the more common one in nature, and probably accounts for all but a few infections. The two-celled conidia are oval, approximately 5 $\mu \times 20 \mu$, and are borne in acervuli. When the perfect stage forms, spermatia 2–3 μ long are first produced in the fruiting body deep within the tissues of fallen leaves. Later, asci and ascospores form, and the previously enclosed fruiting body opens wide at the top and develops into a stalkless apothecium. Individual ascospores are indistinguishable from single conidia.

The pathogen survives the winter as mycelium and conidia in diseased canes and in leaves on the ground. Occasionally, the pathogen survives as mycelium that gives rise to the sexual stage in fallen leaves during the winter and spring. Primary cycles of the fungus may be initiated by rain-splashed conidia or ascospores. Both spores send forth short germ tubes that form appressoria. A penetration peg from the appressorium grows directly through the cuticle and the wall of the underlying epidermal cell, in which a haustorium forms. The fungus then develops a network of subcuticular hyphae. The first symptoms appear in about ten days. Acervuli begin to form in necrotic lesions about three weeks after inoculation. Conidia from these may then be splashed to new infection courts, where secondary cycles occur.

Infection is favored by high humidity and temperatures of 20–25°C. Conidia and ascospores are extruded in viscid masses; splashing rain, therefore, is the most common agent of dispersal. Spores may also be carried to infection courts by crawling insects.

Prevention of Disease

The pathogen cannot be excluded, but can be eradicated to some extent by sanitation: diseased canes should be pruned and fallen leaves collected and destroyed. The only sure way to control black spot is to keep infection courts covered at all times with a good protective fungicide. Plants should be sprayed or dusted every four to seven days with a good organic fungicide, such as maneb, zineb, captan, or folpet.

Differences in susceptibility occur between varieties, but no commercial variety is sufficiently resistant to preclude the necessity of protective spraying or dusting. Thick-leaved varieties and those with a waxy "bloom" exhibit a morphological resistance. Varieties of *Rosa wichuraiana* and *R. multiflora* show a relatively high degree of protoplasmic resistance.

Selected References

BEAN ANTHRACNOSE

Barrus, M. F., 1921. Bean anthracnose. *Cornell Univ. Agr. Exp. Sta. Mem.* (42):97–215.

Dey, P. K., 1919. Studies in the physiology of parasitism, V. Infecion by *Colletotrichum lindemuthianum. Ann. Bot.* (*London*) 33:305–312.

Leach, J. G., 1923. The parasitism of *Colletotrichum lindemuthianum. Minn. Agr. Exp. Sta. Tech. Bull.* (14).

Whetzel, H. H., 1908. Bean anthracnose. *Cornell Univ. Agr. Exp. Sta. Bull.* (255):429–448.

Yerkes, W. D., and M. T. Ortiz, 1956. New races of *Colletotrichum lindemuthianum* in Mexico. *Phytopathology* 46:564–567.

Zaumeyer, W. J., and H. R. Thomas, 1957. A monographic study of bean diseases and methods for their control. *U. S. Dep. Agr. Tech. Bull.* (868).

HELMINTHOSPORUM LEAFSPOTS OF CORN

Drechsler, C., 1925. Leafspot of maize caused by *Cochliobolus heterostrophus,* n. sp., the ascigerous stage of a *Helminthosporium* exhibiting bipolar germination. *J. Agr. Res.* 31:701–726.

Hooker, A. L., D. R. Smith, S. M. Lim, and J. B. Beckett, 1970. Reaction of corn seedlings with male sterile cytoplasm to *Helminthosporium maydis. Plant Dis. Rep.* 54:708–712.

Jennings, P. R., and A. J. Ullstrup, 1957. A histological study of three *Helminthosporium* leaf blights of corn. *Phytopathology* 47:707–714.

Luttrell, E. S., 1958. The perfect stage of *Helminthosporium turcicum. Phytopathology* 48:281–287.

Mercado, A. L., and R. M. Lantican, 1961. The susceptibility of cytoplasmic male-sterile lines of corn to *Helminthosporium maydis* Nish. and Miy. *Philippine Agr.* 45:235–243.

Nelson, R. R., 1959. *Cochliobolus carbonum,* the perfect stage of *Helminthosporium carbonum. Phytopathology* 49:807–810.

Nishikado, Y., and C. Miyake, 1926. Studies on two Helminthosporium diseases of maize caused by *H. turcicum* Pass. and *Cochliobolus heterostrophus* Drech. (*H. maydis* N. and M.). *Ber. Ohara Inst. Landwirt. Forsch. Okayama Univ.* 3:221–266.

Villareal, R. L., and R. M. Lantican, 1965. The cytoplasmic inheritance of susceptibility to *Helminthosporium* leaf spot in corn. *Philippine Agr.* 49:294–300.

APPLE SCAB

Boone, D. M., 1971. Genetics of *Venturia inaequalis. Annu. Rev. Phytopathol.* 9:297–318.

Connor, S. R., and J. W. Heuberger, 1968. Apple scab versus effect of late season application of fungicides on prevention of perithecial development by *Venturia inaequalis. Plant Dis. Rep.* 52:654–658.

Croxall, H. E., B. C. Knight, W. T. Dale, and W. R. Rosser, 1964. A comparison of protective and curative apple scab spray programs. *Plant Pathol.* 13:93–100.

Hirst, J. M., and O. J. Stedman, 1961. The epidemiology of apple scab (*Venturia inaequalis* (Cke.) Winter), I. Frequency of airborne spores in orchards. *Ann. Appl. Biol.* 49:290–305.

———, and ———, 1962. The epidemiology of apple scab (*Venturia inaequalis* (Cke.) Winter). II. Observations on the liberation of ascospores. III. The supply of ascospores. *Ann. Appl. Biol.* 50:525–567.

Keitt, G. W., C. N. Clayton, and M. H. Langford, 1941. Experiments with eradicant fungicides for combatting apple scab. *Phytopathology* 31: 296–322.

Nusbaum, C. J., and G. W. Keitt, 1938. A cytological study of host-parasite relations of *Venturia inaequalis* on apple leaves. *J. Agr. Res.* 56:595–618.

Shay, J. R., E. B. Williams, and J. Janick, 1962. Disease resistance in apple and pear. *Proc. Amer. Soc. Hort. Sci.* 80:97–104.

Wallace, E., 1913. Scab disease of apples. *Cornell Univ. Agr. Exp. Sta. Bull.* (335):545–624.

BLACK SPOT OF ROSE

Aronescu, A., 1934. *Diplocarpon rosae:* from spore germination to haustorium formation. *Mem. Torrey Bot. Club* 61:291–329.

Dodge, D. O., 1931. A further study of the morphology and life history of the rose black-spot fungus. *Mycologia* 23:446–462.

Harris, C. G., 1967. Breeding roses for blackspot resistance. *Gard. Chron.* 162(1):20–21.

Massey, L. M., 1955. Tests with fungicides for rose blackspot. *Amer. Rose Annu.* 40:62–91.

Wolf, F. A., 1913. The perfect stage of *Actinonema rosae*. *Bot. Gaz.* 54: 218–234.

19

Diseases Affecting Photosynthesis —Downy Mildews

The downy-mildew diseases are actually leaf blights induced by certain members of the fungal order Peronosporales of the class Phycomycetes. All are parasitic pathogens that have highly specialized food relations with the plants they attack. The common name, downy mildew, is derived from the fact that the signs of the fungi are masses of mycelia, sporangiospores, and conidia that grow out from the stomata of infected leaves, giving the affected surfaces a downy appearance.

The true downy mildews are members of the family Peronosporaceae, but certain species of the genus *Phytophthora* (family Pythiaceae) are considered downy mildews by many plant pathologists. An important example is *Phytophthora infestans,* the causal agent of late blight of potatoes.

Members of the Peronosporaceae and certain *Phytophthora* spp. have parasitic food relations with the plants that they attack. Their mycelia are usually intercellular, and fingerlike branches, termed **haustoria,** penetrate the cell, invaginate the cytoplast, and absorb nutrients from the living host cell. Sporangia that form on vegetative thalli of *Phytophthora infestans* may germinate either indirectly to form zoospores or directly by means of germ tubes; because sporangia may be disseminated as inocula and then germinate directly to penetrate infection courts, they are often called conidia. In the Peronosporaceae, vegetative spores usually germinate directly to form germ tubes. Most downy mildews survive as oospores, which may germinate in a favorable environment to produce either zoospores or single or multiple germ tubes.

Protective spraying or dusting is usually an effective control measure for most downy mildews, because their inocula must be deposited on above-ground infection courts in order to develop. Because of the specialized food relations between downy mildews and their suscepts, breeding for disease resistance has often been successful.

LATE BLIGHT OF POTATO AND TOMATO

Late blight sporadically becomes epiphytotic and devastates plantings of potatoes and, occasionally, tomatoes. The disease is potentially serious every year, because there is always an abundance of susceptible plants and there is usually sufficient inoculum. The severity of the disease, however, is very much dependent upon weather conditions, and environmental conditions favorable to widespread serious outbreaks of disease occur only once in 5–10 years in most potato-growing regions of the world. Late blight, also known as potato murrain, black blight, dry rot, winter blight, and downy mildew, literally wiped out the potato crop in Ireland in 1845 and in succeeding years, bringing on the great potato famine. With the possible exception of the ergot disease of rye, late blight of potato has caused more human suffering than any other disease of plants.

The symptoms of late blight in the foliage of potato and tomato begin as hydrotic areas with indefinite margins at the tips or on the margins of leaflets. These lesions then become necrotic and turn brown to almost black (Figure 19-1); the necrotic tissue dries and shrivels. Often, a chloranemic border develops around the necrotic areas; this yellow band is widest during wet weather. Necroses extend along the petioles and stems, which are girdled in wet weather. Oval, necrotic lesions develop in diseased stems during dry weather. Necrotic areas on stems are not surrounded by yellow borders. The characteristic, moderately offensive odor that emanates from diseased plants is particularly noticeable when the disease is developing rapidly in the field. Slightly sunken hydrotic areas develop in the outer tissues of infected potato tubers. These areas become necrotic and turn brown

Figure 19-1
Potato foliage affected by late blight of potato.

to purple. Irregularly shaped hydrotic areas form in diseased tomato fruit in the field. Later, these areas become necrotic and brown; they are surrounded by a border that remains green as the unaffected portion of the fruit becomes colored during the ripening process.

During wet weather, the signs of the pathogen—sporangiophores (conidiophores) and sporangia (conidia)—can be seen as a downy growth, particularly on the abaxial surfaces of leaflets.

Etiology

Late blight is caused by the phycomycete *Phytophthora infestans* (Montagne) deBary, a member of the family Pythiaciae. In most potato-growing regions, the fungus survives between crop seasons as mycelium within diseased tubers. Where the sexual stage of the fungus occurs, as in parts of Mexico, the fungus can also survive as oospores in plant debris in the soil. The sexual stage occurs infrequently, if at all, in commercial fields outside of Mexico; oospores, therefore, are unimportant in most potato-producing regions.

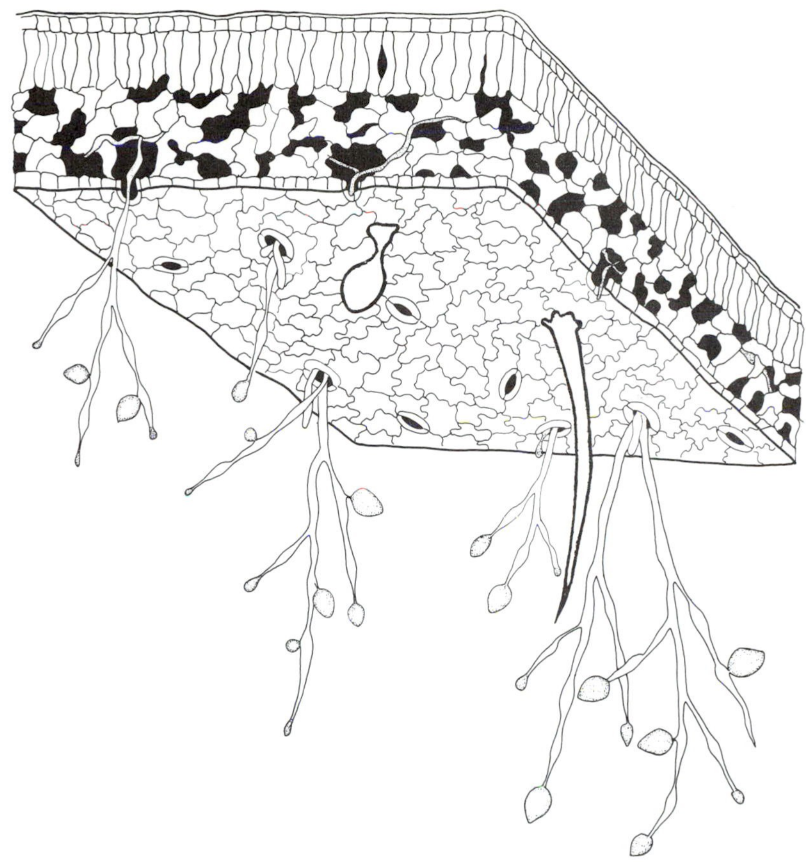

FIGURE 19-2
Blighted potato-leaf tissues, showing sporangia (conidia) of *Phytophthora infestans.*

When infected seed tubers or seed pieces sprout, the fungus grows into the developing shoot to produce sporangiophores and sporangia that protrude through the stomata of the young stem. The sporangia are then splashed by rain or blown by wind to the foliage of young plants, where the primary cycles of disease begin.

Sporangia in the infection court germinate only in a film of water. When the temperature is 20–21°C, two to eight zoospores are released from each sporangium within three hours (Figure 19-2). The zoospores swim about in the water film on the surface of the leaf, lose or withdraw their flagella, and germinate (Figure 19-3) by producing

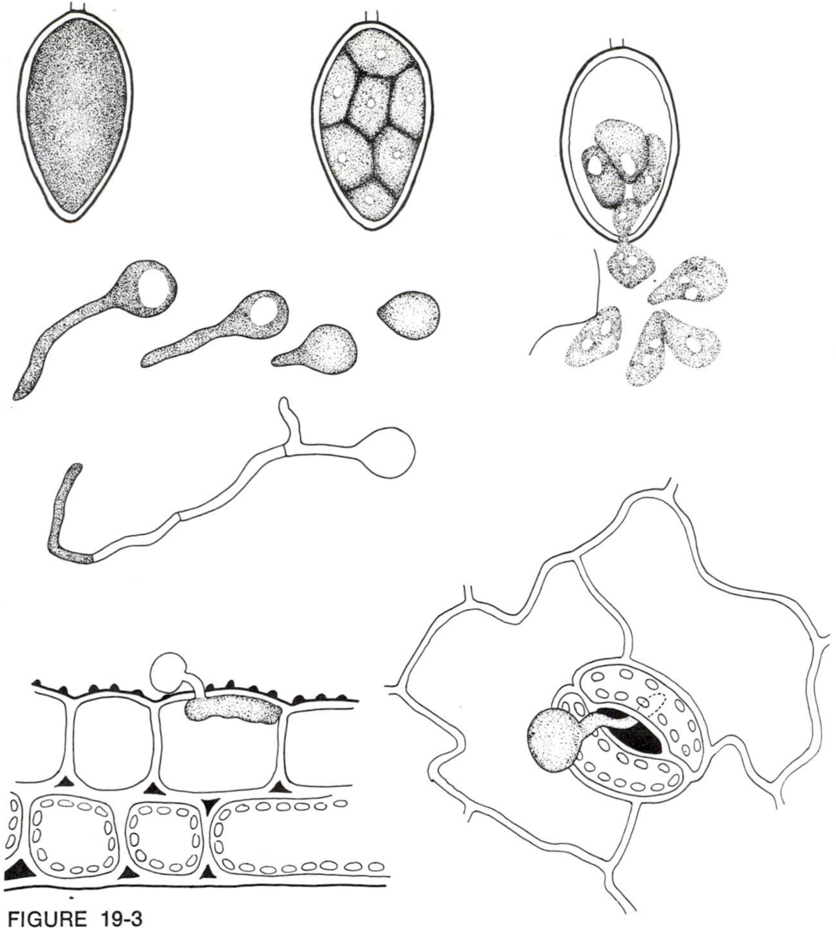

FIGURE 19-3
Phytophthora infestans: sporangia releasing zoospores (*top*); germinating zoospores (*center*); direct penetration (*bottom left*) and stomatal entrance (*bottom right*).

germ tubes. Appressoria form at the tips of elongating germ tubes, and the cuticle of the leaf is penetrated directly by penetration pegs growing from the appressoria.

When the temperature is 22–32°C, a sporangium in a film of water germinates by producing a single germ tube. Thus, the sporangia function as conidia in warm weather. Appressoria form at the ends of germ tubes from sporangia (conidia), and the fungus penetrates di-

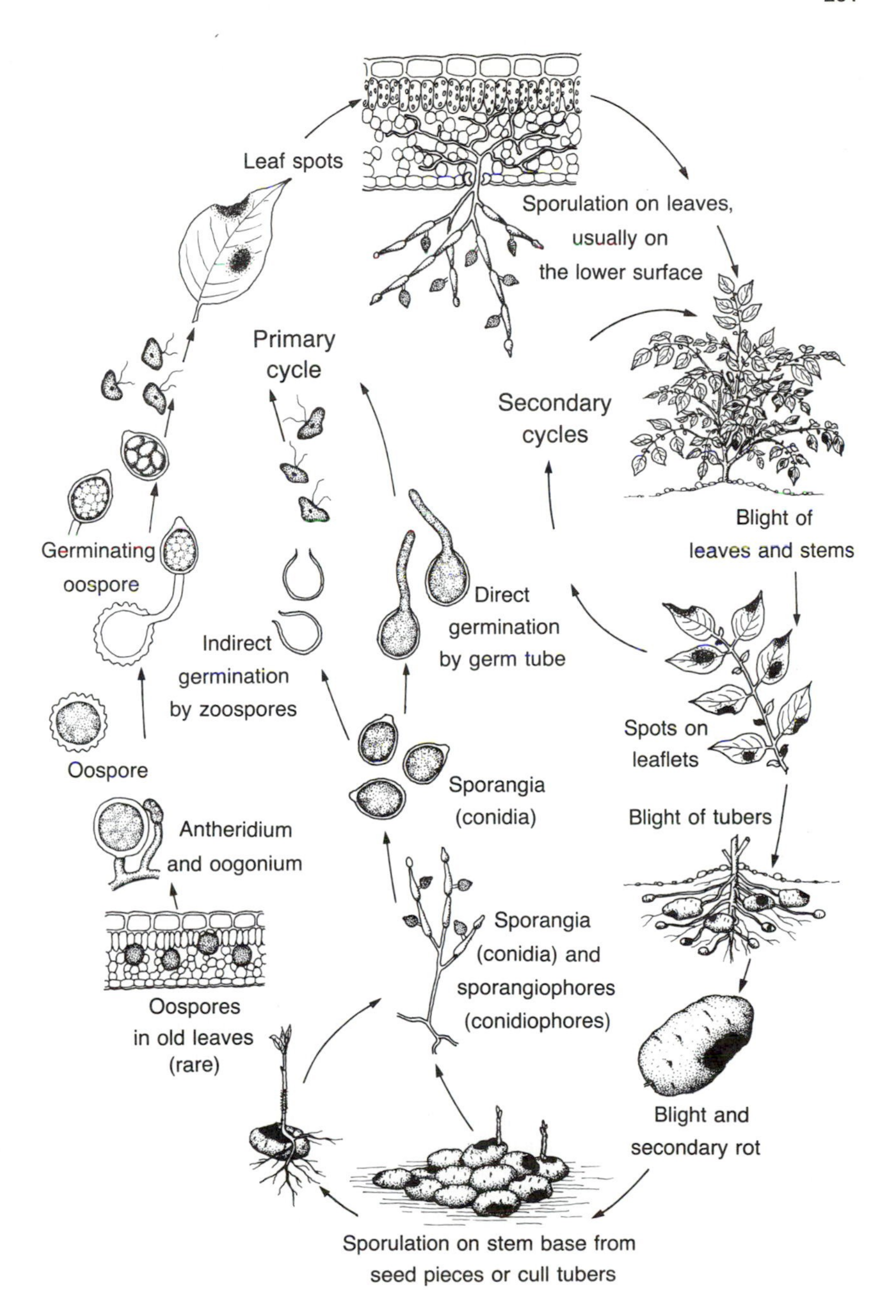

FIGURE 19-4
Late blight of potato.

rectly through the uninjured cuticle beneath the appressoria. Occasionally, branches from germ tubes derived from sporangia enter the leaf through its stomata.

The mycelium develops intercellularly in the diseased leaves of potato and tomato. Fingerlike branches from the intercellular hyphae penetrate cells of the suscept and act as feeding organs. Parasitized host cells die shortly after the tissues have been invaded by the fungus, but symptoms are not visible until two or three days after the onset of infection. The first symptoms are hydrosis and spot. The typical blighting of diseased foliage becomes apparent between five and seven days after ingress. Sporangiophores and sporangia, the signs of the fungus, protrude through stomata approximately one week after infection has begun.

Sporangia are the inocula for secondary cycles of the fungus. The phenomena of inoculation, prepenetration, penetration, and infection proceed as described above. Tubers formed at or just below the surface of the soil are infected late in the season by spores that drop to the soil from diseased leaves. (Figure 19-4).

Epidemiology

Widespread outbreaks of severe late blight occur in potatoes and tomatoes during prolonged periods of cool, wet weather. Usually, there is ample inoculum for an epiphytotic, and most commercial varieties of potato and tomato are susceptible to late blight. Sporangia are produced over a wide range of temperature (3–26°C) when the air is moist. The range of relative humidity favorable to the formation of sporangia is 91–100%. As stated earlier, the sporangia germinate by producing two to eight infective zoospores when the air temperature is 18°C or lower and when there is a film of water in the infection court. Under these conditions, therefore, each sporangium could result in as many as eight new infections. When the weather is warm, however, a single sporangium can give rise to but one infection, because when the temperature is above 18°C, the sporangium behaves as a conidium and germinates by means of a germ tube. Thus, late blight is more severe in cool, wet weather than in warm, wet weather because cool weather is favorable to formation of sporangia and to their germination by means of zoospores.

Prevention of Disease

The number of primary infections by *Phytophthora infestans* can be reduced by planting healthy tubers or seed pieces (exclusion) and by such sanitary practices (eradication) as killing tubers in potato cull piles and destroying volunteer potato plants in the field. These practices and, thus far, the use of resistant varieties do not insure control of late blight. Protective spraying or dusting is the only reliable measure for the control of the disease. In cool-temperate regions, spraying at weekly intervals should be begun when the plants are 4–6 inches high and should be discontinued several weeks before harvest. The disease can be kept in check with such organic fungicides as zineb, maneb, and daconil, unless the sort of weather most favorable to the disease persists for several weeks. When an epidemic is eminent, sprays of fixed copper are recommended.

Late blight is so dependent upon weather for its development in the field that reasonably accurate forecasts of the severity of the disease can be made in some potato- and tomato-growing regions (see The Forecasting of Epidemics, p. 136). Consequently, spray schedules can be modified during the season. Spraying can be discontinued during dry weather, when the disease poses no threat to the crop. Conversely, when an epiphytotic threatens, spraying can be resumed and a potent fungicide, such as a fixed copper, can be used for several weeks.

Solanum demissum, a wild, tuber-forming relative of the potato, is apparently immune to late blight, and has been used in breeding programs to develop varieties of potato that are resistant to the pathogen. Some varieties thus far developed, however, have been susceptible to new races of the pathogen, races that probably have arisen as a result of adaptation. In Mexico, where the sexual stage of the fungus occurs regularly, new races of *Phytophthora infestans* may also arise by hybridization.

DOWNY MILDEW OF TOBACCO

Downy mildew, or blue mold, of tobacco occurs in most tobacco-growing regions of Australia, South America, and the United States; it is also widespread in Europe and the Middle East. Tobacco, certain other species of *Nicotiana,* and a few other solanaceous plants are the only suscepts. In tobacco, the disease is most devastating in the plant bed, but it sometimes causes considerable crop losses in fields of shadegrown tobacco.

When young plants are infected, diseased leaves become cupped or twisted and their tips turn yellow. Later, the leaves become necrotic and show the symptom of blight (Figure 19-5). A rank odor can be detected in beds of plants whose foliage is extensively blighted. The

FIGURE 19-5
Downy mildew (blue mold) of tobacco leaves (*left, center*), and a healthy leaf (*right*) (Photograph by W. B. Tisdale, courtesy of the University of Florida.)

sign of the pathogen is the felt of downy mildew that can be seen on the abaxial surfaces of diseased leaves in wet weather. The mildew is gray or brown, but it may appear violet when viewed obliquely. Blight seldom develops in field-grown plants. Instead, yellow to brown, irregular spots occur on diseased leaves, and the affected tissues finally die and turn brown. Sometimes, the mildew can be seen in wet weather on the adaxial surfaces of the lesions.

Etiology

Downy mildew of tobacco is caused by the phycomycete *Peronospora tabacina* Adam, a typical member of the order Peronosporales. The fungus reproduces sexually in diseased tobacco leaves, forming thick-walled oospores that maintain the fungus between growing seasons. In warm climates, however, the fungus also survives between crop seasons as parasitic mycelium in diseased plants of various species of *Nicotiana*. Conidia produced on these plants, or overwintered oospores, serve as the inoculum for primary cycles of the pathogen. In the United States, conidia are carried northward, in step-wise fashion, and initiate successive cycles of the fungus during the late winter and early spring.

Conidia or oospores will germinate in the infection court only when covered by a film of water. Following germination, the developing hyphae enter the leaves of the suscept by direct penetration of the uninjured cuticle. The mycelium that develops within the diseased leaf grows intercellularly, and parasitizes living cells after producing fingerlike haustoria. The fungus sporulates between four and seven days after the onset of disease, producing conidia on branched conidiophores that emerge through stomata. These spores then initiate secondary cycles when the environment is favorable.

Epidemiology

Like most downy mildews, the one that infects tobacco is favored by cool, wet weather. Conidia are produced most abundantly at 13–15°C, although they will form at any temperature from 3–30°C. Spore germination is also most rapid at approximately 15°C. The disease is checked in warm, dry weather; spots become brittle, and no spores are formed. Consequently, infected plants appear to recover from the disease in dry weather.

Prevention of Disease

The initiation of infections by overwintered oospores can be avoided through the exclusionary practice of using new land for seed beds and through the eradicative practice of fumigating the soil with methyl bromide before seeding. Many growers routinely fumigate the seedbed soil, both to control nematodes and black shank, and to prevent the development of downy mildew from oospores in the soil.

Control measures applied prior to planting cannot guard against infections by conidia subsequently blown in from the outside. Plants in beds must be protected with a fungicide. Presently, most growers apply zineb, a dithiocarbamate, in the form of a dust. Dusting is begun when the plants are first threatened, and is continued at intervals of five to seven days for as long as the weather is cool and wet. The success of zineb as a protective fungicide makes it no longer necessary to fumigate plant beds with such compounds as benzene and paradichlorobenzene.

Commercial varieties of tobacco are susceptible to downy mildew, but *Nicotiana debneyi* is resistant. Perhaps the resistance of this species will be bred into commercially acceptable varieties of tobacco in the near future.

DOWNY MILDEW OF CUCURBITS

Downy mildew of cucurbits occurs throughout the world wherever cucurbits are grown. It is most severe in cucumber, squash, pumpkin, muskmelon, and watermelon.

Yellow spots occur in the leaves of infected plants. The spots, which are somewhat angular, are readily visible on the adaxial surfaces of diseased leaves. Severe infections result in dwarfing and even death of plants. The purple mildew on abaxial surfaces of diseased leaves is the diagnostic sign of the pathogen.

Etiology

Downy mildew of cucurbits is caused by *Pseudoperonospora cubensis* (Berkeley and Curtis) Clinton. It produces vegetative spores, or conidia, upon conidiophores whose tips are branched much like those of the conidiophores of fungi in the related genera *Peronospora* and *Plasmopara*. Vegetative spores often act as conidia and germinate directly by means of germ tubes. They may also germinate indirectly to form biflagellate zoospores.

P. cubensis is not known to reproduce sexually. Oospores, therefore, are not produced as overwintering structures. The fungus survives between growing seasons as mycelium in diseased plants. In cold climates, the pathogen may survive in greenhouse-grown plants; in warm climates, as in the southern United States, the fungus may survive between crop seasons in diseased weeds or volunteer plants of the family Cucurbitaceae.

Conidia from diseased weeds or greenhouse-grown plants are blown to young plants in the field, where the first cycles of the fungus occur. The fungus enters suscepts through stomata, grows intercellularly, and produces ovate, branched haustoria within the parasitized cells. Under favorable conditions, conidia are produced approximately one week after the beginning of the infection. Secondary cycles continue throughout the growing seasons. Conidia are disseminated by wind and by windblown rain.

Epidemiology

Although the optimum temperature for infection, spore germination, and spore formation is approximately 18°C, the disease will develop rapidly in warm weather, provided the relative humidity remains high. Conidia are produced and will germinate when the temperature is as high as 30°C. A film of water in the infection court is essential to spore germination.

Prevention of Disease

Downy mildew of cucurbits is controlled most effectively through protective spraying or dusting and the use of resistant varieties. Dithiocarbamate fungicides, such as zineb or maneb, are usually recommended over fixed copper fungicides, because cucurbits are particularly sensitive to copper.

Selected References

LATE BLIGHT OF POTATO AND TOMATO

Black, W., 1970. The nature and inheritance of field resistance to late blight (*Phytophthora infestans*) in potato. *Amer. Potato J.* 47:279–288.

———, C. Mastenbroek, W. R. Mills, and L. C. Peterson, 1953. A proposal for an international nomenclature of races of *Phytophthora infestans* and genes controlling immunity in *Solanum demissum* derivatives. *Euphytica* 2:173–178.

Cook, H. T., 1949. Forecasting late blight epiphytotics of potatoes and tomatoes. *J. Agr. Res.* 78:545–563.

Cox, A. E., and E. C. Large, 1960. Potato blight epidemics throughout the world. *U. S. Dep. Agr. Handb.* (174).

Crosier, W., 1934. Studies in the biology of *Phytophthora infestans* (Mont.) DeBary. *Cornell Univ. Agr. Exp. Sta. Mem.* (155).

Ferris, V. R., 1955. Histological study of pathogen–suscept relationships between *Phytophthora infestans* and derivatives of *Solanum demissum*. *Phytopathology* 45:546–552.

Hirst, J. M., and O. J. Stedman, 1965. The epidemiology of *Phytophthora infestans*. *Ann. Appl. Biol.* 48:471–517.

Mills, W. R., 1940. *Phytophthora infestans* on tomato. *Phytopathology* 30: 830–839.

Pristou, R., and M. E. Gallegly, 1954. Leaf penetration by *Phytophthora infestans*. *Phytopathology* 44:81–86.

Reddick, D., and W. R. Mills, 1938. Building up virulence in *Phytophthora infestans*. *Amer. Potato J.* 15:29–34.

Van der Plank, J. E., 1966. Horizontal (polygenic) and vertical (oligogenic) resistance against blight. *Amer. Potato J.* 43:43–52.

DOWNY MILDEW OF TOBACCO

Anderson, P. J., 1937. Downy mildew of tobacco. *Conn. Agr. Exp. Sta. Bull.* (405):63–82.

Angell, H. R., and A. V. Hill, 1932. Downy mildew (blue mold) of tobacco in Australia. *Commonw. Aust. Counc. Sci. Ind. Res. Bull.* (65).

Clayton, E. E., 1945. Resistance of tobacco to blue mold (*Peronospora tabacina*). *J. Agr. Res.* 70:79–87.

———, and J. A. Stevenson, 1943. *Peronospora tabacina* Adam, the organism causing blue mold (downy mildew) of tobacco. *Phytopathology* 33:103–113.

Peyrot, J., 1962. Tobacco blue mold in Europe. *FAO (Food Agr. Organ. U. N.) Plant Prot. Bull.* (10):73–80.

Stevens, N. E., and J. C. Ayres, 1940. The history of tobacco downy mildew in the United States in relation to weather conditions. *Phytopathology* 30:684–688.

DOWNY MILDEW OF CUCURBITS

Barnes, W. C., and W. M. Epps, 1950. Some factors related to the expression of resistance of cucumbers to downy mildew. *Proc. Amer. Soc. Hort. Sci.* 56:377–380.

Clinton, G. P., 1905. Downy mildew, or blight. *Peronoplasmopara cubensis* (B. & C.) Clint. of muskmelons and cucumbers. *Conn. Agr. Exp. Sta. Annu. Rep.* 1904:329–362.

Doran, W. L., 1932. Downy mildew of cucumbers. *Mass. Agr. Exp. Sta. Bull.* (283).

Nusbaum, C. J., 1945. The seasonal spread and development of cucurbit downy mildew in Atlantic coastal states in 1944. *Plant Dis. Rep.* 29: 141–143.

20

Diseases Affecting Photosynthesis —Powdery Mildews

The powdery-mildew fungi are so named because of the powdery appearance of their mycelia and chains of conidia that develop profusely on the surfaces of diseased leaves and shoots (Figure 20-1). The powdery mildews are all members of the order Erysiphales of the class Ascomycetes. All known powdery-mildew fungi are obligately parasitic.

The perithecia produced by a powdery mildew on its surface mycelium are spheroidal and have no ostiole; they are termed **cleistothecia.** Appendages that develop on the surfaces of cleistothecia may be mycelial, needlelike, or straight with dichotomously branched ends. The various genera of powdery-mildew fungi are separated on the basis of the morphology of the appendages and whether each cleistothecium contains one ascus or several asci.

Powdery mildews may overwinter in the perithecial stage with ascospores initiating primary cycles in the spring. Some, such as the powdery mildew of winter wheat, liberate ascospores in the autumn after the perithecia have survived through the summer; the fungus survives the winter as mycelium on infected wheat plants. Powdery mildews of woody perennial plants (Figure 20-2) often survive the

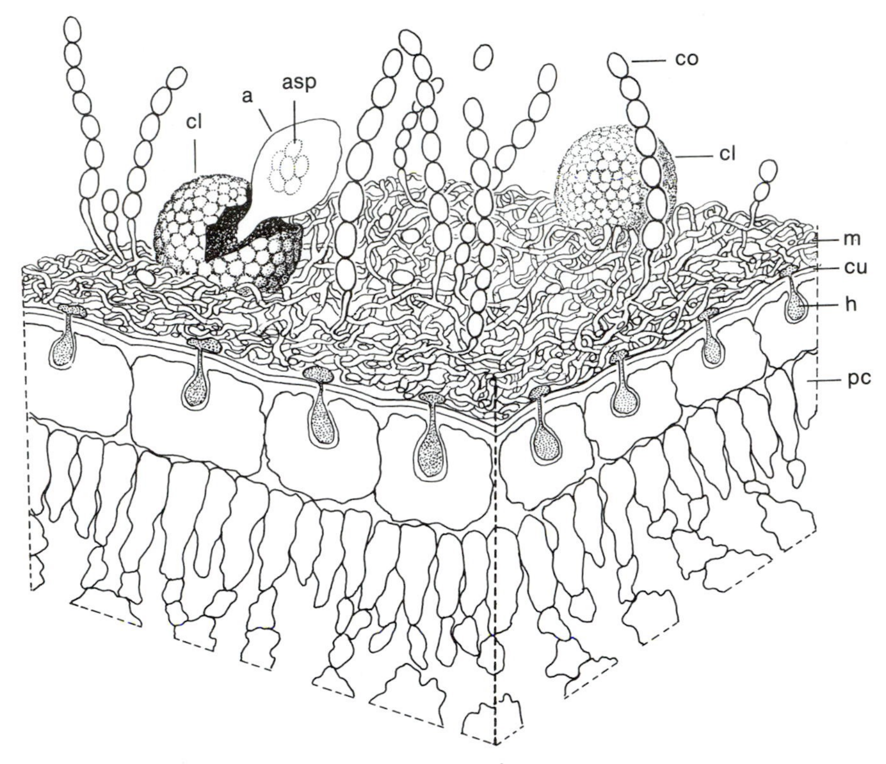

FIGURE 20-1
Model of a leaf infected by a powdery mildew: *co,* conidia; *cl,* cleistothecia; *a,* ascus; *asp,* ascospores; *m,* mycelium; *cu,* cuticle; *h,* haustorium in epidermal cell; *pc,* palisade cell.

winter in dormant buds. Conidia initiate secondary cycles during the growing season.

As with other leaf-infecting fungi, protective spraying and dusting may be effective control measures for powdery mildews. Because most powdery-mildew conidia will not germinate in a film of water, although most germinate best at high relative humidities, those protective fungicides that require solubilization are relatively ineffective. The spores of downy mildews, by contrast, require a film of water for germination.

Sulfur sprays or dusts—which are volatile and thus require no solubilization—are often used to combat powdery mildews. Recently, certain organic fungicides have been used in place of sulfur because they are as effective against the fungi but are not as phytotoxic as

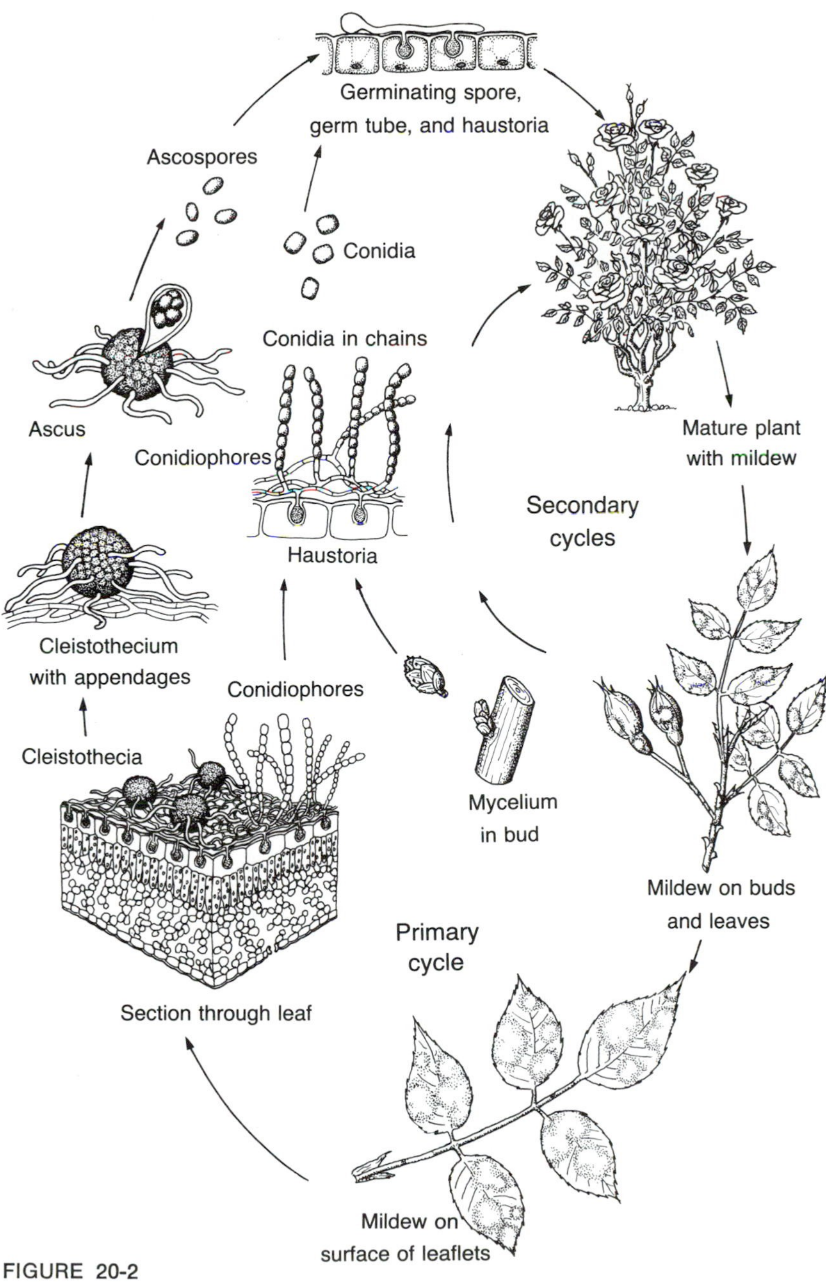

FIGURE 20-2
Powdery mildew of rose.

sulfur. The obligately parasitic nature of the powdery mildews suggests that resistant crop varieties might be selected or developed in breeding programs. Indeed, mildew-resistant varieties of certain forage legumes and cereal grains are used commercially.

POWDERY MILDEW OF CEREALS

Powdery mildew of cereals, which infects almost a hundred species of grains and grasses, is distributed throughout the world, but is most prevalent in the temperate zones. The disease, which occasionally becomes epiphytotic, has been especially severe in barley grown in the southern United States. In New York, the disease caused crop losses of approximately 10–40% in barley over the five-year period beginning in 1937.

The symptoms of the disease are yellowing and curling of leaves and dwarfing of plants. Infected leaves often have islands of green tissue in them; badly infected leaves turn red, then brown, and die. Culms of diseased plants are so weakened that lodging occurs, and the flower heads of severely diseased plants may blast. The diagnostic signs of the pathogen are the cottony to brown weblike mycelial mats, masses of powdery conidia in chains, and carbonous cleistothecia, all of which are produced on the surface of affected tissue (Figure 20-3).

Etiology

Powdery mildew of cereals and grasses is caused by *Erysiphe graminis* De Candolle. The fungus produces conidia in chains from short conidiophores that branch from the mycelium. The basal cells of the conidiophores are distinctly swollen. Cleistothecia are produced late in the summer; when mature, each contains between eight and twenty-five asci, each of which bears four or eight ascospores. Each cleistothecium bears several appendages that resemble hyphal strands.

The fungus usually survives the winter as mycelial mats on perennial grasses or fall-sown cereal grains, but it may also survive as cleistothecia. Ascospores are discharged from the asci that are released when the cleistothecia break upon imbibition of water in the spring. The spores, or conidia produced on mycelial mats in the spring, serve as inocula for the first cycles of the fungus.

Spores blown or splashed on the infection courts germinate best when the relative humidity is approximately 95% and when the tem-

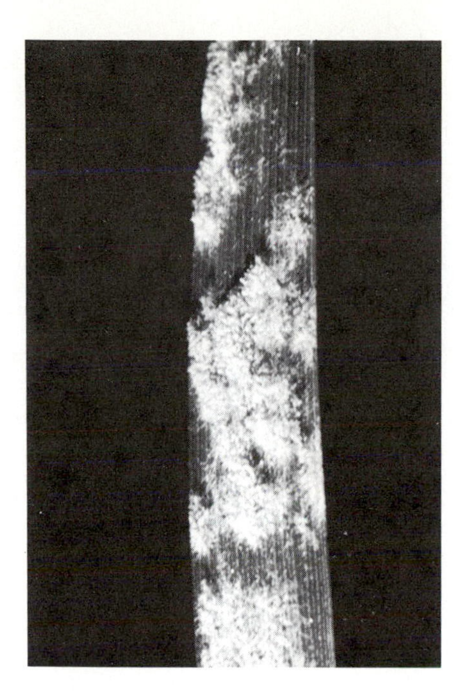

FIGURE 20-3
Powdery mildew of wheat.

perature is 10–15°C. Unlike the spores of most plant-pathogenic fungi, neither the ascospores nor the conidia of the powdery mildews require a film of water for germination. An appressorium forms at the end of each short germ tube, and a penetration peg branches from it to grow through the uninjured cuticle of the leaf. As the wall of the epidermal cell is penetrated, a papilla develops, but this is penetrated by the fungus; the pierced papilla in the wall of the host cell appears as a collar around the penetration peg.

Once within the host cell, an enlargement of the penetration peg develops in two days and invaginates the cytoplasm of the plant cell. This enlargement, the haustorium, finally occupies much of the host cell. In *E. graminis,* the haustorium produces fingerlike extensions on opposite sides along the long axis of the structure (Figure 20-4). Thus, a large surface of the membranous feeding organ of the parasite is placed in intimate contact with the cytoplasmic membrane of the host cell. Nutrients absorbed by the haustorium support the growth of the mycelial felt on the surface of the leaf.

Infection results in increased rates of respiration by infected tissues. Rates of photosynthesis also increase soon after the onset of infection,

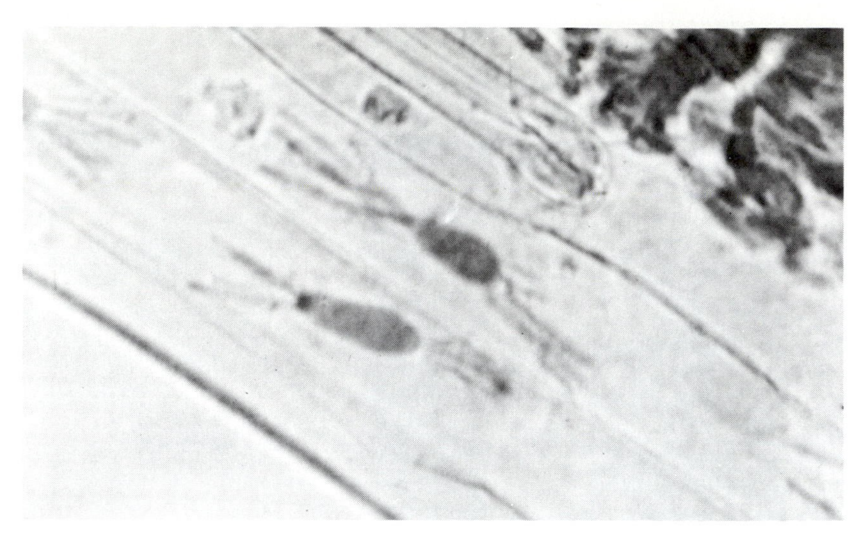

FIGURE 20-4
Haustoria (*center*) of *Erysiphe graminis* in epidermal cells of wheat.

but decrease after about three days. Rates of transpiration are increased 25–70%.

Powdery mildew appears to be most serious in succulent plants. Disease development is favored by cool (10–15°C) moist (90–95% relative humidity) weather. These conditions are best for growth of the mycelium, production of conidia, and germination of conidia and ascospores.

Prevention of Disease

Although the disease could be controlled by eradicative and protective spraying with sulfur, spraying is impracticable in grasses or cereal grains. The disease is controlled through the use of resistant varieties of crop plants. Two kinds of resistance have been observed: In some varieties, the attacking fungus fails to pierce the papilla and dies. In others, penetration occurs, but only small pear-shaped haustoria form; moreover, the cells in which the abortive haustoria form react in a hypersensitive manner, so that only small necrotic flecks form in attacked leaves.

Erysiphe graminis exists in special forms that attack only one species of host plant. Within each special form, there may be physiologic races that can infect only certain varieties of host plants.

FIGURE 20-5
Powdery mildew of cucumber.

POWDERY MILDEW OF CUCURBITS

Powdery mildew of cucurbits occurs wherever cucumbers, melons, squash, and pumpkins are grown; the disease is most severe, however, in places with warm climates. The disease is usually mild, but sporadically becomes epiphytotic, whereupon it may cause the total loss of a crop.

As with other powdery mildews, its diagnosis is dependent upon recognition of the signs of the pathogen. A talclike growth of mycelium and catenulate conidia appears on the adaxial surfaces of diseased leaves (Figure 20-5). Black cleistothecia with appendages that resemble hyphae sometimes form in the mass of superficial mycelium. Symptoms can be detected after the fungus has made some growth on the surface of a leaf. Cloranemic spots develop first, and the tissue later becomes brown and dry. Badly diseased leaves and stems are killed. Mildew rarely occurs on the fruit.

Etiology

Powdery mildew of cucurbits is caused by *Erysiphe cichoracearum* De Candolle. The fungus does not usually overwinter in the perithecial stage, which rarely forms; instead, the fungus survives the winter in warm climates in diseased weeds or volunteer plants of the family Cucurbitaceae. Conidia are blown into cooler regions, in stepwise fashion, during the early part of the growing season.

Wind-blown conidia from diseased plants are the inocula for primary and secondary cycles. Once in the infection court, a conidium will germinate, provided the relative humidity is more than approximately 45%. A short germ tube develops whose tip enlarges into an appressorium. A small penetration tube branches from the center of the appressorium and forces its way through the cuticle of the leaf. A papilla forms in the cell wall, but this is finally penetrated by the fungus. Once within the cell, the fungus enlarges to form a balloonlike haustorium that invaginates the cytoplasm. Parasitized cells do not die, but continue to nourish the fungus for several days. Although none but the epidermal cells are invaded, the fungus adversely affects nearby mesophyll cells, which turn yellow, die, and turn brown. Approximately four to six days after the initiation of disease, signs of the fungus can be seen. By this time, numerous penetration pegs and haustoria have developed from branches of the developing mycelium.

Epidemiology

Powdery mildew of cucurbits is most destructive during warm weather, because the optimum temperature for conidial formation, and for spore germination and penetration, is between 25°C and 30°C. The optimum relative humidity for spore germination is near 100%, but spores will germinate in a much drier atmosphere. Free water in the infection court is deleterious to spore germination.

The severity of the disease is affected not only by the environment, but by the age of the leaf at the time of inoculation. Leaves are most susceptible two to three weeks after unfolding; the very young, folded leaves appear to be immune.

Prevention of Disease

The wind-blown conidia cannot be excluded, but the fungus can be eradicated locally by destroying the weeds and volunteer plants that harbor the pathogen between cropping seasons. Complete eradication is impracticable; consequently, protective sprays or dusts are usually relied upon. Sulfur is phytotoxic to cucurbits, and cannot be used with impunity unless used in amounts too small to provide adequate control. However, a copper–sulfur dust (70% copper to 30% sulfur by weight) has been used safely and with good results on cucumbers. Presently, the organic fungicide Karathane, or Mildex, the common name of which is dinocap, is recommended for melons and all other cucurbits. Some varieties of muskmelon are resistant to powdery mildew.

POWDERY MILDEW OF APPLE

The powdery-mildew disease of apple, which also affects pear, quince, and other pome fruits, occurs wherever these fruits are grown. The disease has been particularly severe at times in the United States, Great Britain, central Europe, Australia, and New Zealand.

Young leaves infected in the bud stage during the preceding season are crinkled, hard and brittle, and somewhat elongated (Figure 20-6). Severely infected leaves are killed. Infected flower buds are shriveled and flowers are blighted; young twigs are also affected. Fleeting infections of young fruit sometimes occur, resulting in a network of russetting (Figure 20-7). As infected fruit enlarge, the skin often cracks in the affected areas. Although the symptoms just described are readily observed, the signs of the pathogen are more conspicuous. All diseased areas are covered with a feltlike growth of mycelium, conidiophores, and oval conidia in chains. Black cleistothecia are sometimes seen in mycelial mats, particularly those on young twigs. Each cleistothecium bears, at its base, short appendages that resemble hyphae; dichotomously branched appendages also grow upward on long stalks from the top of each cleistothecium.

Etiology

Powdery mildew of apple and other pome fruits is caused by *Podosphaera leucotricha* (Ellis and Everhart) Salmon. Each cleistothecium of this fungus bears a single, eight-spored ascus. Although the fungus can survive the winter as cleistothecia, it usually survives as mycelium in diseased leaf or flower buds.

As leaves and flowers emerge in the spring, the external mycelium develops on the affected plant parts. Conidia, which are produced abundantly, are disseminated by wind and water to infection courts on

FIGURE 20-6
Powdery mildew of a young apple leaf.

leaves and young shoots. A germ tube develops from each oval conidium that germinates on a leaf surface, and the cuticle of the leaf is penetrated directly by the tip of the germ tube. No appressoria are formed. Once through the cuticle, the penetration peg forces its way through the wall of an epidermal cell. A haustorium then enlarges within the cell, invaginating the cytoplasm. The fungus has, by this time, established a parasitic relation with the invaded host cell. Nourishment is provided for the external mycelium, which continues to grow beyond the point of ingress. Other penetrations occur at the tips of hyphae, and soon a felty growth of powdery mildew is visible. Conidia that form on this mycelium serve as the inoculum for secondary cycles throughout the growing season.

Epidemiology

Conidia are formed, and germinate most rapidly, in a humid atmosphere when the temperature is near 20°C. The fungus will not grow when the temperature is 33°C or higher. The conidia do not require

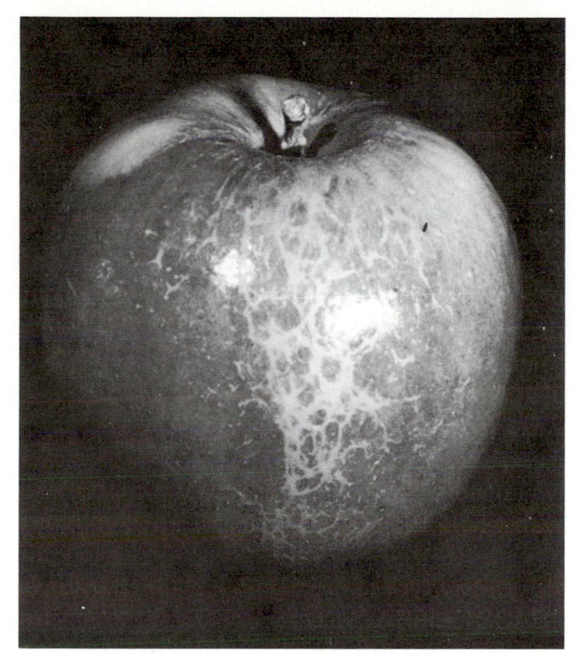

FIGURE 20-7
Cracking and distortion of apple fruit due to infection by powdery mildew.

free water for their germination; in fact, they germinate poorly, if at all, when submerged in water.

Severe outbreaks of disease seem to be correlated with mild winters, particularly when little freezing and thawing occurs in the early spring. Under such conditions, the amount of inoculum surviving in diseased buds is probably high.

Powdery mildew of apple can be forecast accurately if the percentage of infected buds is determined late in the dormant season. A representative sample of bud sticks can be selected from orchard trees late in the winter, and their buds can be forced to open by standing them in water and keeping them in a warm (20–25°C) room. The severity of the mildew in the orchard the following spring can be estimated on the basis of the number of shoots with infected buds.

Prevention of Disease

The wind-blown conidia of *Podosphaera leucotricha* can hardly be excluded, but removing diseased shoots during normal pruning operations will significantly reduce the amount of inoculum. The best method of control, however, is the use of fungicidal sprays or dusts. Sulfur sprays applied at the "pink" and "calyx" stages of blossoming, and twice more at two-week intervals, usually give satisfactory control. Sulfur will also control apple scab. Karathane is recommended for the control of mildew in varieties that are intolerant of sulfur.

Resistant varieties are not usually recommended for the control of apple mildew. The varieties 'Winesap' and 'Rhode Island Greening,' however, are somewhat resistant.

Selected References

POWDERY MILDEW OF CEREALS

Cherewick, W. J., 1944. Studies on the biology of *Erysiphe graminis* DC. *Can. J. Res.* (C)22:52–86.

Reed, G. M., 1909. The mildews of the cereals. *Bull. Torrey Bot. Club* 36:353–388.

White, N. H., and E. P. Baker, 1954. Host pathogen relations in powdery mildew of barley, I. Histology of tissue reactions. *Phytopathology* 44: 657–662.

Yarwood, C. E., 1936. The tolerance of *Ersiphe polygoni* and certain other powdery mildews to low humidity. *Phytopathology* 26:845–859.

———, 1950. Water content of fungus spores. *Amer. J. Bot.* 37:636–639.

POWDERY MILDEW OF CUCURBITS

Markarian, D., and R. R. Harwood, 1967. The inheritance of powdery mildew resistance in *Cucumis melo* L. *Mich. Agr. Exp. Sta. Quart. Bull.* (49):404–411.

Miller, P. A., and J. T. Barrett, 1931. Cantaloupe powdery mildew in the Imperial Valley. *Calif. Agr. Exp. Sta. Bull.* (507).

Reed, G. M., 1907. Infection experiments with the mildew on cucurbits, *Erysiphe cichoracearum* DC. *Trans. Wis. Acad. Sci. Arts Lett.* 15:527–547.

POWDERY MILDEW OF APPLE

Fisher, D. J., 1918. Apple powdery mildew and its control in the arid regions of the Pacific Northwest. *U. S. Dep. Agr. Bull.* (712).

Keil, H. L., and R. A. Wilson, 1960. Evaluation of fungicides for control of powdery mildew. *Plant Dis. Rep.* 44:253–255.

Woodward, R. C., 1927. Studies on *Podosphaera leucotricha* (Ell. & Ev.) Salm. *Trans. Brit. Mycol. Soc.* 12:173–204.

21

Diseases Affecting Photosynthesis —Rusts

Plant diseases now known to be caused by rust fungi have been recognized since the dawn of recorded history. In fact, the Roman god Robigus was supposed to have been responsible for the rusting of cereal grains.

The rust fungi are members of the order Uredinales of the class Heterobasidiomycetes. Their sexual reproduction structures are reminiscent of those of the smut fungi (Chapter 14), but only four sporidia are formed on each promycelium of a rust fungus.

The rusts are pleomorphic: a given species may be capable of producing as many as five spore forms. These spores develop in the following sequence: **pycniospores** or **spermatia** (designated by the symbol 0), **aeciospores** (I), **urediniospores** (II), **teliospores** (III), and **basidiospores** or **sporidia** (IV).

Pycniospores arise from haploid mycelium and are borne in globular sori that are called **pycnia** (or **spermagonia**; see Figure 21-1). The spores accumulate in a slimy mass in the pycnial cavity, from which they are repeatedly extruded. Each pycnium bears spores that are of one mating type; these spores are designated either + or −, and are

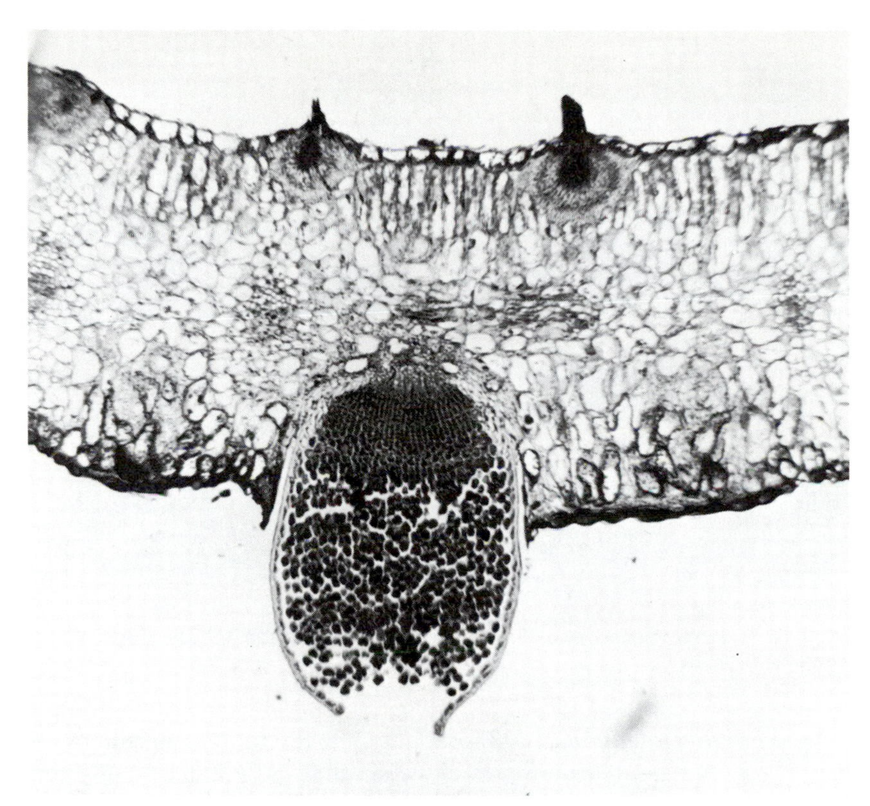

FIGURE 21-1
Pycnia (*top*) and aecia of *Puccinia graminis* in a diseased barberry leaf.

uninucleate. Plasmogamy occurs when pycniospores of one mating type fuse with flexuous hyphae that protrude from the pycnia of the other mating type.

The first spores to develop after plasmogamy, which are binucleate and form on binucleate mycelium, are termed aeciospores. Their sori, or **aecia,** are formed from the same mycelium from which their pycnia had developed earlier. Usually, aecia form on the side of the leaf opposite the pycnia. Morphologically, there are four kinds of aecia, namely **aecidia, caeomae, roestelia,** and **peridermia.** These structures are described in the Glossary.

The urediniospore (**urediospore, uredospore**), is the repeating spore form, and many crops of these spores may be produced successively. Urediniospores are also binucleate; they are formed in sori that are

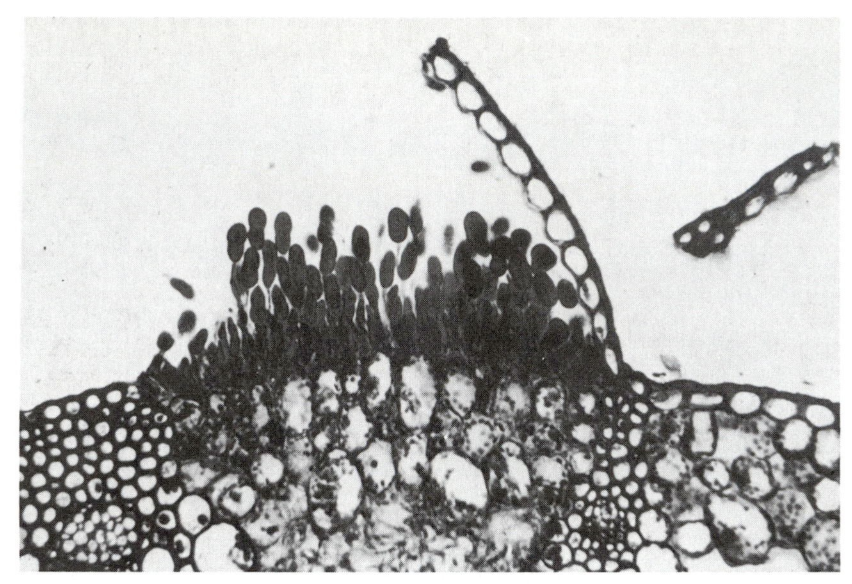

Figure 21-2
Uredinia and urediniospores of *Puccinia graminis* in a diseased wheat leaf.

termed **uredinia** (**uredia, uredosori**), which resemble acervuli (Figure 21-2). Some urediniospores develop unusually thick walls and serve as resting or over-wintering spores; these are called **amphiospores.**

The teliospores (**teleutospores**) develop in sori termed **telia** (**teleutosori;** see Figure 21-3), and, in some forms, the telia replace the uredinia after urediniospores cease to develop. Teliospores are binucleate, thick-walled resting spores, well suited to serve as survival structures. Caryogamy occurs in the mature teliospore, which, upon germi-

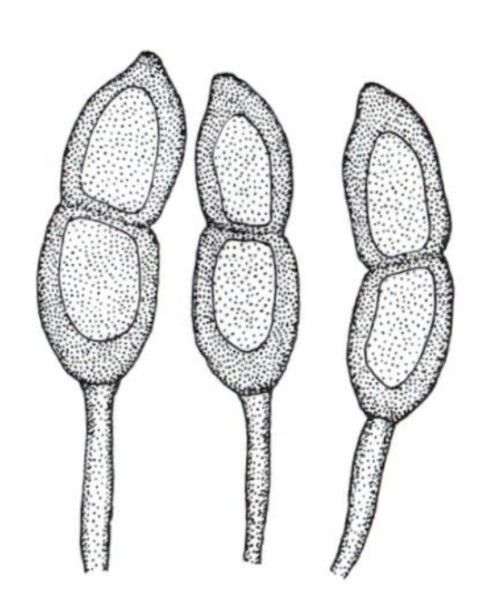

Figure 21-3
Teliospores of *Puccinia graminis*.

nation, puts forth a basidium (promycelium). Meiosis occurs, and the four haploid nuclei move into the basidium, where they are separated by septa. A sterigma develops from each cell of the basidium, and a nucleus moves into each basidiospore (sporidium) before its abstriction from the sterigma. Basidiospores germinate by germ tubes, which may penetrate an appropriate host; subsequently, pycnia form in the host.

Many rust fungi produce all of the five spore stages discussed above —pycnia (0), aecia (I), uredinia (II), telia (III), and basidia (IV)— and are termed **long-cycle (macrocyclic) rusts.** The same term is used, even though some spore stages are lacking, as long as there is one binucleate spore stage in addition to the telial stage. A rust which has only one binucleate spore stage, and that the telial stage, is termed a **short-cycle (microcyclic) rust.**

Some rusts produce all of their spore forms on one plant species, and are termed **autoecious.** Others, however, produce some spore forms on one plant and others only on another plant species; these are **heteroecious** rusts. In heteroecious rusts, pycnia and aecia occur on one host, and telia alone, or uredinia and telia, occur on another. The host that bears telia is the **primary host;** the one that bears pycnia and aecia is the **alternate host.**

Three rust diseases are discussed below.

STEM RUST OF CEREALS

Stem rust occurs on many grasses and cereal-grain crops, including wheat, barley, rye, wild barley, brome grass, *Agropyron* spp., and *Elymus* spp. The disease is distributed throughout the world wherever susceptible plants are grown. In diseased cereals (Figure 21-4), chloranemic flecks develop in leaves, leaf sheaths, and stems. Soon after the beginning of disease, the fungus produces sori that bear urediniospores, which are red in mass. The stalked urediniospores break through the cuticle in the infected areas. Later, the sori become black, owing to the production of teliospores, which have dark, thick walls. The leaves of diseased plants die prematurely; diseased plants are dwarfed, therefore, and their production of grain is reduced.

The fungus that causes stem rust of cereals also infects the common barberry (*Berberis vulgaris*) and various species of *Mahonia,* a related genus. Yellow spots with purple borders become apparent on the adaxial surfaces of diseased leaves; flask-shaped pycnia (sperma-

FIGURE 21-4
Stem rust of wheat.

gonia), which bear pycniospores (spermatia), develop in the necrotic areas. Later, clusters of cup-shaped, yellowish aecia develop on the abaxial surfaces of infected leaves (Figure 21-5). The masses of cluster-cups are 0.5–1.0 cm in diameter; each cup is barely distinguishable with the unaided eye.

From the foregoing account, it is apparent that diagnosis of stem rust is largely dependent upon recognition of the signs produced in diseased tissue. Indeed, the name "rust" was given not to the disease, but to the pathogenic fungus itself, whose uredinial sori give the ap-

FIGURE 21-5
Barberry leaves infected by *Puccinia graminis.*

pearance of iron rust when they are abundant on diseased wheat plants.

Etiology

Stem rust is caused by *Puccinia graminis,* a heteroecious rust with its pycnial (0) and aecial (I) stages on the barberry and its uredinial (II) and telial (III) stages on cereal grains or grasses. The fungus occurs in special forms, each of which is capable of infecting only certain species of suspectible plants. Thus, *P. graminis* f. sp. *tritici* attacks only wheat (*Triticum*), *P. graminis* f. sp. *secalis* attacks only rye (*Secale*), *P. graminis* f. sp. *phlei-pratensis* attacks only timothy (*Phleum pratense*), and so on.

In cold climates, the fungus survives the winter as teliospores, which undergo a period of rest in the summer or autumn and a period of cold-induced dormancy in the winter. As the weather becomes warm in the spring, each of the two cells of the teliospores may germinate

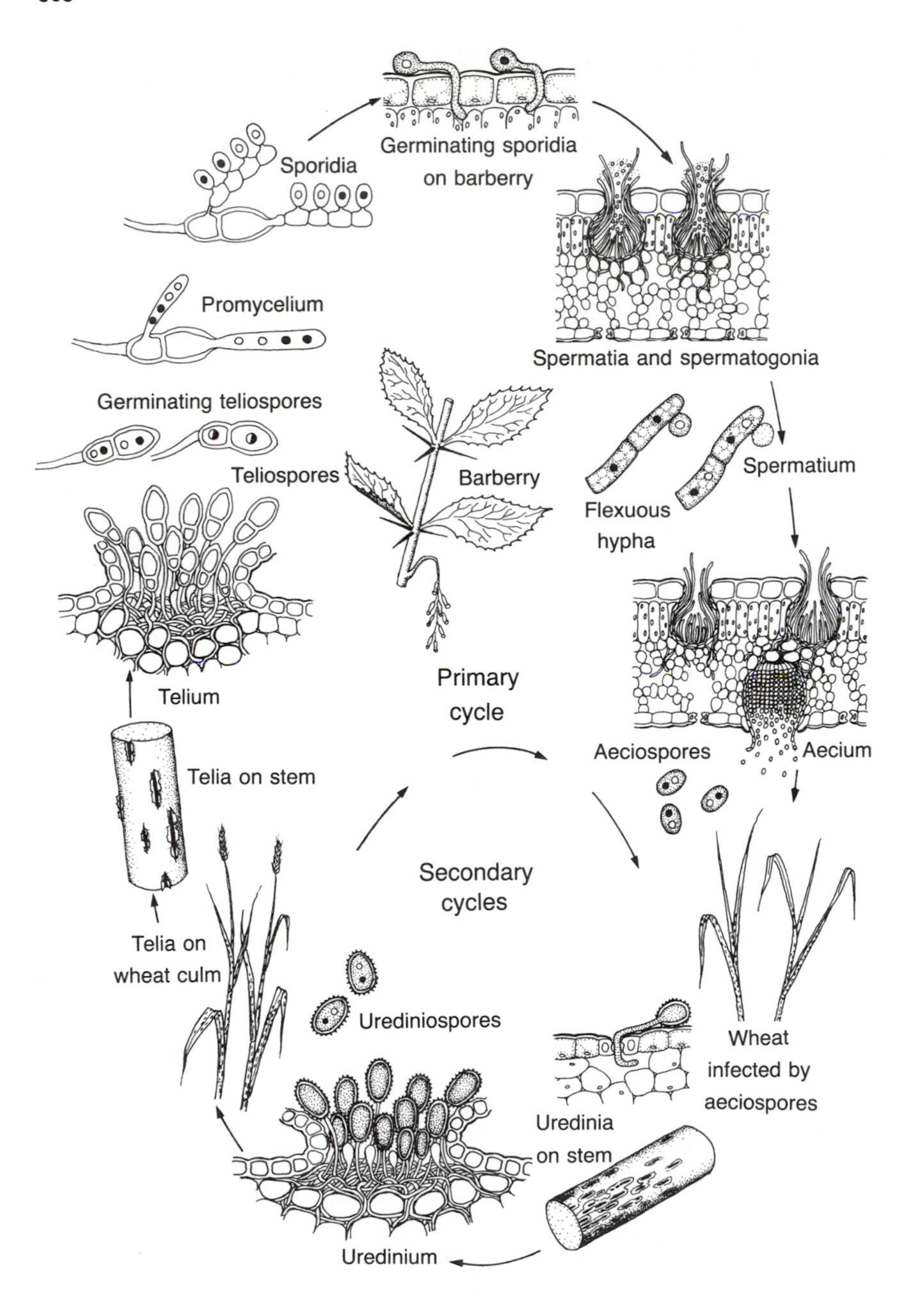

FIGURE 21-6
Stem rust of cereals.

to form a promycelium. Caryogamy occurs in the mature teliospore, and reduction takes place in the young promycelium. Three septa form in the promycelium, separating it into four cells, each of which contains a single haploid nucleus. A sterigma develops from each cell, a nucleus moves to the tip of each sterigma, and each tip enlarges to form a sporidium. The mature sporidium becomes a ballistospore and is forcibly discharged. Sporidia are wind-blown, and survive only if they are deposited upon the young, tender leaves of the barberry.

Sporidia germinate on the surfaces of barberry leaves if a film of water is present. A short germ tube is produced, and a penetration peg branches from its tip and penetrates the cuticle. Parasitic mycelium develops in the barberry leaf, producing haustoria that penetrate the walls of host cells and invaginate their protoplasts. Pycnia soon develop in diseased tissues; these are dark, flask-shaped structures filled with unicellular spores in a matrix of amber-colored liquid. Ostioles are erumpent, and strands of flexuous hyphae emerge from them.

Plasmogamy occurs during the pycnial stage when pycniospores of one mating type fuse with cells of flexuous hyphae of the opposite mating type. Following plasmogamy, strands of parasitic mycelium become dicaryotic, presumably by the passage of nuclei through the hyphae. Meanwhile, aecial primordia have developed at the abaxial surfaces of diseased leaves. The "cluster cup," or aecidial kind of aecium, has taken form by the time the hyphae in the base of each aecium have become dicaryotic. These dicaryotic hyphae grow in the cavities of the aecia, producing chains of binucleate aeciospores that fill each aecium. Mature aeciospores are released to the wind, and survive only if blown to the cereal host. Aeciospores cannot reinfect the barberry.

Aeciospores germinate on the leaves of the cereal host and germ tubes enter through the stomata. Host cells are parasitized by intercellular hyphae that send forth haustoria into individual cells. Several days after infection by aeciospores, the fungus sporulates, producing urediniospores. These reinfect wheat, and secondary cycles develop throughout the growing season. Finally, the two-celled teliospores are produced in sori that had previously borne urediniospores, and the cycle is complete (Figure 21-6).

In relatively warm climates, the fungus can survive the winter as urediniospores and perpetuate cycles in wheat. Thus, the barberry, while essential to the completion of the sexual cycle, is not required for continued development of the fungus in cereals. Urediniospores are blown long distances from the southern United States to the north in the spring and summer; stem rust occurs, therefore, regardless of whether barberries are present. The severity of rust in cereals is usually greater, however, in regions where the barberry grows than in regions where it does not.

Stem rust develops most rapidly in wet, cool (18–20°C) weather; severe disease results under these conditions if there is an abundance of inoculum in the area when plants have emerged from the boot stage but have not yet entered the flowering stage. Heavy applications of nitrogen, by delaying maturity and increasing the density of stands, often result in severe epiphytotics of rust.

Prevention of Disease

Exclusionary measures of disease control are ineffective against stem rust of cereals because inoculum can be blown into an area from hundreds of miles away. One eradicative measure, eradication of the alternate host, has resulted in a reduction in the amount of inoculum available for primary cycles; but barberry eradication does not result in satisfactory control because wind-borne inoculum may be disseminated into regions where the barberry has been eradicated. Theoretically, the disease in wheat could be controlled through the use of protective fungicides, but such measures have not proved to be practicable.

Stem rust is controlled largely through the use of rust-resistant varieties of cereals. Many such varieties show protoplasmic or plasmatic defensive reactions, whereas others are hypersensitive and show necrogenic defensive reactions. A few varieties of wheat exhibit what has been termed functional resistance, wherein ingress is impeded by closed stomata, which fail to open until late in the morning and which close early in the afternoon. Thus, the infection court is relatively dry during the only time that ingress can occur.

Resistance of cereal grains is inherited monogenically, and resistance is dominant over susceptibility. Once resistant breeding lines have been found, it is relatively easy, through crossing and back-crossing programs, to develop agronomically acceptable varieties that are also resistant to stem rust. Only a slight genetic change in a segment of the population of the pathogen, however, may result in the development of a more virulent line of the pathogen. Thus, different races of *Puccinia graminis* are constantly being detected in commercial fields planted with grain varieties resistant to previously recognized races. Some 250 different races of *P. graminis* have been identified.

BLISTER RUST OF WHITE PINES

Blister rusts, which are caused by fungi of the genus *Cronartium*, affect most species of pine. All produce their pycnial and aecial stages in some species of pine and their uredinial and telial stages in some species of dicotyledonous plant. The white-pine blister rust (*Cronartium ribicola* Fischer), which is representative of all blister rusts of pine, will be discussed in this account.

Blister rust occurs in all white pines wherever they are grown in Europe, Asia, and North America. The disease apparently originated in Asia, and reached Europe in the middle of the nineteenth century, more than a hundred years after the eastern white pine of North America (*Pinus strobus*) had been introduced into Europe. The disease was introduced into the United States and Canada early in the twentieth century. Seedlings of *Pinus strobus* grown in Europe were used for reforestation projects in the United States because they could be purchased more cheaply than locally grown seedlings. Unfortunately, some of the imported seedlings were infected by *Cronartium ribicola*. These seedlings were planted in forests, and blister rust was soon established in North America.

The pine needles are first infected in late summer or early autumn by sporidia; yellow to brown lesions appear in the affected needles. During the following season, the bark of twigs bearing diseased needles becomes yellowish and slightly swollen. Pycnia first appear in the bark of swollen twigs in the summer following the onset of infection or in the early spring of the following year. These signs of the pathogen form amber-colored blisters that rupture to release pycniospores in a sugary liquid. During the following spring, aecia—the second sign of the pathogen—develop in the old pycnial scars (Figure 21-7). Meanwhile, additional pycnia form at the periphery of the swollen portion of the stem. Aecia are white elongate blisters that rupture in the spring to expose masses of orange-yellow aeciospores that are borne in chains. The tumefaction on the stems of pine consists of a central portion with dead bark, surrounded by a zone of aecia; pycnia occur only between the aecia and the advancing margin of the infection.

In gooseberry and currant (*Ribes* spp.), in which primary infections are initiated by aeciospores blown from the infected pines, uredinial pustules develop mostly on the abaxial surfaces of cotyledons and true leaves. Small twigs, young buds, and flower parts are sometimes

FIGURE 21-7
White pine infected by the blister-rust fungus. (Courtesy of R. N. Campbell, University of California, Davis.)

affected. Later, hairlike columns of teliospores develop in the areas that had borne the uredinia (Figure 21-8). Yellow to brown flecks are often visible on the adaxial surfaces of leaves above the uredinia or telia. Obviously, with the exception of the tumefaction in diseased pines, blister rust is diagnosed largely on the basis of the signs of the fungus.

FIGURE 21-8
Leaves of currant infected by the blister-rust fungus.

Etiology

White-pine blister rust is caused by *Cronartium ribicola,* a heteroecious, long-cycle rust fungus. The pycniospores of *C. ribicola* are pear-shaped, thin-walled, and roughly only 3–5 $\mu \times$ 6–8 μ; aeciospores, however, are approximately 20 $\mu \times$ 30 μ, and have thick walls with warty surfaces. The urediniospores are about the same size as the aeciospores, but have tiny spines on their walls. The teliospores, which are 10 $\mu \times$ 50 μ, are packed together in cylindrical hairlike columns.

The fungus commonly overwinters in diseased pine stems, but can also survive as mycelium in diseased twigs of *Ribes.* Aeciospores borne on pine trees in the spring are dispersed, mostly by wind, to *Ribes* plants. In the presence of a film of water in the infection court, the spores germinate to form germ tubes, the tips of which enter the plants through their stomata. Once inside, the fungus enlarges in the substomatal cavity and sends branches between the cells of the mesophyll. Haustoria penetrate cell walls and invaginate the cytoplasm therein. Urediniospores form within a few days to a week; these, upon maturity, are blown or splashed to other foliage of *Ribes* plants, where they initiate infections in much the same way as the aeciospores. Secondary cycles of the pathogen develop in *Ribes* as a consequence of reinfections by urediniospores. Telial columns emerge from diseased leaves in the summer and early autumn. Teliospores germinate in place to produce sporidia, which are discharged forcibly; they may then be carried by the wind to pines, whose needles may be entered through their stomata. The fungus parasitizes the pine needle, and the mycelium grows into the twig, where, one to three years later, pycnia and aecia will develop in the gall tissue.

Epidemiology

Blister rust develops most rapidly during moist seasons. Water is essential to the infection of pines by sporidia and the infection of *Ribes* by the aeciospores or urediniospores. At least 48 hours with a saturated atmosphere and a temperature near 20°C are needed for production and germination of sporidia and for the subsequent infection of pine.

The pathogen is spread long distances by wind-blown aeciospores, which may remain viable after having been blown for hundreds of miles. The sporidia, however, will die if blown farther than a quarter of a mile.

Prevention of Disease

Blister rust could have been excluded from North America, but now that it has become established, long-distance dissemination of aeciospores by wind precludes exclusion as a means of control. The disease in pines has been controlled best by removing plants of the genus *Ribes;* these may be removed by hand or killed with such chemical weed killers as 2,4-D and 2,4,5-T. In regions where large-scale *Ribes*-eradication programs are impractical, the disease is controlled by isolating pine plantations from the primary host, by using resistant varieties of currants and gooseberries in the commercial production of those fruits, and by controlling the disease in currants by means of protective spraying. Tests with antibiotics—Actidione (cycloheximide), for example— were initially encouraging, but, because of the erratic results obtained in controlled experiments, widespread use of such chemicals has not developed.

CEDAR–APPLE RUST

Were it not for the fact that the name "cedar–apple rust" has been firmly established by usage, this disease might better be called "juniper–apple rust," because the name "cedar," which properly refers to members of the genus *Cedrus*, is here employed to designate members of the genus *Juniperus*. *Gymnosporangium juniperi-virginianae*, the causal agent of cedar–apple rust, is representative of the species of that genus, which induce the formation of galls or blisters in species of *Juniperus* and cause leafspots of various pome fruit. *G. juniperi-virginianae*, *G. clavipes*, and *G. globosum* are prevalent on *Juniperus virginiana* (commonly known as eastern red-cedar); *G. juniperi-virginianae*, however, attacks apple, whereas the other two species attack *Crataegus* spp. (hawthorn) in the United States.

Symptoms in apple begin as yellowish spots on leaves (Figure 21-9) and, occasionally, on young twigs and immature fruit. The lesions turn orange, and bright red bands finally develop around the orange cen-

FIGURE 21-9
Apple leaves infected by the cedar–apple rust fungus.

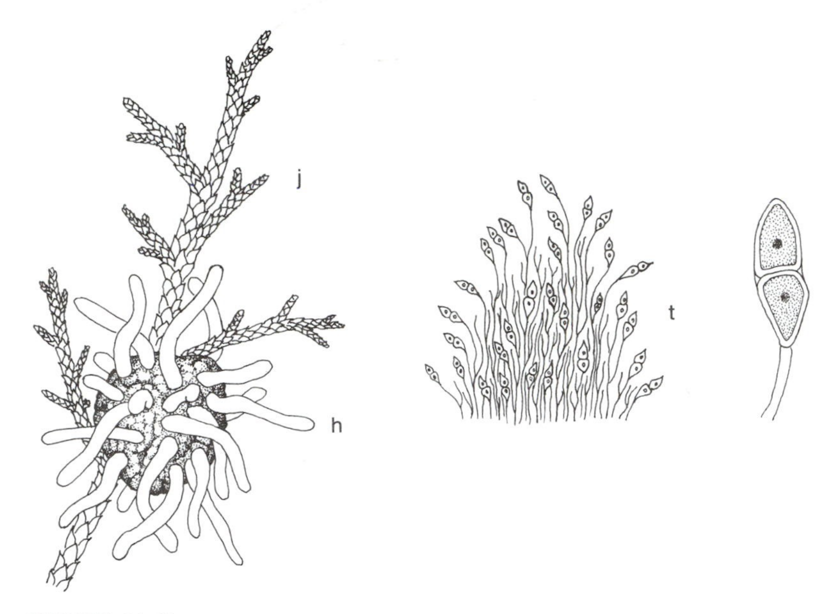

FIGURE 21-10
Telial horns (*h*) from gall of juniper (*j*) infected by the cedar–apple rust fungus, and the tip of a telial horn with an enlarged teliospore (*t*).

ters. A yellow fluid containing pycniospores exudes from the black, flask-shaped pycnia that form as signs of the pathogen on the adaxial surfaces of diseased leaves. Hyperplasia occurs on the abaxial surfaces of infected foliage, and roesteloid aecia are produced from the small blisters. The outer covering (peridium) of each aecium is elongate and cylindrical at first. Later, it splits longitudinally, the aeciospores are released. Young aecia can be seen as hairlike projections from the leaf; older aecia with split peridia appear fuzzy. Diseased fruit are distorted, and badly diseased twigs are dwarfed and sometimes killed.

In infected junipers, leaf galls—having begun as small swellings on the adaxial surfaces of infected needles—enlarge to form brown spongy masses. The telia form in the spring as gelatinous, yellowish columns that emerge from depressions in the leaf gall (Figure 21-10).

Etiology

Cedar–apple rust is caused by *Gymnosporangium juniperi-virginianae*, which produces pycnia and aecia on apple and telia on juniper. The fungus does not produce uredinia.

The mycelium survives the winter in the galls in junipers. Teliospores are formed in the early spring, and are packed together in the brightly colored, gelatinous columns that emerge from the galls. Each teliospore is two-celled, oval, stalked, and approximately 18 $\mu \times 50 \mu$. Spores germinate in place, producing promycelia and sporidia. The sporidia are discharged forcibly and are blown by the wind to apple trees.

The sporidium produces a short germ tube on the young apple leaf that directly penetrates its surface. The fungus grows between the cells of the leaf and sends haustoria into parasitized cells. Pycniospores are produced in pycnia on the upper surfaces of diseased leaves, and plasmogamy occurs as pycniospores of opposite mating types mingle and fuse. Binucleate mycelium is seen first in the aecial initials on the lower surfaces of leaves. Two-celled aeciospores develop in the summer and reach maturity in August. The ovoid aeciospores have thick walls with small warty projections; they measure approximately 20 $\mu \times 25 \mu$.

Owing to the uncertainty that exists about whether aeciospores undergo a period of rest, it is not known whether infection of the junipers occurs in the autumn or spring; it seems most likely that they are infected in the late summer or early autumn. The chances that aeciospores overwinter in fallen apple leaves and are wind-blown to junipers in the spring are slim.

Juniper leaves are entered by the fungus, presumably through their stomata, and are parasitized by an intercellular mycelium from which haustoria develop. Although hyperplasia begins soon after infection, the galls in juniper do not reach full size for another year; the telia form during the second spring after the initiation of disease in the juniper.

Prevention of Disease

Probably the most effective method of controlling cedar–apple rust would be to select sites for apple orchards at least a mile from the nearest junipers. In Virginia, there is a law against planting junipers within three miles of commercial apple orchards. In regions where eastern red-cedar is already common in the field, the disease in apples is controlled very effectively by means of protective fungicides. Ferbam is currently added to the spray mixtures used on apples in the spring. The immature foliage, twigs, and fruit should be kept covered with the fungicide, but, because mature leaves cannot be penetrated by the fungus, ferbam need not be used after the leaves are fully expanded and cuticularized.

Certain varieties of apple and some lines of eastern red-cedar are relatively resistant to rust. Resistance in apple is monogenic and dominant. How long a resistant variety of apple will remain so under field conditions cannot be predicted. The fungus apparently exists in several physiological races, and the chances are good that a race virulent to a new resistant variety would soon make itself known. Thus, the use of resistant varieties of apple is of dubious value as a control measure, because an appropriately virulent race of the pathogen probably exists and because breeding and selection programs for the development of new varieties of tree fruits is necessarily slow.

Selected References

STEM RUST OF CEREALS

Allen, R. F., 1923. A cytological study of infection of Baart and Kanred wheats by *Puccinia graminis tritici*. *J. Agr. Res.* 23:131–152.

Hart, H., 1929. Relation of stomatal behavior to stem-rust resistance in wheat. *J. Agr. Res.* 39:929–948.

Rowell, J. B., 1971. Chemical control of the cereal rusts. *Annu. Rev. Phytopathol.* 6:243–262.

Stakman, E. C., 1914. A study in cereal rusts. Physiological races. *Minn. Agr. Exp. Sta. Bull.* (138).

———, D. M. Stewart, and W. Q. Loegering, 1962. Identification of physiologic races of *Puccinia graminis tritici*. *U. S. Dep. Agr. Agr. Res. Serv. Rep.* (E617).

Williams, P. G., K. J. Scott, J. L. Kuhl, and D. J. Maclean, 1967. Sporulation and pathogenicity of *Puccinia graminis* f. sp. *tritici* grown on an artificial medium. *Phytopathology* 57:326–327.

BLISTER RUST OF WHITE PINES

Colley, R. H., 1918. Parasitism, morphology, and cytology of *Cronartium ribicola*. *J. Agr. Res.* 15:619–660.

Hirt, R. R., 1956. Fifty years of white pine blister rust in the Northeast. *J. Forest.* 54:435–438.

Mielke, J. L., 1943. White pine blister rust in western North America. *Yale Univ. Sch. Forest. Bull.* (52).

Moss, V. D., 1961. Antibiotics for control of blister rust on western white pine. *Forest Sci.* 7:380–396.

Van Arsdel, E. P., A. J. Riker, and R. F. Patton, 1956. The effects of temperature and moisture on the spread of white pine blister rust. *Phytopathology* 46:307–318.

CEDAR–APPLE RUST

Bliss, D. E., 1933. The pathogenicity and seasonal development of *Gymnosporangium* in Iowa. *Iowa Agr. Exp. Sta. Res. Bull.* (166):339–392.

Hotson, H. H., and V. M. Cutter, 1951. The isolation and culture of *Gymnosporangium juniperi-virginianae* Sehw. upon artificial media. *Proc. Nat. Acad. Sci. U. S. A.* 37:400–403.

Palmiter, D. H., 1952. Rust diseases of apples and their control in the Hudson Valley, *N. Y. Agr. Exp. Sta. Geneva Bull.* (756).

Reed, H. S., and C. H. Crabill, 1915. The cedar rust disease of apples caused by *Gymnosporangium juniperi-virginianae. Va. Agr. Exp. Sta. Tech. Bull.* (9).

22

Diseases Affecting Translocation—Viral and Mycoplasmal Diseases

Although viral and mycoplasmal diseases often adversely affect the processes of growth and photosynthesis, they also impair translocation as a consequence of their movement in, and infection of, phloem tissue. Because of the etiological similarities of most viral and mycoplasmal diseases of plants, it seems logical to group all of them together.

Generalizations applicable to plant viruses and viral diseases, and to those diseases caused by mycoplasmalike organisms, have been made in earlier chapters and need not be repeated here. Examples of the "mosaic" diseases and the "yellows" diseases are discussed below.

CUCUMBER MOSAIC

Cucumber mosaic, one of the most important diseases of the cultivated cucumber, is caused by a plant virus that also infects celery, spinach, banana, tobacco, tomato, gladiolus, lilies, and other plants. The symptoms of the disease are mosaic in young foliage, mottle of mature leaves, necrotic lines in older leaves, and mosaic and distortion of fruit (Figure 22-1). Also, the internodes of affected plants are shorter

FIGURE 22-1
Cucumber fruit infected by cucumber-mosaic virus.

than those of healthy plants. Diseased plants often die near midseason. Distorted fruit frequently have wartlike areas that are dark green. Albication occurs in some severely affected fruit.

Etiology

The disease is caused by cucumber-mosaic virus, which is a spheroidal (perhaps polyhedral) plant virus. It has a thermal inactivation point (10 minutes) of approximately 60°C, and is inactivated within a few days in dried leaf tissue.

The virus survives between growing season in perennial weeds, greenhouse-grown plants, and in the seeds of *Echinocystis lobata*, a species of wild cucumber. It is transmitted from diseased plants by aphids (Chapter 3). Secondary cycles also occur as a result of mechanical transmission of the virus by workers during early harvesting procedures. Once inside host plants, the virus somehow mobilizes the metabolic potential of the infected cell so as to achieve its own replication at the expense of the host.

Widespread outbreaks of cucumber mosaic result when there is an abundance of overwintered inoculum in the vicinity of commercial fields of susceptible plants. Symptoms in diseased plants are most severe as the temperature increases from approximately 15°C to 28°C.

Prevention of Disease

Screening the ventilators of greenhouses excludes viruliferous aphids, and thereby protects the plants within from cucumber-mosaic virus. Eradication of the virus through the removal of its weed hosts, while often impracticable, has been used successfully in the control of the disease in celery grown in Florida. In this instance, only one weed host, *Commelina* sp., was prevalent, and it grew almost exclusively in hedgerows near the fields; thus, most of the inoculum could be destroyed by eradication of a single, readily accessible alternative host.

Resistant varieties of spinach and cucumber are used in the commercial control of cucumber-mosaic virus. Resistance in spinach is dominant and monogenic, that in cucumber is polygenic. Because the virus is still able to replicate itself in resistant varieties of spinach and cucumber, the resistance is perhaps more accurately termed "tolerance." In spinach, little replication of the virus occurs below 28°C, and symptoms rarely occur in field-grown plants of tolerant varieties of spinach. Symptoms do not develop in the tolerant varieties of cucumber. Tolerant varieties of both hosts originated in the Orient.

TOBACCO MOSAIC

The mosaic disease of solanaceous plants that is caused by tobacco mosaic virus (TMV) can be a serious disease of tobacco and tomato wherever those plants are grown. Most other species of solanaceous plants are also susceptible, as are at least 73 species in 14 other families of plants.

The diagnostic symptom in tobacco plants is a mosaic of the leaves, in which dark green areas and the adjacent yellowish areas are separated by sharp lines of demarcation (see Figure 3-2). The green areas are often raised as blisters and are usually darker green than healthy tissues. These symptoms develop in young leaves that have previously shown some chlorosis and distortion (downward-curled margins). Sometimes, spots of bright yellow appear in older leaves. Invariably, these spots yield a mutant of the virus, the "aucuba" strain, so called because of the resemblance of the yellow-spotted leaves of affected plants to those of the gold-dust aucuba, *Aucuba japonica variegata.*

In tomato, the diagnostic symptom is a chlorotic mottle, in which areas of light green to yellow and green occur in a diffuse pattern in the leaves. In mottle, the boundaries between areas of different colors are not sharp as they are in mosaic: the light green and darker green tend to blend into each other. Fruits from TMV-infected plants often have brown vascular discolorations; in diseased fruits of some varieties of tomato, small, sunken, yellow spots may develop. Certain strains of the virus cause necroses along the stems and in the leaves and fruits of tomato. When tomatoes are simultaneously infected by TMV and the potato-mottle virus, a severe necrotic-streak disease results. Symptoms in doubly infected plants are much more severe than would be expected on the basis of the severity of symptoms in singly infected plants; thus, the two viruses are synergistic in tomato.

TMV causes local necrotic lesions in the inoculated leaves of several host plants, notably *Nicotiana glutinosa* (see Figure 3-3) and the pinto bean. As with other plant viruses, the positive correlation between the numbers of lesions and the concentration of active virus in the inoculum makes the biological assay of TMV possible.

Etiology

Mosaic of tobacco and tomato is caused by mechanically transmissible plant virus. The intact particles, which are rigid rods that measure approximately 15 mμ $\times$ 300 mμ (see Figure 3-3), consist of a core of infectious ribonucleic acid and a covering of protein. TMV is, perhaps, the most heat-stable of all plant viruses: in crude juice pressed from infected plants, it will withstand heating at 90°C for 10 minutes. It is not inactivated in dried plant tissue, even after many years, nor is it completely inactivated during the processing of tobacco products. Oddly enough, TMV is not transmitted by aphids; the reason for this anomaly remains a mystery. Much of the basic research on plant viruses has been done with TMV; its stability and high concentration in diseased plants make it an ideal tool for research. Despite the unique properties of TMV, valid generalizations about plant viruses have been made from specific information obtained from experiments with TMV.

Although the virus is seed-borne in petunia and tomato, it usually survives between growing seasons in plant residues or plant products. The primary cycle of the pathogen is usually initiated when transplants are handled by workers whose hands have been contaminated by the virus. The virus is literally rubbed into the epidermal cells of plants that are injured slightly while being handled. Virus that has overwintered in crop residues in the soil can initiate disease in the slightly injured roots of transplants. Secondary cycles occur after transmission during the "suckering" of tobacco and tomatoes and during other cultural practices.

Once within the protoplast of a host, TMV particles are deproteinized and the infectious ribonucleic acid begins to replicate itself and to coat itself with protein. Soon, the infected protoplast becomes filled with intact particles of TMV. Two kinds of viral inclusions form in diseased cells: one is amorphous and protoplasmic, whereas the other is crystalline. The latter have been shown to consist of layers of viral particles.

Prevention of Disease

Tobacco or tomato plants should not be set in soil containing crop residues that might harbor the virus. Land on which no diseased plants have grown for two or three years is probably safe, however. Perhaps the most important control measure is to make certain that the hands of the workers are not contaminated with TMV. Before transplanting, suckering, or staking operations, workers should wash their hands thoroughly with soap and water; and of course, workers must neither handle nor use tobacco products while handling plants or working in the fields.

TMV-resistant varieties of tobacco and tomato have been developed. Most resistant varieties are hypersensitive, although some varieties of tobacco have resistance of the 'Ambalema' type. 'Ambalema' tobacco is really tolerant; the virus replicates slowly and induces only mild symptoms in such plants. In the hypersensitive varieties, resistance is inherited as a dominant monogenic characteristic.

LEAFROLL OF POTATO

The leafroll disease is one of the most important diseases of potato in Europe, the United Kingdom, and North America. During the nineteenth century, and the first quarter of the twentieth century, it contributed heavily to the "running-out" or degeneration diseases of po-

FIGURE 22-2
Leafroll of potato.

tato, diseases now known to be caused by mixed infections by the leafroll virus and potato viruses X, Y, A, and S.

The disease often goes undetected during the season in which plants are inoculated. If symptoms form at all, they consist only of an upward curling of the laminar tissues of the leaflets near the top of the plant. The diagnostic symptoms of leafroll are seen in most varieties, however, when the plants have been grown from infected seed pieces or seed tubers: the lower leaves curl upward at the margins, plants are dwarfed and have an abnormally erect habit of growth, leaves become thickened and leathery (Figure 22-2), and there is a noticeable rattle of the foliage when one brushes the plants while walking through the field. A brown discoloration occurs in the phloem tissues of stems and tubers of some varieties, but this symptom is by no means diagnostic; many severely affected varieties do not show the symptom of phloem necrosis.

Etiology

Leafroll of potato is caused by a circulative (persistent), aphid-borne virus. It is one of the few circulative viruses transmitted by aphids, which usually transmit stylet-borne (nonpersistent) viruses. The leafroll virus does not replicate within its insect vector.

The virus survives between growing seasons in the buds of infected tubers. As shoots develop from diseased tubers, the virus moves systemically throughout the plant, apparently being restricted to the phloem tissue. The green peach aphid, *Myzus persicae*, is the chief agent of inoculation. It acquires virus as it feeds upon the phloem of diseased plants, and transmits it to other plants on which it may feed.

Once injected into the phloem of a plant by an aphid, the virus replicates and moves systemically. Starch accumulates in the leaves of diseased plants, causing the leaves to be stiff. Most of the starch accumulates in the spongy parenchyma; there is little accumulation of excess starch in the closely packed cells of the palisade parenchyma, and the swelling of the starch-gorged spongy parenchyma results in an upward curling of the laminar tissue. Failure of starch to move out of diseased leaves may be partly due to malfunctioning of the phloem, at least in those varieties of potato in which phloem necrosis occurs. This does not account for the accumulation of starch in those varieties in which there is no necrosis of the phloem. Possibly, the viral infection somehow prevents the conversion of starch to sugar, thereby preventing the translocation of manufactured food materials out of the diseased leaves.

Epidemics of leafroll occur when relatively large numbers of infected tubers are planted and when there is an abundance of the winged forms of the green peach aphid.

Prevention of Disease

Exclusion of the virus through the use of healthy seed tubers is by far the most effective means of controlling leafroll. Several states of the United States and a number of countries have cooperative inspection programs through which one can obtain seed tubers certified to be almost completely free of leafroll and other potato viruses. Plants in fields up for certification are inspected in midseason and just before harvest, and only those fields having a low percentage of infected plants will be certified. A certain tolerance (that is, a maximum allowable percentage of infected plants) is agreed to by all concerned. Diseased plants discovered in the first inspection can be rogued, thereby eradicating much of the virus in the field. In addition to passing inspections in the field, sometimes a third test must be passed before seed tubers are certified: Samples of tubers are taken, their rest period is broken, and they are planted during the winter in the greenhouse or in a field where the climate is mild. Batches of seed tubers whose samples give rise to more diseased plants than tolerated cannot be certified. Many of the certified seed fields will be found in cold climates or at high elevations, where the aphid vectors of potato viruses are not prevalent. Thus, seed tubers grown in Scotland, Ireland, Switzerland, Canada, and the northern United States are likely to be reasonably free of potato viruses.

ASTER YELLOWS

The pathogen of aster yellows infects some 200 species of plants in approximately 40 families, and is particularly destructive in aster, lettuce, and carrot. The disease has been so devastating that asters are no longer grown extensively in much of the northern United States. The recent discovery of mycoplasmalike organisms (Chapter 3) in yellows-infected plants suggests that aster yellows is caused by a mycoplasma rather than a virus.

Early symptoms of the disease caused by the aster-yellows pathogen are clearing of the veins and general chlorosis of the young leaves. The internodes of the main stem are shorter than normal, and axillary buds sprout, giving rise to spindly, yellow shoots (Figure 22-3). The entire diseased plant takes on the appearance of a witches' broom. Flowers develop color unevenly; some petals are green, whereas others develop the characteristic color of the variety. Also, there is frequently a necrosis of the main stem near its apex. Diseased lettuce plants show the typical symptoms and also bolt prematurely, producing seed in place of leaves or heads. The pathogen of aster yellows causes a stunting of potatoes and a twisting of the upper leaflets. Anthocyanescence occurs in the twisted leaflets, hence the name "purple top" of potato. Tubers of diseased potato plants become flaccid and produce weak, spindly sprouts.

FIGURE 22-3
Aster plant infected by the pathogen of aster yellows; *inset,* flower from a healthy plant.

Etiology

The disease is caused by the aster yellows mycoplasma, which is transmitted naturally only by various species of leafhoppers, although it may also be transmitted by grafting. The pathogen occurs in two strains: the eastern strain, which does not infect celery, and the western strain, which does.

The pathogen survives between growing seasons in perennial hosts. The six-spotted leafhopper, *Macrosteles divisus,* and several other species of leafhoppers are the agents of inoculation. The pathogen is acquired from the phloem of an infected plant by the insect, which is not infective until approximately ten days after acquiring the pathogen. During this time, the pathogen is replicating within the insect and circulating through its system (see Chapter 3). Once the insect becomes infective, it remains so for the rest of its life, often 100 days or more. The pathogen is transmitted to previously healthy plants that are fed upon by infective leafhoppers. The pathogen is injected directly into the phloem; it then moves systemically throughout the plant, but remains confined to the phloem. The physiological effects of the pathogen include, among other things, the impairment of growth-regulating mechanisms: apical dominance is broken—there is a dwarfing effect upon the main stem—while axillary shoots appear to be stimulated.

Prevention of Disease

Aster yellows can be controlled in part by the exclusionary practice of planting susceptible crops more than 200 feet from weeds that are alternative hosts. Eradication of weed hosts is also recommended.

Probably the most successful way to control aster yellows is to grow susceptible plants in beds protected by insect-proof shade cloth; because the leafhopper vectors do not usually fly very high off the ground, considerable protection is afforded plants that are enclosed within a six-foot high, insect-proof screen. Some degree of control has been achieved by repeated applications of such insecticides as DDT or mixtures of sulfur with pyrethrum or rotenone.

Little progress has been made in the development or selection of varieties resistant to aster yellows.

Selected References

CUCUMBER MOSAIC

Doolittle, S. P., 1920. The mosaic disease of cucurbits. *U. S. Dep. Agr. Bull.* (879).

Munger, H. M., and A. G. Newhall, 1953. Breeding for resistance in celery and cucurbits. *Phytopathology* 43:254–259.

Pound, G. S., 1948. Strains of cucumber mosaic virus pathogenic on crucifers. *J. Agr. Res.* 77:1–12.

Sill, W. H., Jr., W. C. Burger, M. A. Stahmann, and J. C. Walker, 1952. Electron microscopy of cucumber virus 1. *Phytopathology* 42:420–422.

Wellman, F. L., 1934. Identification of celery virus 1, the cause of southern celery mosaic. *Phytopathology* 24:695–725.

TOBACCO MOSAIC

Allard, H. A., 1914. The mosaic disease of tobacco. *U. S. Dep. Agr. Bull.* (40).

Esau, K., and J. Cronshaw, 1967. Relation of tobacco mosaic virus to the host cells. *J. Cell Biol.* 33:665–678.

Grant, T. J., 1934. The host range and behavior of the ordinary tobacco mosaic virus. *Phytopathology* 24:311–336.

Holmes, F. O., 1932. Symptoms of tobacco mosaic disease. *Contrib. Boyce Thompson Inst. Plant Res.* 4:323–357.

———, 1960. Inheritance in tobacco of an improved resistance to infection by TMV. *Virology* 12:59–67.

Markham, R., J. Hitchborn, G. Hills, and S. Frey, 1964. The anatomy of the tobacco mosaic virus. *Virology* 22:342–359.

LEAFROLL OF POTATO

Bawden, F. C., 1932. A study on the histological changes resulting from certain virus infections of the potato. *Proc. Roy. Soc. London B Biol. Sci.* 111:74–85.

Smith, K. M., 1929. Insect transmission of potato leafroll. *Ann. Appl. Biol.* 16:209–229.

Whitehead, T., 1934. On the respiration of healthy and leafroll infected potatoes. *Ann. Appl. Biol.* 21:48–77.

ASTER YELLOWS

Davis, R. E., and R. F. Whitcomb, 1971. Mycoplasmas, rickettsiae, and chlamydiae: possible relations to yellows disease and other disorders of plants and insects. *Annu. Rev. Phytopathol.* 9:119–154.

Doi, Y., M. Teranaka, K. Yora, and H. Asuyama, 1967. Mycoplasma- or PLT group-like microorganisms found in the phloem elements of plants infected with mulberry dwarf, potato witches' broom, aster yellows, or paulownia witches' broom. *Ann. Phytopathol. Soc. Jap.* 33:259–266.

Kunkel, L. O., 1926. Studies on aster yellows. *Amer. J. Bot.* 13:646–705.

Maramorosch, K., 1952. Direct evidence for the multiplication of aster yellows virus in its insect vector. *Phytopathology* 42:59–64.

Shikata, E., K. Maramorosch, and K. C. Ling, 1969. Presumptive mycoplasma etiology of yellows diseases. *FAO* (*Food Agr. Org. U. N.*) *Plant Prot. Bull.* (17):121–127.

23

Noninfectious Diseases of Plants

Plants often show conspicuous symptoms of disease even though no transmissible pathogen can be detected. Such a plant may be exhibiting the symptoms of a disease caused by a noninfectious agent. Such agents include deficiencies and excesses of nutrients, biologically produced toxicants, improper soil relations, adverse meteorological conditions, and air pollutants. These agents, and some of the diseases they cause, are discussed in this chapter.

NUTRIENT DEFICIENCIES

Deficiencies of essential mineral nutrients are among the most prevalent noninfectious agents of disease in green plants. Twelve such elements, and the symptoms of plants deficient in them, are discussed below.

Nitrogen

Nitrogen is a constituent of such essential compounds as chlorophyll, hormones, amino acids, and important proteins, including enzymes and vitamins. Nitrogen-deficient plants are stunted and woody, and their foliage is light green or yellow. The ratio of root tissue to shoot tissue in nitrogen-deficient plants is high, and sugar accumulates in the tissues because nitrogen becomes limiting for respiration. Symptoms of nitrogen deficiency are similar in all affected plants. The condition is corrected by side dressings of nitrogen-rich fertilizer—for example, one containing ammonium nitrate.

Phosphorus

Phosphorus is found in nucleoproteins and in lipoids, or phosphatized fats. It is essential to carbohydrate transformations and respiration; several of the intermediates in anaerobic respiration are phosphorylated sugars and phosphorylated 3-carbon compounds. Phosphorus is also involved in high-energy bonding, as in the conversion of adenosine diphosphate (ADP) to adenosine triphosphate (ATP). The element is essential to timely differentiation and maturation of plant tissues.

Phosphorus-deficient plants grow and mature slowly. Sugar accumulates, and anthocyanescence often occurs. Phosphorus-deficiency anthocyanescence is so striking in the common weed lamb's quarters (*Chenopodium album*) that this plant is often used as a biological indicator of phosphorus deficiency in the soil. Cotton is particularly susceptible to phosphorus-deficiency disease. Affected plants are dwarfed and their foliage is dark green. Other highly susceptible plants are peach, citrus fruits, and flax. The disease is readily controlled by applying phosphate fertilizers.

Potassium

Although not a constituent of organic compounds detected in the living plant cell, potassium is required in large amounts by growing plants. The element is essential as a catalyst, certainly, in nitrate reduction and in reactions that convert starch to sugar; it is also essential to photosynthesis and to cambial activity.

"Potash hunger" is particularly important in potatoes, tobacco, cotton, forage legumes, and other crop plants that require uncommonly large amounts of the element. Potassium deficiency often results in the yellowing of leaves, particularly at their margins. This is usually followed by a yellowing of interveinal tissue and, finally, by the bronzing and necrosis of previously yellowed tissues. Potash hunger in alfalfa and clovers is characterized by white necrotic spots in laminar tissues. Specific diseases resulting from a deficiency of potassium include cotton rust and white spot of alfalfa. The deficiency is quickly and easily controlled through soil application of a potassium salt in the form of a commercial fertilizer.

Boron

Boron increases the mobility of sugars and calcium and is important in cell division and protein synthesis. Boron is essential to pollination, and it affects flower formation, fruit set, and seed production.

Growing points die, roots are thick and stunted, cambial activity is impaired, stems are brittle, internodes are shortened, and seed production is reduced in boron-deficient plants. Symptoms of boron deficiency appear in most fruit and vegetable crops when its concentration in the plants is less than 20 ppm on a dry-weight basis. Specific diseases caused by a deficiency of boron include cracked stem of celery, heart rot of beet, internal cork of apple (Figure 23-1), brown heart of cabbage and turnip, internal brown spot of sweet potato, terminal-bud breakdown of tobacco, fruit pitting and dieback of olive, and rosetting and anthocyanescence of alfalfa. Boron deficiencies can be corrected by the application of 8–10 pounds of boron per acre: fritted boron can be applied to the soil, or borax or fertilizer–borate sprays can be applied either to the soil or directly to the plants.

FIGURE 23-1
Internal cork in boron-deficient apple fruit.

FIGURE 23-2
Blossom-end rot of tomato.

Calcium

Calcium is often the limiting factor in cell-well formation, and, consequently, is indirectly involved in cell division. As calcium pectate, it is perhaps the principal constituent of the middle lamellae. Calcium ions also function in permeability phenomena.

Plants deficient in calcium are stunted, and their foliage curls and becomes brittle because of the accumulation of starch. Necroses also develop in some tissues of calcium-deficient plants. Blossom-end rot of tomato (Figure 23-2), black heart of celery, and withertip of flax are a few specific diseases associated with a deficiency of calcium. Apples, peaches, tobacco, and peas are also among the crop plants that seem particularly sensitive to a scarcity of calcium.

Copper

Copper catalyzes certain reactions in respiration and is a constituent of some enzymes. It is indirectly involved in the synthesis of chlorophyll, and it is important both in reproduction and in the metabolism of carbohydrates and proteins. Of all the essential plant nutrients, only molybdenum is required in smaller amounts than copper. Whereas symptoms of deficiency of most trace elements begin to show when the concentrations of those elements in plant tissues is less, on the average, than 15–50 ppm, copper-deficiency symptoms are not seen until copper concentrations drop below 2–10 ppm.

Copper-deficient plants bear chlorotic foliage, and their twigs often

FIGURE 23-3
Symptoms of copper deficiency in citrus. (Photograph by A. S. Rhoads, courtesy of the University of Florida.)

show the necrotic symptom known as dieback. Cupping of the young leaves is not unusual; in citrus, gum pockets form in the twigs, which also bear multiple buds (Figure 23-3). Specific plant diseases attributed to copper deficiency include dieback of citrus, the "reclamation disease" of oats, and withertip of apple. Among the crop plants most likely to suffer from copper deficiency are grasses, legumes, fruit and nut trees, and solanaceous crops.

Iron

Iron, although not a constituent of chlorophyll, is essential to the synthesis of that compound. It is also an electron carrier in the oxidation and reduction reactions of respiration, and is a constituent of certain enzymes and other proteins. Iron is not readily mobile within the plant.

Deficiency of iron is characterized by a chlorosis of young leaves in which the laminar tissue becomes chlorotic while the veins remain green. The leaves of grasses suffering from a deficiency of iron have a striped appearance; in citrus, the iron-deficiency disease has been termed "green netting" (Figure 23-4). Iron deficiency often occurs

FIGURE 23-4
Iron-deficiency chlorosis of a grapefruit leaf. (Photograph by A. S. Rhoads, courtesy of University of Florida.)

in calcareous soils, where the iron present in the soil is largely unavailable to the plant, owing to its tendency to form insoluble compounds with calcium. Iron-deficiency chlorosis is pronounced in azaleas and camellias grown in alkaline soils. Although iron is usually available in acid soils, its availability is decreased in such soils when excessive amounts of phosphorus are applied.

Iron-deficiency symptoms can be corrected with foliar or soil applications of iron chelates or ferrous salts.

Magnesium

Magnesium, the metallic constituent of chlorophyll, also catalyzes reactions important in oxidative respiration. Unlike iron, magnesium is readily mobile in plants.

The leaves of magnesium-deficient plants become chlorotic in blotchy areas between the veins (Figure 23-5). Older leaves are affected first, but young leaves become chlorotic also if the condition is not corrected. Necroses often occur within the yellow blotches and, as in potatoes, along the margins of leaves or leaflets. In tobacco, the

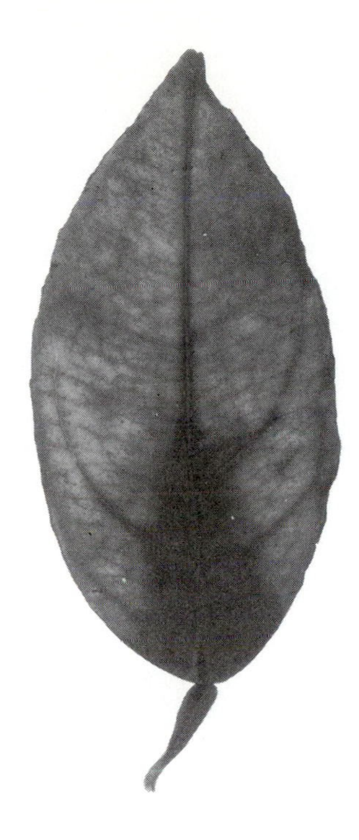

FIGURE 23-5
Leaf of sweet orange showing symptoms of magnesium deficiency. (Photograph by A. S. Rhoads, courtesy of the University of Florida.)

disease "sand drown" has been attributed to a deficiency of magnesium.

Certain soils in the eastern United States are deficient in magnesium. The deficiency can be corrected by applying magnesium sulfate or dolomitic limestone to the soil.

Manganese

Manganese is required in respiratory reactions and in the synthesis of the vitamins riboflavin and ascorbic acid. It is also essential to the reduction of carbon dioxide in photosynthesis.

Interveinal foliar chlorosis develops in manganese-deficient plants (Figure 23-6). The fact that symptoms appear first on the older leaves of plants with manganese-deficiency chlorosis distinguishes this disease from iron-deficiency chlorosis. Chlorosis due to a deficiency of manganese is regular, not blotchy as it is in magnesium-deficient leaves.

The gray-speck disease of oats and marsh spot of garden peas are due to deficiencies of manganese. Numerous other crop plants have been reported to suffer from manganese deficiency. Symptoms are

FIGURE 23-6
Leaf of grapefruit showing symptoms of manganese deficiency. (Photograph by A. S. Rhoads, courtesy of the University of Florida.)

corrected by applying manganese—either as inorganic salts or as organic compounds—either to the soil or directly to the foliage of affected plants.

Molybdenum

Of all the elements essential for the growth of green plants, molybdenum is required in the smallest amounts: a concentration of less than 0.5 ppm is required. Molybdenum is important in protein synthesis and is a constituent of an enzyme that is responsible for the reduction of nitrates to nitrites. It is essential to symbiotic fixation of nitrogen, as in the nodules formed on the roots of legumes infected with *Rhizobium* spp.

Plants deficient in molybdenum are stunted, have pale yellow foliage, and often have cupped and curled leaves with marginal chlorosis. In citrus trees deficient in the element, the leaves are marked with circular chlorotic spots that occur in rows on each side of the midrib.

The whiptail disease of cauliflower, which is characterized by malformed leaves and irregular flower formation, is a molybdenum-deficiency disease. Symptoms of molybdenum deficiency have also been recognized in corn, legumes, sugar beets, flax, tomatoes, and other crop plants.

The liming of acidic soils often corrects molybdenum-deficiency symptoms in plants by releasing the molybdenum that remains chemically bound in an acid medium. Symptoms may also be corrected by applying ammonium molybdate or sodium molybdate to the soil or to the foliage of deficient plants.

Sulfur

Sulfur is a constituent of amino acids and proteins. It also functions in the formation of chlorophyll. Sulfur-deficiency symptoms rarely occur, but leaves of affected plants become pale green and then turn yellow. A dwarfing and yellowing disease in cotton has been attributed to a deficiency of sulfur, and a sulfur deficiency in citrus has also been described. The condition can be prevented by using commercial fertilizers that have been supplemented with sulfur.

Zinc

Zinc is required for the production of growth-regulating substances (hormones) and is a catalyst in oxidation reactions in green plants. It is also important in the formation of chlorophyll and in photosynthetic activity.

Zinc-deficient plants are stunted because the internodes do not elongate normally. Sometimes, as in apple, the internodes are so repressed that rosetting occurs. The foliage of affected plants becomes chlorotic and bronze-colored. Specifically, zinc deficiency results in such diseases as white bud of corn, rosette of fruit trees, bronzing of tung, little leaf of stone fruit, and mottle leaf of citrus. Pear, avocado, grape, alfalfa, forage grasses, and many other crop plants have also been reported to suffer from zinc deficiency. Applications of organic zinc compounds or inorganic zinc salts either to the soil or directly to growing plants have been recommended to control zinc-deficiency diseases.

NUTRIENT EXCESSES

Plant nutrients, particularly those required only in trace amounts, sometimes accumulate to toxic concentrations in plant tissues. Internal bark necrosis of apples, for example, is apparently due to excessive amounts of magnesium and iron.

An excess of boron often results in a blotchy chlorosis that starts at the tips of the leaves of affected plants and extends thence along the leaf margins. Necrotic spots develop in the chlorotic areas, and badly affected leaves drop away. The toxicity usually develops after injudicious use of boron to correct a previously existing deficiency of the element. Boron is leached from the soil by rain or heavy irrigation, and the soil can be limed to make the boron less readily available to plants. Fruit and nut trees and common beans are highly sensitive to boron, cereals and tomatoes are moderately sensitive, and crucifers and beets are boron-tolerant.

Many of the alkaline soils in the United States have concentrations of soluble salts that are high enough to be toxic to plants. For example, the so-called black alkaline soils contain excessive sodium, which must be leached, after calcium has been added to the soil, to make them suitable for agriculture. The white alkaline soils, however, contain salts of calcium and magnesium in excess. Some plants are extremely chlorotic when grown on calcareous soils—for example, grapes grown on certain soils in France and pineapples in Puerto Rico. Excessive magnesium has been associated with fasciation of stems and with leaf crinkling. Large amounts of calcium in the soil may result in increases in calcium in plant tissues, or may have an indirect effect upon the health of the plant by influencing the availability of other elements—for example, iron and zinc deficiencies are common in apple trees grown in alkaline soils. Excess calcium and magnesium can be leached from the soil by rain or heavy irrigation.

Copper and manganese have been known to accumulate in the sandy soils of certain citrus orchards in Florida because of excessive fertilization, causing the citrus trees to become unthrifty and chlorotic. Roots have been killed by excess copper in the soil, and the acid conditions in such a soil can prevent the uptake of necessary iron. Excess zinc is less commonly a problem, but it has been shown to cause chlorosis in citrus seedlings, apparently by preventing absorption of iron by the roots.

When plants are grown in the presence of excessive amounts of soluble salts, they may suffer a necrosis of the foliage that usually begins at the tips of leaves and progresses thence along the leaf margins. Such necrotic tissues are often invaded by saprophytic fungi.

BIOLOGICALLY INDUCED TOXICANTS

Exudates from the roots and fallen leaves and hulls of English walnut trees contain juglone, which has long been known to be toxic to herbaceous plants. Tomatoes, for example, do not survive in soils near living walnut trees or in soils from which walnut trees have been recently removed.

Saprophytic microorganisms that live in the soil may also produce metabolic by-products that are toxic to green plants. For example, *Aspergillus wentii*, when growing in the vicinity of plant roots, produces a toxin that causes such diseases as frenching of tobacco and strap-leaf of chrysanthemum and other plants. Apparently, other nonparasitic fungi, and possibly bacteria, produce toxins that adversely affect growth-regulating substances in plants. Diseased plants usually have extremely narrow, leathery leaves whose margins curl downward.

ADVERSE METEOROLOGICAL CONDITIONS

Insufficient and Excessive Amounts of Water

Plants subjected to drought may suffer chlorosis, anthocyanescence, or, if the drought is severe, permanent wilting of the foliage; leaves and fruit of woody plants may drop prematurely. Firing and scorch are common leaf symptoms of plants that do not receive sufficient moisture; the necroses that develop in affected foliage and fruit probably result from an internal buildup of certain nutrients to toxic levels. Certain effects of drought upon woody plants may be observed the following season, when twigs and branches die back from the tips.

Sudden changes in soil-moisture levels were once thought to cause blossom-end rot of tomato fruit and black heart of celery. Both diseases are presently thought to result mainly from a deficiency of calcium, but improper water relations contribute to their severity: the diseases are most likely to occur when the soil becomes drier than usual for a period and then suddenly is subjected to excessive rainfall or irrigation.

Concentrations of toxic compounds, such as nitrites, build up in waterlogged soils. Plants grow poorly in soils with excessive water and typically show a general yellowing of foliage. Woody plants lose their leaves, and dieback of shoots occurs. Coupled with decreased rates of transpiration, excessive soil moisture contributes to edema, a disease characterized by intumescences in the epidermis of leaves. Epidermal cells enlarge in groups to form small, pimplelike blisters.

Excessive soil moisture results ultimately in the wilting of herbaceous plants, whose roots, meantime, have become blackened and decayed. Doubtless, the death of the roots is due partly to the accumulation of nitrites and other toxic compounds produced by anaerobic organisms in the waterlogged soil, and partly to the direct actions of saprophytic bacteria and fungi, which readily enter damaged root tissues and assist in their breakdown.

Low- and High-Temperature Effects

Every factor that affects the growth of living organisms can be described in terms of cardinal quantities. For example, there is a minimum temperature below which a plant will not grow, an optimum at which it will grow best, and a maximum above which it will not grow. Cardinal temperatures vary widely among species; thus, certain plants grow only in the torrid zone, others only in the temperate zones, and so forth. Indeed, the minimum temperature for a given species is usually the factor that determines the geographical distribution of that species north or south of the equator.

Low temperatures may adversely affect green plants in one of two ways. First, temperatures near the minimum for plant growth will limit the rates of metabolic reactions, which then proceed too slowly for good growth; if prolonged, such low temperatures may cause early death. Second, temperatures below 0°C may kill plants because of intercellular and intracellular ice formation.

Freezing or frost damage causes necrosis in the laminar tissues of leaves and often results in distortion of the foliage that grows from cold-damaged buds. Necrotic areas of frosted leaves usually appear papery and bleached; they are seldom brown. The new growth of perennial plants is sometimes so severely damaged by cold that twigs die and affected leaves fall away. Perennial plants are also subject to winter killing, which is usually most severe after periods of alternate freezing and thawing.

Low-temperature damage in living plant tissues is due to the formation of ice crystals in the intercellular spaces and, ultimately, in the cytoplasm or the vacuoles of the cells themselves. The formation of ice crystals between plant cells can cause mechanical damage, but most harm to cells probably results from their desiccation: water is lost rapidly from the protoplasts through plasma membranes, whose permeability to water is increased at low temperatures. Most damage occurs, however, when the temperature is low enough for sufficient time that ice crystals are induced to form in the protoplast itself. Upon thawing, the ice crystals release water, which accumulates in the intercellular spaces. In plant tissues exposed suddenly to exceedingly low temperatures—for example, in those quick-frozen with liquid nitrogen—ice crystals do not form. If such tissues are thawed rapidly, the formation of ice crystals can be prevented, and tissues so treated are capable of functioning normally when placed in an appropriate environment. The rapid freezing of cells with little or no free water permits the long-term preservation of such organisms as bacteria and fungi. The "freeze-dry" method has also been used successfully to maintain plant viruses in storage.

Potatoes may be adversely affected by exposure to low temperatures. In potato tubers stored at approximately 5°C, the starch is converted to sugar more rapidly than the sugar can be catabolized. This increased sugar content not only makes the tubers unpalatable, but it makes them especially unsuitable for potato chips: the tissues actually caramelize during the cooking processes. Saccharified potatoes can be made usable again by raising the temperature of the storage area to 10°C for a week or more. Potato tubers stored for prolonged periods at low temperatures break down internally: blotchy, ringlike, or netted necroses develop within them.

Diurnal variations in winter temperatures can lead to the development of trunk cankers in trees. Freezing damage—particularly to cambial, phloem, and cortical cells—often occurs during winter nights in the temperate and frigid zones. When cold-damaged tissues are heated, during the day, by direct rays of the sun, their temperature rises rapidly. This results in such rapid expansion of the warmed parts of the damaged bark that it cracks open and a canker develops. Wood-destroying pathogens often enter the stems of woody plants through these winter sunscald cankers.

The leaves of plants exposed to excessively high temperatures usually become desiccated: leaf tissues die—first at the leaf tips, then along the margins—and plants wilt rapidly because of increased rates of transpiration. Heat cankers develop in the stems of flax plants subjected to high temperatures. Symptoms of scald and internal browning develop in the fruits of deciduous trees when they are exposed to high temperatures.

High temperature, under anaerobic conditions, brings about the black-heart disease of potato tubers (Figure 23-7). Anaerobic conditions may occur in poorly ventilated storages or in water logged soils, and potato tubers in either environment can develop black heart. The rate of anaerobic respiration is greatly increased, whereas that of aerobic respiration is actually decreased. Metabolic products that accumulate deep within the tubers cause autolytic breakdown of the central tissues, which turn black upon death.

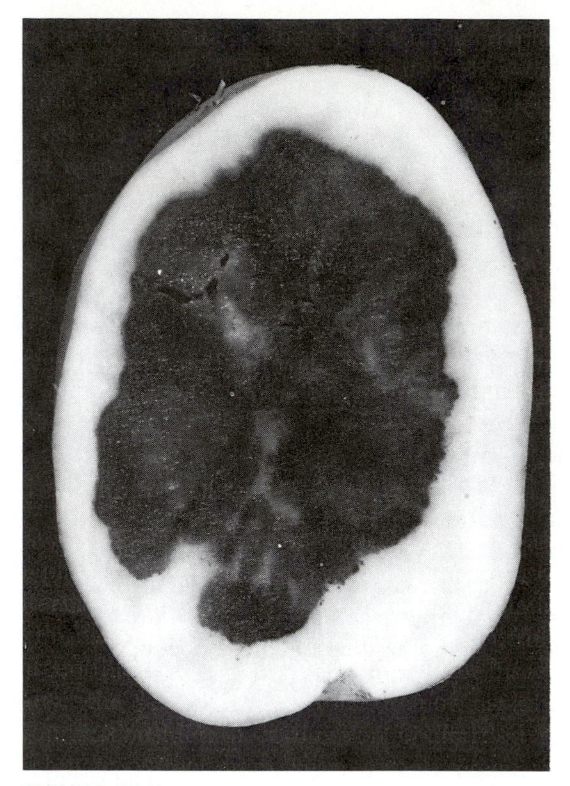

FIGURE 23-7
Black heart of potato tubers.

AIR POLLUTANTS

The air at the surface of the earth is approximately 78% nitrogen, which is considered inert. It is certainly unavailable as a plant nutrient, except in plants capable of fixing atmospheric nitrogen through the activities of symbiotic root-infecting bacteria. Oxygen, which is used by green plants in aerobic respiration, accounts for some 21% of the atmosphere at the earth's surface; and carbon dioxide, which is essential to photosynthesis—the reaction that is ultimately the source of all energy and food used by living organisms—accounts for only about 0.03% of the volume of the air.

The 0.07% of the air that is not accounted for by nitrogen, oxygen, and carbon dioxide comprises a variety of other compounds: some of these are inert, but certain others continuously impair the functions of green plants. These air-polluting chemicals are, therefore, noninfectious agents of plant diseases.

Chemicals that damage plants have always existed in the air, but heavy concentrations of disease-inducing substances have become increasingly destructive in recent years in highly industrialized regions. Atmospheric pollution is beginning to receive the serious attention of plant pathologists, some of whom have reviewed the subject (American Phytopathological Society, 1968).

Air pollutants can be separated into four groups: (1) those that arise from a specific combustion source; (2) those that arise as the products of photochemical reactions between combustion products and naturally occurring chemicals in the air; (3) those that occur naturally in the atmosphere but that may reach damaging concentrations under certain meteorological conditions (ozone is apparently the only known number of this group); and (4) those toxic gases (exemplified by ethylene) that are released into the air by living plants.

Pollutants from Specific Combustion Sources

Sulfur dioxide, as a constituent of smoke from industrial plants, has been recognized for almost a century as a noninfectious agent of plants disease. The laminar tissues of badly affected leaves are killed; the necrotic areas are tan at first, but turn brown later. Less severely affected foliage shows a yellowing of interveinal tissues, but the veins remain green. Damage by sulfur dioxide is common near smelter plants; the sulfur content of the leaves of plants grown nearby may be increased fourfold. Alfalfa and cotton are highly sensitive to damage by sulfur dioxide.

Fluorides may be emitted in plant-toxic concentrations (Treshow, 1971) from metal smelters, aluminum plants, ceramics factories, and plants that produce phosphate fertilizers. Symptoms of fluoride damage are scorch in dicotyledonous plants and leaf-tip necrosis in monocotyledonous plants. Necrotic areas become reddish-brown and brittle; they sometimes fall away, giving the leaf a tattered appearance.

Products of Photochemical Reactions

The silver-leaf disease of plants growing near Los Angeles, California, once thought to be the effect of a plant-pathogenic organism, was shown, in the 1940s, to be associated with "smog." The disease is characterized by the silvery or bronze appearance of the abaxial surfaces of the leaves of affected plants. The lower epidermis collapses, the adjacent mesophyll cells shrink, and the space at the lower surface of the leaf fills with air.

It was soon discovered that silver leaf could not be induced by any single one of the fifty detectable constituents of the polluted air in the Los Angeles area, but the symptoms could be reproduced by exposing plants to mixtures of ozone and vapors of certain hydrocarbons. Organic peroxides—reaction products from mixtures of nitrogen dioxide and certain hydrocarbons—were also shown to cause the symptoms of silver leaf; their instability, however, indicated that these peroxides probably did not cause widespread outbreaks of silver leaf under field conditions. It now seems most likely that silver leaf is usually caused by another photochemical reaction product that is prevalent in "smog." This product, peroxyacetyl nitrate (PAN), is but one of a homologous series of phytotoxic peroxyacyl nitrates.

Ozone may also be a product of photochemical reactions, but because it may occur in plant-toxic concentrations under certain entirely natural conditions, it will be discussed in the section below.

Naturally Occurring Phytotoxic Air Pollutants

Ozone, at concentrations of only 0.03–0.05 ppm, adversely affects chlorophyll-rich tissues; its effects, therefore, are most noticeable on the adaxial surfaces of leaves. It causes stippling, flecking, and mottling, all due mainly to damaged chloroplasts, particularly those in the palisade parenchyma (Figure 23-8).

Just as ozone may arise from photochemical reactions involving combustion, it may also arise from photochemical reactions with plant products. For example, the ozone-induced needle blight of white pine apparently develops after aerial chemical reactions that involve the terpenes that volatilize naturally from the pine trees. But ozone also reaches toxic concentrations under natural conditions: it is released during thunderstorms, and is evidently brought down from the upper atmosphere during temperature inversions and during conditions that lead to anticyclonic flows of air at the surface of the earth.

Phytotoxic Gases Released by Green Plants

Green plants often release ethylene from diseased or mechanically damaged tissue. Excessive amounts of ethylene cause epinasty, pre-

FIGURE 23-8
Ozone-damaged tobacco leaf (*right*) compared with a healthy leaf (*left*).

mature leaf abscission, abscission of petals or flowers, and poor floral development. Ethylene is particularly damaging to cut flowers in transit or storage.

Prevention of Diseases Caused by Air Pollutants

The prevention of plant diseases caused by air pollutants is presently more a challenge than a reality. Some progress has been made, however: the control of pollutants at the sources of combustion of petroleum products, while far from satisfactory, is being accomplished to some degree with systems for improved combustion and filtration. That individual plants of a cultivar or species may show different sensitivities to pollutants indicates that the development of toxicant-resistant varieties of crop plants is a distinct possibility. The protection of growing plants with sprays or dusts designed to minimize air-pollutant damage is presently too costly for commercial use, but the possibility of controlling these diseases cheaply with protective chemicals deserves further attention.

Selected References

American Phytopathological Society, 1968. Symposium on trends in air pollution damage to plants. Invitational papers presented at the fifty-ninth annual meeting of the American Phytopathological Society, Washington, D. C., 20–24 August, 1967. *Phytopathology* 58:1075–1113.

Curtis, O. F., and D. G. Clark, 1950. *An Introduction to Plant Physiology,* pp. 361–403. New York: McGraw-Hill.

Darley, E. F., and J. T. Middleton, 1966. Problems of air pollution in plant pathology. *Annu. Rev. Phytopathol.* 4:103–118.

Heald, F. D., 1933. *Manual of Plant Diseases,* pp. 58–247. New York: McGraw-Hill.

Rich, S., 1964. Ozone damage to plants. *Annu. Rev. Phytopathol.* 2:253–266.

Sprague, H. B., ed., 1964. *Hunger Signs in Crops* (3rd ed.). New York: David McKay.

Treshow, M., 1971. Fluorides as air pollutants affecting plants. *Annu. Rev. Phytopathol.* 9:21–44.

Wallace, T., 1961. The diagnosis of mineral deficiencies in plants (2nd ed.). New York: Chemical Publishing Co.

24

Plant Pathology—
Some Agricultural Perspectives

Plant pathologists seek to enhance crop production by reducing losses from plant diseases caused by infectious and noninfectious agents. They achieve their objective by sensible application of the principles of plant-disease control, which, in turn, relate directly to the principles of plant pathogenesis (Table 1-1). Although they have long been aware of the environmental and biological factors that affect disease development in fields of plants, plant pathologists have seemingly been more concerned with the fate of the individual plant. Recently, however, they have become increasingly concerned with the subject of disease in *populations* of plants. This has led to a long-overdue synthesis of analytical research on programs of disease control through the prevention of epidemics, in which the "public health" of plant populations is the main concern. Thus, epidemiology has taken its place alongside etiology as a scientific basis for the effective control of plant diseases. This new approach has been called the "modern phytopathology" (Zadoks, 1974).

The purpose of this chapter is to expand the principles of plant-disease control and place them in harmony with the concepts of "modern phytopathology." This will allow us to view the science of plant pathology from an agricultural perspective. First, however, we must treat several recent developments that have contributed to our present-day concept of plant pathology.

RECENT ADVANCES IN PLANT PATHOLOGY

Plant pathology has developed rapidly in several of its branches, particularly in those dealing with the science and art of disease control. One important development is that the increased use of systemic fungicides has added a new dimension, therapy or cure, to complement the more familiar preventive practices (Chapter 9). Others are that tissue-culture techniques are now being used on a commercial scale to propagate pathogen-free plants; that the breeding and selection of crop varieties for disease resistance have taken on new meanings as scientists have become increasingly aware of the genetic vulnerability of the species of plants that man manages; and that the science of epidemiology (Chapter 8) has been recognized as a necessary foundation for effective plant-disease control. These topics are treated in this chapter; but our knowledge of several groups of primary agents of plant disease has also burgeoned during the past few years, and some of the recent advances in our comprehension of etiology also deserve attention here.

Plant Viruses

Although the details of viral replication in host-plant cells remain unknown (page 44), it now seems reasonable that events proceed in an orderly sequence, at least for the majority of plant viruses containing ribonucleic acid (RNA). First, the protein coat or **capsid** (which is composed of protein subunits termed **capsomeres**) is shed from the infectious RNA within. Second, the viral RNA, with the aid of the enzyme RNA polymerase (replicase), replicates itself using raw materials from the infected cells. Third, the coat protein forms; and, finally, complete viral particles, often called **virions,** are assembled from new coat-protein units and new viral genomes. The sites of all these activities in host cells have not yet been ascertained, but it seems likely that host-cell membranes, and vesiculations in them, play important roles. For example, viruses may enter protoplasts by **pinocytosis,** a process in which particles are engulfed in small vesicles or vacuoles formed by deep invagination of plasma membranes. Removal of coat protein could occur in pinocytic vesicles of the plasma membrane. Other membranes, such as those enclosing nuclei and chloroplasts, may well play important parts in RNA replication and in the encapsidation or final assembly of new virions. The subject of replication of plant viruses has been reviewed by Hamilton (1974).

From the foregoing account, it is easy to see how genetic recombination might occur among different plant viruses simultaneously infecting a single host (page 44). During the multiplication of two different viruses in the same host cell, the final assembly of virions could result in as many as six combinations of nucleic acids (genomes) and protein coats (capsids): in two of the possible combinations, the new particles would be identical to either of the two parents; in two others, each of the two genomes would be encapsidated with the protein of the other; and in the last two, each of the two genomes would be enclosed in a capsid consisting of proteins from both parents (Figure 24-1). Rochow (1972) refers to reproduction of the parental type as **homologous encapsidation.** The other combinations are examples of **heterologous encapsidation:** when the genome of one virus type is encapsidated by protein derived wholly from the other, it is referred to as **transcapsidation** or **genomic masking;** when the protein coat is derived partly from one virus and partly from the other, it is termed **phenotypic mixing.** Strong evidence for heterologous encapsidation among plant virus is provided by research on two serologically distinct circulative viruses of yellow dwarf of barley, one of which depends on the presence of the other for its transmission by an aphid vector (Rochow, 1969, 1970). Also, Dodds and Hamilton (1971) reported transcapsidation of the nucleic acid of tobacco-mosaic virus with the protein of barley stripe-mosaic virus in barley plants simultaneously infected with both viruses.

Our understanding of the nature of plant viruses has been complicated somewhat by research done in the late 1960s and early 1970s on plant viruses having more than one nucleoprotein component. Some of these viruses, though by no means all of them, have their genetic information divided among those components. This subject was reviewed by van Kammen (1972), who arranged the plant viruses with divided genomes into several types: first, the satellite virus (of tobacco-necrosis virus), which cannot replicate itself but depends on an association with the replicative tobacco-necrosis virus for its own reproduction; second, the tobacco-rattle virus, which has two components (one that can replicate by itself and another that cannot), both of which are required to produce virions; third, the cowpea-mosaic type, each of whose members has two components,

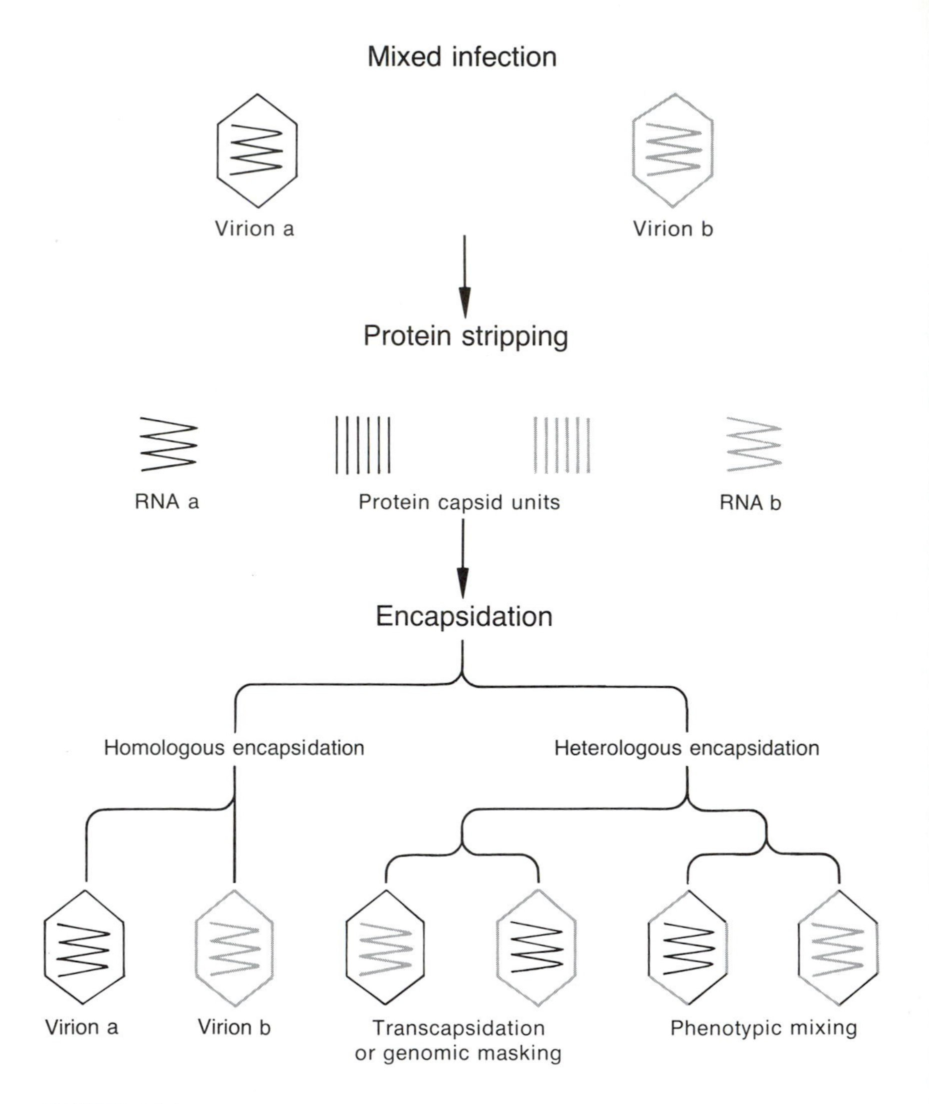

FIGURE 24-1
Some possible recombinations of nucleic acids and protein capsids from two plant viruses that could result from simultaneous infection of one plant by both viruses. (Adapted from a diagram by Rochow, 1972. Printed with the permission of Annual Reviews, Inc., Palo Alto, California.)

both of which are required to produce virions and neither of which can replicate individually; and fourth, the alfalfa-mosaic virus, in which three nucleoprotein groups are necessary to produce virions.

Recently acquired knowledge of plant viruses has occasioned some progress toward an acceptable classification of them. Though viral classification remains a subject of considerable controversy, information now at hand has permitted the tentative arrangement of some of the plant viruses into 16 groups. This classification scheme, which is the product of the Plant Virus Subcommittee of the International Committee on Nomenclature of Viruses (Harrison et al., 1971), owes much to the earlier work of Brandes and Wetter (1959) and Brandes and Berks (1965). The viruses are grouped on the basis of similarities and differences among them with respect to as many as 56 characters relative to their behavior in their host plants, their vector relations, the properties of their particles, and their composition. Eight of the important characters of a plant virus are customarily presented in a kind of "short-hand," the cryptogram, which usually appears immediately after the name of the virus in taxonomic treatments. The cryptogram, which was suggested by Gibbs et al. (1966), has been refined and standardized symbols have been assigned (Anonymous, 1970). For example, the cryptogram for cucumber mosaic virus is R/1: 1/18: S/S: S/Ap. This means that the genome consists of ribonucleic acid (R) that is single-stranded (1); that the nucleic acid has a molecular weight of 1 million (1) and that it comprises 18% of the particle (18); that the particle is spherical or nearly so (S) and that the outline of the protein coat is also spherical (S); and that the hosts of the virus are seed plants (S), with the virus being transmitted by aphids (Ap). Although Harrison et al. (1971) suggested names for each of the 16 groups of plant viruses, individual viruses are usually identified by their common names (Hopkins, 1957; Martyn, 1958, pp. 63–64). An abbreviated classification of plant viruses is set forth in Table 24-1. Descriptions of plant viruses are being published regularly by the Commonwealth Mycological Institute, Kew, Surrey, England. More than a hundred plant viruses have been described, each by a recognized expert conducting research with the particular virus.

Plant Viroids

The several plant-infecting entities that apparently exist as naked RNA (page 44) include the agents of spindle tuber of potato (Figure 24-2), exocortis of citrus, and chlorotic mottle of chrysanthemum (Figure 24-3). These agents, in the words of Diener (1972), repre-

TABLE 24-1

A partial classification of plant viruses. (The cryptograms, which are given in brackets after the names of the viruses, can be interpreted by means of the key at the end of the table.)

POLYTYPIC GROUPS

TOBRAVIRUS Group

Type member: tobacco-rattle virus [R/1: 2.3/5 + 0.9/5: E/E: S/Ne]

Other member: pea early-browning virus

TOBAMOVIRUS Group

Type member: tobacco-mosaic virus [R/1: 2/5: E/E: S/0]

Other members: tomato-mosaic virus, ribgrass-mosaic virus, cucumber green-mottle virus, odontoglossum-ringspot virus, Sammon's opuntia virus, sunn-hemp mosaic virus

POTEXVIRUS Group

Type member: potato virus X [R/1: 2.1/6: E/E: S/(Fu)0]

Other members: clover yellow-mosaic virus, white-clover mosaic virus, cactus virus X

Possible members: cassava common-mosaic virus, cymbidium-mosaic virus, papaya-mosaic virus

CARLAVIRUS Group

Type member: carnation-latent virus [R/1: */6: E/E: S/Ap]

Other members: chrysanthemum virus B, pea-streak virus, potato virus M, potato virus S, red-clover vein-mosaic virus

POTYVIRUS Group

Type member: potato virus Y [*/*: */*: E/E: S/Ap]

Other members: bean common-mosaic virus, bean yellow-mosaic virus, potato virus A, tobacco-etch virus

Possible members: celery-mosaic virus, lettuce-mosaic virus, papaya-ringspot virus, sugarcane-mosaic virus, tulip-breaking virus, turnip-mosaic virus, watermelon-mosaic virus

CUCUMOVIRUS Group

Type member: cucumber-mosaic virus [R/1: 1/18: S/S: S/Ap]

Other member: tomato-aspermy virus

Possible member: peanut-stunt virus

TYMOVIRUS Group

Type member: turnip yellow-mosaic virus [R/1: 1.9/34: S/S: S/Cl]

Other members: wild-cucumber mosaic virus, eggplant-mosaic virus

COMOVIRUS Group

Type member: cowpea-mosaic virus [R/1: 1.5/24 + 2.6/33: S/S: S/Cl]

Other members: bean-pod mottle virus, radish-mosaic virus, squash-mosaic virus

NEPOVIRUS Group

Type member: tobacco-ringspot virus [R/1: 2.2/40: S/S: S/Ne]

Other members: tomato-ringspot virus, grapevine-fanleaf virus, tomato black-ring virus

BROMOVIRUS Group
Type member: brome-mosaic virus [R/1: 1/22: S/S: S/(Ne)]
Other members: broad-bean mottle virus, cowpea chlorotic-mottle virus

TOMBUSVIRUS Group
Type member: tomato bushy-stunt virus [R/1: 1.5/17: S/S: S/*]
Other members: carnation Italian-ringspot virus, pelargonium leaf-curl virus

CAULIMOVIRUS Group
Type member: cauliflower-mosaic virus [D/2: 4.5/16: S/S: S/Ap]
Other members: carnation etched-ring virus, dahlia-mosaic virus

MONOTYPIC GROUPS

ALFALFA-MOSAIC VIRUS [R/1: 1.3/18 + 1.1/18 + 0.9/18: U/U: S/Ap]

PEA ENATION-MOSAIC VIRUS [R/1: */29: S/S: S/Ap]

TOBACCO-NECROSIS VIRUS [R/1: 1.5/19: S/S: S/Fu]

TOMATO SPOTTED-WILT VIRUS [R/*: */*: S/S: S/Th]

KEY TO THE CRYPTOGRAMS

First element: type of nucleic acid (*R,* RNA; *D,* DNA)/strandedness of the nucleic acid (*1,* single; *2,* double)
Second element: molecular weight of the nucleic acid (millions)/nucleic-acid fraction of the viral particle (percent)
Third element: outline of the viral particle (*S,* spherical; *E,* elongated; *U,* elongated with rounded ends)/outline of the protein coat (*S,* spherical; *E,* elongated; *U,* elongated with rounded ends)
Fourth element: host (*S,* seed plant)/vector (*Ap,* aphid; *Ne,* nematode; *Fu,* fungus; *Th,* thrips; *Cl,* beetle; *Au,* leafhopper; O, spreads without a vector)
A parenthetical symbol represents doubtful or unconfirmed information; an asterisk means that the property in question is not known; and a second element consisting of two or more parts separated by plus signs indicates that the virus has two or more nucleoprotein components.

Source: Harrison et al. (1971), Anonymous (1970), and Matthews (1970).

sent a "novel class of subviral pathogens, the viroids . . . characterized by the apparent absence of a dormant phase (virions) and by genomes . . . much smaller than those of known viruses." The viroidal RNA of spindle tuber of potato, for example, has a molecular weight of approximately 50,000, 40 times smaller than that of the RNA (and 800 times smaller than that of the intact virion) of tobacco-mosaic virus. Although some objections have been raised against the use of the term viroid for such pathogens as that of spindle tuber of

FIGURE 24-2
Healthy potato tuber (*left*) and one infected by the potato spindle-tuber viroid (*right*).

potato (McKinney, 1973), we agree with Diener (1973) that ". . . redefinition of *viroid* to encompass nucleic-acid species with properties similar to those of PSTV [potato spindle-tuber virus] is appropriate and serves a useful function."

Mycoplasmalike Organisms, Rickettsiae, and Bacteria

Mycoplasmalike organisms (pages 42 and 43), though associated with plant diseases once thought to be of viral origin, more nearly resemble bacteria than viruses (see Box 3-2). Consequently, Breed et al. (1957) classified the mycoplasmalike organisms as a monogeneric family and order of the bacteria. Later, Buchanan and Gibbons (1974) treated them as members of a separate class of organisms (Mollicutes) closely related to the class Bacteria (Table 24-2). It remains to be seen whether the plant-infecting mycoplasmalike organ-

FIGURE 24-3
Leaf of a chrysanthemum plant infected by the chlorotic-mottle viroid.

isms belong to the genus *Mycoplasma,* whose members infect some animals, or whether they contain members sufficiently characteristic to justify erection of new genera (Saglio et al., 1973), a new family, or even a new order. Recently, Markham et al. (1974) fulfilled Koch's Postulates (pages 16 and 17) with respect to a mycoplasmalike organism, thereby showing it to be the cause of the citrus little-leaf disease.

Another group of bacterialike organisms, the rickettsiae, have only recently been incriminated as agents of disease in plants (Giannotti et al., 1970; Goheen et al., 1973; Hopkins and Mollenhauer, 1973). Structurally, the rickettsiae more nearly resemble bacteria than do the mycoplasmalike organisms, because the rickettsiae have distinct cell walls (the outer margins of which usually appear scalloped). The rickettsiae are presently viewed as members of a unique class of organisms related to the bacteria (Buchanan and Gibbons, 1974). Those that cause plant diseases (for example, Pierce's disease of grapevines) are obligately parasitic and occupy the tracheary elements of diseased plants (Figure 24-4). The discovery of rickettsiae

TABLE 24-2
Classification of plant-pathogenic bacteria and related organisms

- Kingdom PROCARYOTAE [Monera] (organisms with procaryotic nuclei)
  - Division SCOTOBACTERIA (procaryotes indifferent to light)
    - Class BACTERIA
      - GRAM-NEGATIVE AEROBIC RODS AND COCCI
        - Family PSEUDOMONADACEAE
          - [examples: *Pseudomonas solanacearum* (of brown rot of tomato and other solanaceous plants) and *Xanthomonas phaseoli* (of common bacterial blight of bean)]
        - Family RHIZOBIACEAE
          - [example: *Agrobacterium tumefaciens* (of crown gall of various plants)]
      - GRAM-NEGATIVE FACULTATIVELY ANAEROBIC RODS
        - Family ENTEROBACTERIACEAE
          - [example: *Erwinia carotovora* (of soft rot of fleshy tissue)]
      - ENDOSPORE-FORMING RODS AND COCCI
        - Family BACILLACEAE
          - [examples: *Bacillus* spp. (of potato seed-piece rots)]
      - ACTINOMYCETES AND RELATED ORGANISMS
        - CORYNEFORM BACTERIA
          - [example: *Corynebacterium insidiosum* (of alfalfa wilt)]
        - Order ACTINOMYCETALES
          - Family STREPTOMYCETACEAE
            - [example: *Streptomyces scabies* (of potato scab)]
    - Class MOLLICUTES
      - Order MYCOPLASMATALES (pleuropneumonialike organisms)
        - Family MYCOPLASMATACEAE
          - [examples: Mycoplasmalike organisms presumed to cause many of the "yellows" diseases of plants. The agent of citrus "stubborn" disease has been named *Spiroplasma citri* by Saglio et al. (1973), and is presently treated as a genus of uncertain affiliation.]
    - Class RICKETTSIAE
      - Order RICKETTSIALES
        - Family RICKETTSIACEAE
          - [example: Organism that causes Pierce's disease of grapevines (Goheen et al., 1973; Hopkins and Mollenhauer, 1973)]

Source: Buchanan and Gibbons (1974).

in the xylem of diseased plants once thought to be affected by viruses disproves the notion that a plant virus might occupy the tracheary elements of its host. All viruses that infect their plant hosts systemically move in the sieve elements of the phloem, and some plant viruses are completely restricted to the phloem.

The families of plant-infecting bacteria (Chapter 3) are classified in Table 24-2 to illustrate their taxonomic relations with the two "new" groups of plant pathogens, the mycoplasmalike organisms and the rickettsiae.

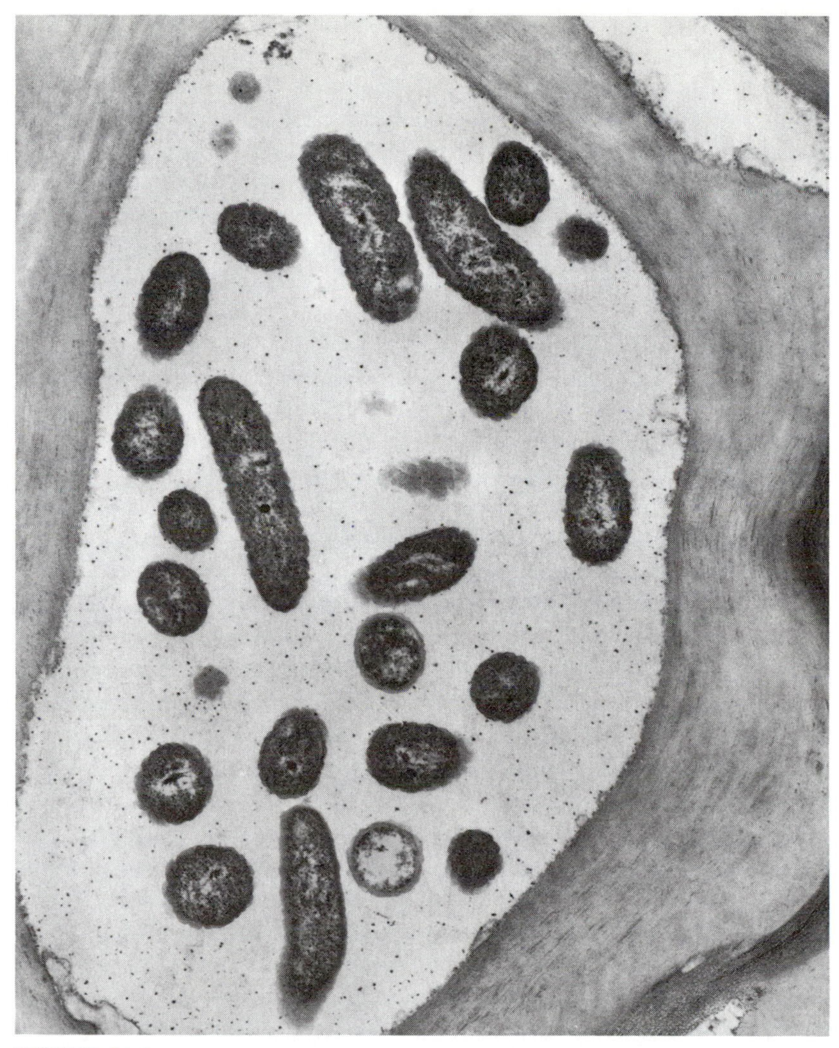

FIGURE 24-4
Electron micrograph of rickettsialike organisms in a xylem vessel of a grapevine, X 24,367. (Courtesy of D. L. Hopkins.)

Pathogen-Free Propagative Stock from Tissue Cultures

Treatments of the propagative parts of plants with heat or chemicals have long been used as exclusionary and eradicative measures in the control of plant diseases (pages 144 and 145). Propagative materials free of plant pathogens can also be obtained by another procedure, tissue culture. (We use the term "tissue culture" here in a broad sense: we include such diverse procedures as the culturing of protoplasts, embryos, anthers, microspores, root tips, lateral and adventitious buds, and the internal meristematic cells of cambial regions; we exclude, however, the familiar horticultural practices of vegetative propagation—the rooting of cuttings, layering, grafting, planting of modified stem or root organs, and so forth.)

Murashige (1974) lists four areas for the possible commercial application of tissue-culture methods: (1) the collecting of pharmaceuticals and other natural products of plants; (2) the genetic improvement of crop-plant varieties; (3) the recovery of disease-free clones from diseased plants and the preservation of valuable germ plasm in a healthy condition; and (4) the rapid clonal multiplication of selected varieties. Neither the first nor the fourth of these possibilities has been fully realized, though there is now some commercial clonal multiplication of a few ornamental plants by tissue-culture methods; moreover, present-day methods of tissue culture are not yet commercially practicable for woody plants. The second and third possibilities numbered above, however, have already been turned to account in commercial agriculture. Haploid plants may be developed from cultures of anthers or microspores, and protoplasts cultured from one plant may be fused with those from another that is genetically different. Tissue-culture methods thus offer new opportunities for the genetic improvement of crops.

Of perhaps the greatest immediate interest to plant pathologists is the propagation of pathogen-free stock obtained from diseased "mother plants." It is possible to obtain pathogen-free tissues for propagation from systemically infected plants if they are not uniformly invaded by the infecting pathogens. Commercial growers of chrysanthemums practice propagation by the culture of ordinary tip cuttings to obtain plants free of the vascular-wilt pathogen, *Verticillium albo-atrum,* which usually does not invade the vascular tissues near the tips of shoots (Dimock, 1943). Also, virus-free cuttings and budwood can often be obtained from diseased plants following heat therapy. Perhaps the greatest use of tissue culture in the control of plant diseases has been in the field of plant virology. Ever since the pioneering work of Morel and Martin (1952), researchers have contributed to our knowledge of growing virus-free plants from diseased ones by techniques variously termed "meristem culture" (Quak, 1957), "tip culture" (Stone, 1963), or "meristem-tip culture" (Hollings, 1965). By whatever name they are called, these techniques consist of excising apical meristem tissues a few millimeters long, growing them aseptically in a culture medium, and transplanting the resulting plantlets into soil, where they can grow to maturity.

The greatest value of tissue culture as a method of propagating pathogen-free plants has accrued to the plant breeders: with tissue-culture techniques, they have often been able to free their breeding lines of pathogens—particularly viruses—that have infected their propagative materials and that might have been perpetuated in their stocks inadvertently through ordinary vegetative propagation.

Resistance of Plants to Pathogens and of Pathogens to Chemicals Used for Their Control

Resistance of plants to plant pathogens is of two kinds: structural or functional resistance to penetration by the pathogen, or protoplasmic resistance of the penetrated plant tissues to infection (colonization) by the pathogen (Chapter 9). We are concerned here with protoplasmic resistance, which, in epidemiological terms, can be divided into two categories, horizontal resistance and vertical resistance (page 156).

The concepts of horizontal and vertical resistance were derived mathematically from studies of plant epidemiology (van der Plank, 1963, 1968). For example, if we were to prepare a bar graph showing the degree of resistance of a crop-plant variety on the ordinate and the several races of a pathogen on the abscissa, vertical resistance would be represented by long bars above certain of the races, indicating almost absolute resistance to those races. Horizontal resistance would be represented by a series of bars of various lengths above all races of the pathogen. Horizontal resistance thus amounts to a generalized resistance to a pathogen expressed not in absolute terms but in varying degrees of resistance to the various races of the pathogen. Most crop-plant varieties have at least some horizontal resistance to most plant pathogens because repeated selection—saving seed, for example, from only the "best" plants—has preserved many genes that confer this kind of resistance. When the plant breeder improves such a variety by incorporating into it a gene (or a few genes) conferring vertical resistance, the newly developed variety may then possess almost absolute resistance to one or more races of the pathogen in addition to whatever degree of horizontal resistance it may have had initially. Vertical and horizontal resistance of a hypothetical crop-plant variety to a hypothetical pathogen are depicted in Figure 24-5.

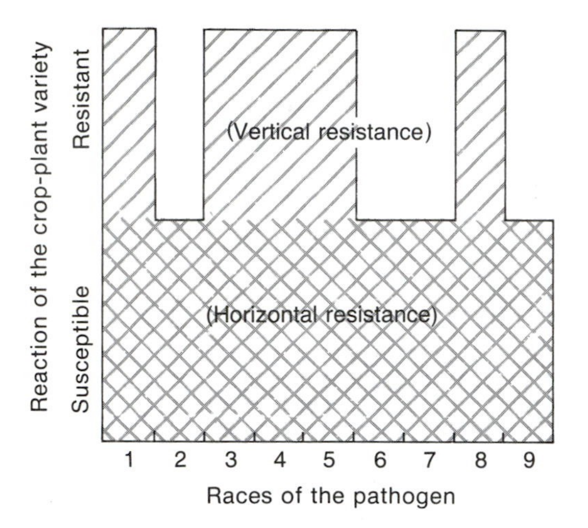

FIGURE 24-5
Diagram depicting vertical and horizontal types of disease resistance. The hypothetical crop-plant variety is horizontally resistant to all races of the pathogen, vertically resistant to races 1, 3, 4, 5, and 8.

Vertically resistant varieties are relatively easy to develop, assuming a suitable source of the resistance. Resistant progenies from controlled crosses are easy to select because they are clearly resistant—often hypersensitive (page 125)—whereas susceptible progenies are clearly susceptible. Because vertical resistance to a particular pathogen is conferred by only one gene or a very few genes, it is easy to achieve; but vertical resistance, however valuable it may be, has one innate weakness with which the plant breeder must always contend: Vertical resistance may break down, often very quickly, as races of the pathogen to which the plant is not resistant increase in number.

Vertical resistance breaks down because of the severe pressure that a resistant variety exerts on the population of a pathogenic species. If a population of an obligately parasitic pathogen encounters a population of vertically resistant plants, the pathogen becomes an "endangered species," because it is deprived of its source of sustenance. The pathogenic species would become extinct, were it not for one important ability: that to exist in a variety of genetic forms. In nature, of course, long-term survival of a pathogenic species is not threatened by a new vertically resistant variety of a suscept species because, among the common strains of the pathogen, there are usually a small number of individual organisms that, because of genetic differences, are unaffected by the resistance that has thwarted those of the common strains. Freed by the pressure of vertical resistance from the competition of its now effectively nonpathogenic sister strains, the virulent strain increases in number until finally, within a few years, it causes widespread disease in the new variety. This breakdown of vertical resistance is to be expected, because a difference in a single gene in a strain of the pathogen species may make the difference between a virulent strain and an ineffective one. Horizontal resistance, which is conferred by many genes, does not break down similarly. It may "erode" (Robinson, 1969), if a few of the genes contributing to horizontal resistance are somehow lost in later generations. In effect, horizontal resistance treats all races of the pathogen about the same, permitting the common or "wild-type" strains to maintain their relative positions of prominence in the population.

The increased resistance of parasites to selective chemicals used for their control is strikingly similar to the increased numbers of virulent strains of some pathogens in the presence of plants with vertical resistance. Familiar examples of the resistance of certain strains of organisms to chemicals include resistance of bacteria to antibiotics and the resistance of insects to certain insecticidal chemicals. Only recently has the problem of resistance of plant-pathogenic fungi to fungicides become serious, and this problem sprang from the solution of another. The solution to the problem of curing plants of certain diseases by chemotherapy entailed the use of potent chemicals with highly specific modes of action. Erwin (1973) stated the problem this way: "As fungicides become more specific, one would expect fungi to become resistant, as biochemical specificity or selectivity probably is due to interruption of only one genetically controlled event in the metabolism of the fungus. Thus, either by single-gene mutation or by selection of resistant individuals in a population, a resistant population could arise quickly." Whereas the earlier fungicides (copper- and sulfur-containing compounds, the dithiocarbamates, and captan) possess generalized protoplasmic toxicity, the antibiotics and such systemic fungicides as benomyl possess very specific metabolic toxicities. Consequently, little or no selection pressure for resistant strains was exerted by the earlier fungicides, and great selection pressure is exerted by the newer systemic ones.

It seems, then, that concepts of disease resistance in plants are applicable, at least in an epidemiological sense, to the concepts of chemical control of plant diseases. Zadoks (1974) has stated the relations between these two classes of concepts in terms of a simile: "Some systemics happen to exert a strong selective pressure on fungal populations, just like different genes for [vertical] resistance do."

A REEXAMINATION OF THE PRINCIPLES OF PLANT-DISEASE CONTROL

Earlier (on pages 14 and 15, and in Chapter 9), we presented four principles of plant-disease control. In this section, we expand that number to six, accepting new categories to accommodate control measures developed (or now more thoroughly understood) as a consequence of recent advances in plant pathology.

Epidemiology as the Basis for Plant-Disease Control

Epidemiology (Chapter 8) is the science of epidemics, which are widespread outbreaks of disease. The population of a pathogen explodes, as it were, and sometimes the entire population of a suscept is destroyed by the disease that pathogen causes. An epidemic in a crop plant, however, becomes a reality only if all of the following six conditions are fulfilled: (1) there must be an abundance of crop plants, and (2) the varieties grown must be susceptible; (3) there must be an abundance of the species of pathogen, and (4) the predominating strains of the pathogen must be virulent in the suscept species; (5) the environment must be favorable to disease development, and (6) the favorable environment must persist throughout most of the growing season (pages 10, 11, 132, and 133). The plant pathologist, in quest of his goal of preventing epidemics, seeks to make one or more of these six factors limiting for disease development. He thus bases his practice of disease-control principles on the science of epidemiology. "One may control disease in two general ways: one may either reduce the initial inoculum from which an epidemic starts or reduce the rate at which infection builds up during the epidemic. One may, of course, also do both" (van der Plank, 1968).

The Principles of Plant-Disease Control

Expansion of the four classical principles presented earlier (pages 14 and 15) seems warranted because of recent emphasis on the

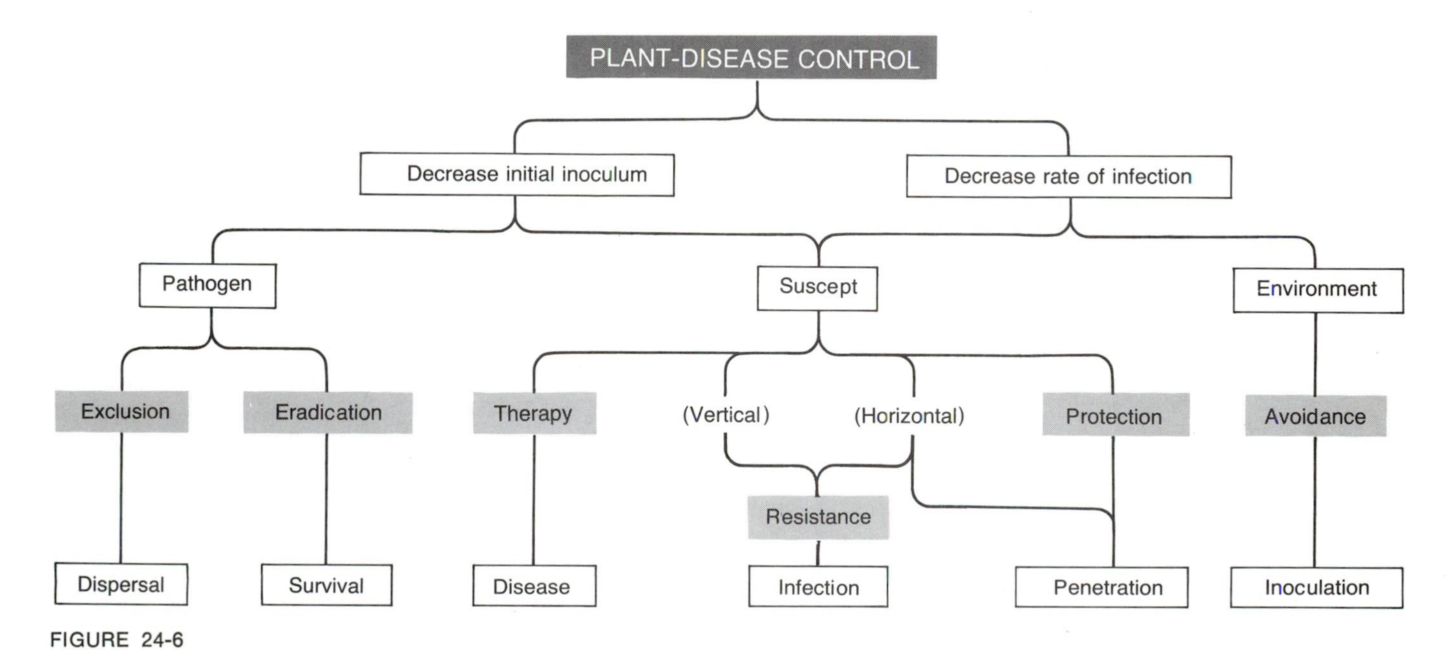

FIGURE 24-6
Interrelations among the principles of plant-disease control, epidemiological approaches, the objects or "targets" of disease-control measures, and events in the life cycles of plant pathogens interrupted by the control measures.

science of epidemiology and in view of the different "targets" of disease-control efforts (National Academy of Sciences, 1968; G. C. Kent, 1971, personal communication). Accordingly, curative measures involving the application of heat or chemicals, formerly labeled eradicative, are considered here under a new principle, **therapy.** Also, cultural practices once considered protective are considered under a new category, **avoidance.** Both "new" principles are unique: therapy because it is curative rather than preventive (as are all other disease-control principles), and avoidance because it results from something done, not to the pathogen and not so much to or with the suscept, but to and with the environment. For example, late planting of corn in warm soils results in the avoidance of *Fusarium*-induced seedling blight, which is severe only in cool, moist soils. As van der Plank (1968) points out, the development of disease resistance depends upon two very different phenomena, vertical and horizontal resistance, each of which we treat here as a subdivision of "development of resistance." Vertical resistance has the effect of reducing inoculum, whereas horizontal resistance has the effect of slowing the rate of production of inoculum by populations of the pathogen in fields planted with susceptible crops.

The six principles that characterize the modern concept of plant pathology and plant-disease control are now termed: **exclusion, eradication, therapy, development of resistance, protection,** and **avoidance.** They can be viewed from three standpoints: (1) whether they involve reduction in the initial inoculum or the rate of disease development, (2) whether they primarily involve control of the population of the pathogen, the cure or defense of the suscept, or involve the environment as it relates to disease, and (3) whether they involve interruption of dispersal, survival, inoculation, penetration, infection, or the actual course of disease. These interrelations are illustrated in Figure 24-6.

THE PLACE OF PLANT PATHOLOGY IN AGRICULTURE

An abundance—even a sufficiency—of food, fiber, and shelter from the plants that man manages depends not only on soil fertility, good tillage, and favorable weather, but also on the control of the natural enemies of crop plants, those noxious organisms collectively termed **pests.** In the so-called developed countries, plant pests (vermin, insects, weeds, and plant pathogens) probably take at least a third of our plant products before harvest (Bierne, 1967). Who knows what the average annual losses amount to in the less developed countries?

More important, perhaps, than average yearly losses are the potential losses that could be suffered if pest-control measures were ineffective. For example, average losses to all plant diseases in the United States may not amount to more than about 15% of the crops affected, but southern leaf blight of corn alone took that much of the corn crop in 1970 (see Boxes 1-1 and 8-1). Moreover, many plant pathogens presently limited in their geographic distribution would threaten crop production elsewhere if exclusionary practices were to fail (Thurston, 1973); and the introduction of new crop-plant varieties into one region from another may expose those varieties to pathogens and other pests to which they had never before been exposed (see Box 9-2). Nevertheless, the wheat varieties of "the Green Revolution," which were bred and selected in Mexico, have performed well in Asia and Africa for almost a decade (Saari and Wilcoxson, 1974). True, many of these varieties have proved susceptible to certain races of rust fungi found in Asia and Africa, but this is to be expected, for their vertical resistance to races of those obligately parasitic pathogens exerted a strong selection pressure for "new" and virulent races. Rust resistance is now being incorporated into "Green Revolution" wheats by means of continuing wheat-breeding programs in India, Pakistan, Turkey, and Egypt (Saari and Wilcoxson, 1974). Such programs are commendable because they constitute a significant step toward the solution of one of the problems of world hunger. In the long run, most solutions will be found in the less developed countries themselves, where new varieties and new pest-control procedures must be adapted to local conditions. Although the principles of plant-pest control are universal in scope, the application of those principles must vary from region to region and from time to time.

The science of plant pathology, we believe, is presently placing increased emphasis upon the relation of the environment to pathogen–suscept interactions in plant disease. With epidemiology as its foundation, and knowledge of population dynamics—the rise and fall of (pathogen and suscept) populations with time—as its central theme, plant pathology plays a major role in the realm of applied ecology. Plant pathologists, now capable of treating multiple variables through computer technology, can begin to deal more scientifically and more effectively with the numerous factors affecting plant-disease development. Until recently, the artful farmer has had to deal with many of these factors by intuition.

Accuracy in disease forecasting is enhanced as scientists begin to study the interrelations between such diverse factors as weather, innate varietal susceptibility, and the production, release, and dispersal of inoculum. For example, Berger (1973a) showed that counting the number of spores of *Cercospora apii* trapped daily and keeping track of weather conditions can permit accurate forecasting of early blight of celery. Recommendations for spray applications can be adjusted so that sprays can be applied only when needed. Berger (1973b) has also worked out a practical basis for forecasting epidemics of the northern leaf blight of corn, which is caused by *Helminthosporium turcicum*. As with early blight of celery, knowledge of spore "flights," as well as knowledge of favorable environmental conditions, is essential to useful forecasting. Berger found that sprays of maneb applied on the days of peak spore occurrences are effective, but those applied only one day earlier or one day later are not.

Perhaps the most sophisticated approaches to understanding the multifaceted aspects of a plant-disease epidemic are those using computer-simulation models of the epidemic. "EPIDEM" (Waggoner and Horsfall, 1969), "EPIMAY" (Waggoner et al., 1972), and "BLITECAST" (Krause et al., 1975) are well-known examples. The computer simulation takes into account all the discernible factors—qualitative and quantitative—that relate to pathogen, suscept, and environment, as well as those that determine disease severity over a period of time. It also provides the basis for predicting the influence of changing weather on the future course of the epidemic (Waggoner and Horsfall, 1969).

Finally, we believe that the science of plant pathology has "come of age" and stands ready to contribute its fair share to world crop production. Related sciences have also made significant advances, and the time is right for a thorough integration of our knowledge. Now that man can cope scientifically with a multitude of variables at one time, there is every reason to believe he can accomplish this monumental task. The plant pathologist, it seems, should first make whatever contributions he can to the complete integration of pest-control procedures. Then, with his colleagues in other pest-control sciences and in the horticultural sciences, he can work toward the integration of pest-control procedures into the broad area of crop-plant management and production. The plant pathologist can thereby assist in protecting crop plants against the constant dangers mentioned at the beginning of Chapter 1: "the buffeting of the elements, competition from weeds, plagues of insects, and the ravages of disease."

Selected References

Anonymous, 1970. Editorial note. *Neth. J. Plant Pathol.* 76:227–228.

Beirne, B. P., 1967. *Pest Management.* Cleveland, Ohio: CRC Press.

Berger, R. D., 1973a. *Helminthosporium turcicum* lesion numbers related to numbers of trapped spores and fungicide sprays. *Phytopathology* 63:930–933.

———, 1973b. Early blight of celery: analysis of disease spread in Florida. *Phytopathology* 63:1161–1165.

Brandes, J., and R. Berks, 1965. Gross morphology and serology as a basis for classification of elongated plant viruses. *Advan. Virus Res.* 11:1–24.

———, and C. Wetter, 1959. Classification of elongated plant viruses on the basis of particle morphology. *Virology* 8:99–115.

Breed, R. S., E. G. D. Murray, and N. R. Smith, eds., 1957. *Bergey's Manual of Determinative Bacteriology* (7th ed.). Baltimore: Williams & Wilkins.

Buchanan, R. E., and N. E. Gibbons, eds., 1974. *Bergey's Manual of Determinative Bacteriology* (8th ed.). Baltimore: Williams & Wilkins.

Diener, T. O., 1972. Viroids. *Advan. Virus Res.* 17:295–313.

———, 1973. Virus terminology and the viroid: a rebuttal. *Phytopathology* 63:1328–1329.

Dimock, A. W., 1943. Coming up to date on *Verticillium* wilt and *Septoria* leafspot of chrysanthemums. *Chrysanthemum Soc. Amer. Bull.* 11:3–10.

Dodds, J. A., and R. I. Hamilton, 1971. Evidence for possible genomic masking between two unrelated plant viruses. *Phytopathology* 61:889–890.

Erwin, D. C., 1973. Systemic fungicides: disease control, translocation, and mode of action. *Annu. Rev. Phytopathol.* 11:389–442.

Gibbs, A. J., B. D. Harrison, D. H. Watson, and P. Wildy, 1966. What's in a virus name? *Nature* 209:450–454.

Giannotti, J., C. Vago, G. Marcoux, G. Devauchelle, and J. L. Duthoit, 1970. Infection de plante par un type inhabituel de microorganisme intracellulaire. *C. R. Hebd. Seances Acad. Sci.* (D)271:2118–2119.

Goheen, A. C., G. Nyland, and S. K. Lowe, 1973. Association of a rickettsialike organism with Pierce's disease of grapevines and alfalfa dwarf and heat therapy of the disease in grapevines. *Phytopathology* 63:341–345.

Hamilton, R. I., 1974. Replication of plant viruses. *Annu. Rev. Phytopathol.* 12:223–245.

Harrison, B. D., J. T. Finch, A. J. Gibbs, M. Hollings, R. J. Shepherd, V. Valenta, and C. Wetter, 1971. Sixteen groups of plant viruses. *Virology* 45:356–363.

Hollings, M., 1965. Disease control through virus-free stock. *Annu. Rev. Phytopathol.* 3:367–396.

Hopkins, D. L., and H. H. Mollenhauer, 1973. Rickettsia-like bacterium associated with Pierce's disease of grapes. *Science* 179:289–300.

Hopkins, J. C. F., ed., 1957. Common names of plant-virus diseases used in the *Review of Applied Mycology*. *Rev. Appl. Mycol.* 35(Suppl.).

Krause, R. A., L. B. Massie, and R. A. Hyre, 1975. BLITECAST: A computerized forecast of potato late blight. *Plant Dis. Rep.* 59:95–98.

Markham, P. G., R. Townsend, M. Bar-Joseph, M. J. Daniels, Audrey Plaskitt, and Brenda M. Meddins, 1974. Spiroplasmas as the causal agents of citrus little-leaf disease. *Ann. Appl. Biol.* 78:49–57.

Martyn, E. B., 1968. *Plant Virus Names*. Kew: Commonwealth Mycological Institute.

Matthews, R. E. F., 1970. *Plant Virology*. New York: Academic Press.

McKinney, H. H., 1973. Comments on virus terminology and the "viroid." *Phytopathology* 63:438.

Morel, G., and C. Martin, 1952. Guérison de dahlias atteints d'une maladie à virus. *C. R. Hebd. Seances Acad. Sci.* (D)235:1324–1325.

Murashige, T., 1974. Plant propagation through tissue cultures. *Annu. Rev. Plant Physiol.* 25:135–166.

National Academy of Sciences, 1968. *Plant Disease Development and Control* (Nat. Acad. Sci. Publ. 1596). Washington, D. C.: National Academy of Sciences.

Quak, F., 1957. Meristeem cultuur, gecombineerd met warmtebehandeling vor het verkrijgen van virus-vrije anjerplanten. *Tijdschr. Plantenziekten* 63:13–14.

Robinson, R. A., 1969. Disease resistance terminology. *Rev. Appl. Mycol.* 48:593–606.

Rochow, W. F., 1969. Biological properties of four isolates of barley yellow dwarf virus. *Phytopathology* 59:1580–1589.

———, 1970. Barley yellow dwarf virus: phenotypic mixing and vector specificity. *Science* 167:875–878.

———, 1972. The role of mixed infections in the transmission of plant viruses by aphids. *Annu. Rev. Phytopathol.* 10:101–124.

Saari, E. E., and R. D. Wilcoxson, 1974. Plant disease situation of high-yielding dwarf wheats in Asia and Africa. *Annu. Rev. Phytopathol.* 12:49–68.

Saglio, P., M. L'Hospital, D. La Fléche, G. DuPont, J. M. Bové, J. G. Tully, and E. A. Freundt, 1973. *Spiroplasma citri* gen. and sp. n.: mycoplasma-like organism associated with "stubborn" disease of citrus. *Int. J. Syst. Bacteriol.* 23:191-204.

Stone, O. M., 1963. Factors affecting the growth of carnation plants from shoot apices. *Ann. Appl. Biol.* 52:199–209.

Thurston, H. D., 1973. Threatening plant diseases. *Annu. Rev. Phytopathol.* 11:27–52.

van der Plank, J. E., 1963. *Plant Diseases: Epidemics and Control.* New York: Academic Press.

———, 1968. *Disease Resistance in Plants.* New York: Academic Press.

van Kammen, A., 1972. Plant viruses with a divided genome. *Annu. Rev. Phytopathol.* 10:125–150.

Waggoner, P. E., and J. G. Horsfall, 1969. EPIDEM. A simulator of plant disease written for a computer. *Conn. Agr. Expt. Sta. Bull.* (698).

———, ———, and R. J. Lukens, 1972. EPIMAY. A simulator of southern corn leaf blight. *Conn. Agr. Expt. Sta. Bull.* (729).

Zadoks, J. C., 1974. The role of epidemiology in modern phytopathology. *Phytopathology* 64:918–923.

APPENDIX 1

Techniques for Diagnosis of Plant Diseases

Accurate diagnosis of plant disease depends upon Koch's postulates or rules of proof of pathogenicity (see Chapter 1). Only the most rudimentary techniques basic to the study of plant diseases are described below; more complicated methods are discussed in standard works (see Selected References at the end of this chapter). Special techniques are described in articles published in such periodicals as *The American Journal of Botany, Annals of Applied Biology, Transactions of the British Mycological Society,* and *Phytopathology.*

DETECTING PATHOGENS ASSOCIATED WITH DISEASES

Constant association of a microorganism with a given symptom picture is the first evidence for the pathogenicity of that organism. Detection of pathogenic bacteria, fungi, and nematodes requires both the use of optical equipment and, quite often, special methods of preparing the material for viewing.

Optical Equipment

The Hand Lens. This instrument is an ordinary magnifying glass or simple microscope. Signs of plant pathogens, including bacterial ooze, asexual and sexual fruiting bodies of fungi, some fungal spores, and certain structures of nematodes, may be detected by observation through a hand lens that magnifies 10–20 diameters. The lenses most useful in field diagnoses are 10 ×, 12 ×, or 15 ×. Although it is possible to detect the presence of signs through proper use of the hand lens, details of their structure cannot be viewed with such low magnification.

The Monobjective Microscope. Whereas the hand lens is a simple microscope, the monobjective instrument is a compound microscope, having two separate lens systems: The objective lens system near the specimen magnifies it a definite amount and the ocular lens system in the eyepiece magnifies the subject image projected by the objective lenses. Thus, the total magnification of the specimen is the product of the magnifications provided by the two lens systems. For most plant-pathology work with the monobjective microscope, 10 × oculars are used with 10 × or 44 × objectives. When bacteria or fine structures of nematodes are to be viewed, an oil-immersion objective (approximately 97 ×) is used.

The Stereoscopic Dissecting Microscope. This instrument is actually two compound microscopes combined so that there is one for each eye. The nosepiece carries paired objectives inclined at an angle of approximately 12°. Thus, the specimen is viewed from separate angles by each eye, and stereoscopic perception of depth is achieved. With paired objectives, however, resolution is not increased at very high magnification, and stereoscopic microscopes provide workable magnifications only up to approximately 144 diameters. Usually, the nosepiece carries three pairs of objectives; some modern stereoscopic microscopes, however, provide for viewing at any magnification within a wide range.

Stereoscopic microscopes are used in plant pathology for moderately close observation of diseased specimens. Because the image is not inverted, the stereoscopic dissecting model is particularly valuable for dissections. For example, fungal structures to be viewed later at higher magnifications can be dissected from diseased tissue under the stereoscopic microscope. Also, single nematodes can be picked out readily from specimens with the aid of this microscope.

Preparation of Plant Pathogens for Microscopy

Bacteria. Bits of diseased plant tissue containing bacteria are chopped thoroughly in a drop of water on a clean microscope slide. After about 5 minutes has been allowed for the diffusion of bacterial cells into the water, the pieces of tissue are scraped away and the bacterial suspension in water is spread thinly over the surface of the slide. The water is allowed to evaporate, and the bacteria are fixed to the slide with gentle heating. A drop of 1% methylene blue in ethanol is applied and allowed to stand for a few minutes, and the excess stain is washed away with water. After the rinse water has evaporated, the stained bacteria can be viewed with the microscope, using the oil-immersion objective. They appear blue on a colorless background.

Bacteria can also be viewed after a background stain has been applied. Tissue containing bacteria is chopped in a drop of 2% Congo red in water. The stain and bacteria are smeared thinly over the slide after bits of tissue have been removed, and the bacteria are fixed to the air-dried slide with gentle heating. The slide is then flooded with 95% ethanol to which has been added concentrated hydrochloric acid (1 drop/10 ml); this makes the red stain turn blue. After the alcohol has evaporated, the mount can be viewed with the oil-immersion objective. The bacteria appear colorless on a blue background.

Fungi. The media in which fungal structures are most often mounted on microscope slides include water, mineral oil, and a glycerine–water solution. Plain tap or distilled water is the most convenient and useful medium for making temporary mounts. Water for microscope mounts should be kept free of all debris; if kept in a dropping bottle, it should be renewed frequently and the bottle and its dropping pipette should be washed and rinsed each time the water is renewed. A drop of water may be deposited on the microscope slide with a pipette, glass rod, looped transfer wire, or with the tip of a finger. Mineral oil is useful when living conidiophores and their conidia are to be viewed *in situ* after having been plucked from the substrate with needle or forceps. A solution of glycerine in water (50% by volume) may be used when the mount is to be kept for several days.

Such fungal structures as conidiophores and conidia and, occasionally, other fruiting bodies and mycelium can be picked from diseased tissues with forceps or can be scraped off the surface with a scalpel. Such structures can be deposited in a drop of the mounting medium on a slide, covered with a cover glass, and viewed immediately. If the specimen to be examined has been dried, it is best to soak the tissue in 2% potassium hydroxide before the fungal tissues are scraped off. Excess potassium hydroxide solution should be blotted from the tissue before it is scraped.

Preliminary microscopic studies of fungal structures that are imbedded in diseased plant tissues can often be made most satisfactorily from crushed mounts, which are prepared as follows: A drop of mounting medium is placed on a clean slide, and a small piece of tissue containing the fungal structure (pycnidium, for example, or perithecium) is submerged in the drop. The tissue is then chopped with a scalpel and crushed with the flat side of the instrument. The triturated tissue in the mounting medium is then covered with a cover slip and viewed.

Scrapings and crushed mounts result in torn and distorted fungal structures and suscept tissues. Thin sections through diseased tissues, however, enable the plant pathologist to view fungi as they exist in diseased tissues. The procedures required for preparation of permanent mounts of microtome sections are too laborious and time-consuming to be of use in routine examinations; instead, thin sections are cut freehand with a sharp razor and are mounted directly for viewing. With practice, any person of average dexterity can learn to cut excellent freehand sections.

If thin leaves or slender stems are to be sectioned, the material must be supported for sectioning. A piece of dry or alcohol-soaked pith stick is split to a depth of about 2 cm and the piece of tissue to be sectioned is inserted into the slit and oriented with a portion protruding. The end of the pith stick is then squared off with the razor. To cut the sections, hold the pith firmly between the thumb and index finger, rest the flat side of the razor on the finger and slice through the pith and plant material. Move the razor diagonally across the pith, cutting the sections as thinly as possible. After several sections have built up on the top side of the razor, wipe them off into water in a watch glass, and select the thinnest sections for viewing. These can be lifted with a needle or forceps into the mounting fluid on a slide. Firm tissues, such as stems, can be cut without the pith support.

A saturated aqueous solution of chloral hydrate is often used as a mounting medium for crushed mounts or for freehand sections; fungal hyphae stand out sharply in the cleared plant tissue. When large leaf lesions are to be viewed to detect the distribution of mycelium and fruiting bodies *in toto* and *in situ,* the lesions can be soaked 1–7 days in chloral hydrate or in 80% ethanol to remove the chlorophyll, rinsed in water, soaked 1–7 days in 5% potassium hydroxide, rinsed, and mounted in glycerine or chloral hydrate. Soaking tissues in potassium hydroxide solution (2–10%) removes brown oxidation products in necrotic lesions, making imbedded fungal structures readily visible.

Several stains are satisfactory for staining hyaline mycelium and other fungal structures—notably, either 1% cotton blue or 1% acid fuchsin (red). Specimens may be crushed or immersed in a drop of stain on a slide, covered, and viewed immediately. It is best to place freehand sections or bits of tissue in the stain in a watch glass, and, after a few minutes, to rinse the tissues in water to remove excess stain; stained and rinsed sections can then be mounted in glycerine or another medium.

Nematodes. After plant tissues containing living endoparasitic nematodes have been kept for a while in water, the nematodes will migrate from the diseased tissues into the water, from which they can be lifted singly and prepared for viewing with the microscope. Some nematodes —*Meloidogyne* spp., for example—can be viewed *in situ* with the stereoscopic microscope; swollen females can be easily teased out of the tissue with dissecting needles, after which they can be mounted on slides.

A fine splinter of bamboo, carefully sharpened with a razor while being viewed with the stereoscopic dissecting microscope, is used to lift individual vermiform nematodes from water and to transfer them to a drop of water on a slide. The stereoscopic microscope is used to locate an individual and to trap it on the sharp, somewhat rough point of the splinter. After the nematode has been transferred to water on a slide, it is killed by gentle heating. The nematode is ready for viewing after a cover glass has been applied. A few strands of glass wool should be mounted under the cover glass to keep it from crushing the nematode.

PURE-CULTURE METHODS AND TECHNIQUES FOR ISOLATING PLANT PATHOGENS

Culture Media

Koch's second rule of proof of pathogenicity—isolation and growth of the suspected pathogen in pure culture—is applicable only to bacteria and fungi, because plant-pathogenic viruses and nematodes are obligately parasitic. Bacteria and fungi can be grown in a wide variety of media, both solid and liquid. Only several of the familiar media will be described here.

Potato–Dextrose Agar. The most commonly used solid medium for culturing phytopathogenic fungi and many bacteria is potato-dextrose agar (PDA). This medium is made with 200–500 g potatoes, 15–20 g dextrose (D-glucose), 15–20 g agar, and sufficient distilled water to make one liter of medium. The potatoes are washed, peeled, sliced, and cooked in 500 ml water until soft; the potatoes can be either boiled in a pan over a burner or prepared in a steamer or an autoclave—in any case, the cooking time is approximately 30 minutes. The agar is melted in 500 ml distilled water. The water in which the potatoes were cooked is strained through cheesecloth into the melted agar, and the glucose is added and dissolved. The volume of the medium is then increased to one liter with distilled water, and the hot medium is filtered through cheesecloth into tubes or flasks. It is most convenient to pour the medium into a large funnel supported on a ring stand; a short length of rubber tubing attached to the funnel and equipped with pinch clamp is used as a delivery tube. Filled flasks and tubes are stoppered with plugging cotton and autoclaved at 120°C and 15 lb/in^2 for 20–30 minutes. Sterilized medium in flasks can be poured into petri dishes; slants may be made in the tubes by allowing the medium to cool and solidify while the tubes are kept in a tilted position.

Water Agar. Water agar is made by dissolving 15–20 g of agar in sufficient tap or distilled water to make one liter of medium. This nonnutrient medium is useful both for isolating such fungi as *Pythium* spp. and for obtaining hyphal-tip transfers. Hyphae grow in length on this medium, but they do not branch as profusely as they do on PDA.

Nutrient Agar. Many plant-pathogenic bacteria grow better in nutrient agar than in PDA. The formula is 3 g beef extract, 10 g peptone, and 15–20 g agar, all dissolved in sufficient distilled water to make one liter of medium.

Isolating Plant Pathogens

Although plant pathogens are often isolated from diseased tissues, some of them can be isolated from the soil. Whatever the substance from which a plant pathogen is to be isolated, it is essential that the pathogen is placed in a new environment that favors its development over that of any one of its saprophytic competitors; this is usually more difficult to accomplish with soil-infesting pathogens than with those infecting the tissues of living plants.

Isolating Pathogens from the Soil. Only three of the many techniques described for isolating plant pathogens from soil will be mentioned here, namely (1) the direct-plating method, (2) the dilution method, and (3) the trapping method.

1) In the direct-plating method, sterilized culture medium is melted over boiling water and—after the mouth of the container has been flamed—poured into sterilized petri dishes, where it is allowed to harden. Pinches of soil suspected of harboring the pathogen are placed on the medium with the tip of an alcohol-flamed spatula or scalpel. The inoculated medium is then incubated and observed at intervals. After colonies of bacteria and fungi have grown from the sample of soil, they can be cut out of the agar with a flamed needle and transferred individually to other medium-filled petri dishes or to test-tube slants. Also, microscope mounts can be made for the purpose of identifying the organisms observed.

Original platings of soil are usually made on a nonnutrient medium, such as water agar: fungal growth is less luxuriant on water agar than on PDA, and it is easier to transfer parts of individual colonies from the former than from the latter. When the pathogen is a fungus, the growth of contaminating bacteria can be retarded by adding one or two drops of 50% lactic acid to each petri dish just before the melted agar medium is poured into it. Fungi are usually much more tolerant of an acid substrate than are bacteria. When the pathogen to be isolated is a bacterium, nutrient agar, PDA, or another special medium will produce more satisfactory results than water agar.

2) In the dilution method, soil suspected of being infested with the pathogen is thoroughly mixed with sterilized water and allowed to settle. One volume of the supernatant is then mixed with nine volumes of water, giving a 1/10 dilution. Several additional serial dilutions are made, and a small volume of each is added to a sterilized petri dish. Then melted agar medium that has been cooled to approximately 37°C is poured into the dishes. Colonies of microorganisms can be picked out individually a day or two after the plates have been inoculated.

3) In the trapping method, seeds are planted—or plant parts are placed—in soil infested with the pathogen. Seeds planted in soil infested with damping-off fungi will give rise to seedlings that are

diseased; the pathogen can then be isolated from the diseased seedlings, as described in the next section. Phycomycetous pathogens in soil can be trapped in green apples, and the pathogen can then be isolated from the apple tissue: A small amount of soil is placed in the bottom of a moist chamber (a preparation dish, for example); a green apple, surface-sterilized with 2% mercuric chloride or with 15% commercial bleach, is punctured with a flamed needle to a depth of 1–2 inches several times at the calyx end; the punctured end is pressed firmly into the soil; and the moist chamber is kept closed for several days. Such pathogens as *Pythium debaryanum* and certain species of *Phytophthora* will grow into the flesh of the fruit at the wounds, and will cause a rot. After the apple has been opened, bits of diseased tissue can be transferred with a needle directly to hard PDA or water agar in petri dishes.

Thielaviopsis basicola, the cause of black root-rot of tobacco and other plants, can be trapped in discs of carrot root tissue by a technique described by Yarwood (1946). Carrot discs are placed on moist filter paper in a moist chamber (in this case, a petri dish), and bits of infested soil are placed on the discs. After 24 hours, the soil is washed from the discs, which are then replaced in the moist chamber. Within a few days, the discs will be covered with masses of black chlamydospores and with endoconidia of the fungus. These can be transferred aseptically to an agar medium. Carrot discs can also be used to trap the crown-gall bacterium from infested soil.

Isolating Pathogens from Diseased Plant Tissues. Tissues sampled during the active stage of an infection are likely to have within them only the pathogen responsible for the infection; the surfaces of such tissues, however, are usually contaminated with saprophytic organisms. Thus, surface disinfestation of tissues selected for isolation work is prerequisite to successful isolation. Many surface contaminants can be removed by washing diseased tissues with soap or detergent and then rinsing them for 1–2 hours in running tap water. Tender tissues, such as seedling stems containing damping-off fungi, are cut into small pieces and planted directly on acidified PDA immediately after they have been rinsed in running tap water.

Additional surface disinfestation of less tender tissues is usually accomplished with 0.5–1.0% calcium hypochlorite (10–20% commercial bleach) or with a 0.1% aqueous solution of mercuric chloride. Tissues treated with bleach can be planted directly on PDA; those treated with mercuric chloride, however, must be rinsed in sterilized water before they are planted, lest residual mercury ions inhibit the growth of the pathogen.

To isolate plant-pathogenic fungi by the tissue-plating method, the following procedure is recommended: Pieces of root, stem, leaf, and other diseased tissue showing distinct symptoms are washed and placed in a beaker, which is then covered with cheesecloth and left in running water for 1–2 hours. Squares of tissue approximately 0.5 cm across are cut from the advancing margins of lesions placed in 15% commercial bleach; after 5–10 minutes, or when green tissues begin to bleach, the small pieces of tissue are lifted with alcohol-flamed forceps onto hardened agar medium in petri dishes. Later, the predominating organism can be transferred to fresh agar slants for growth in pure culture.

When sclerotia are associated with diseased tissue, they can be surface-sterilized, cut in half, and planted on solidified PDA.

Bacteria (and fungi that sporulate in or on diseased tissue) can be isolated by a dilution method: Bits of diseased tissue are chopped with a flamed scalpel in a few drops of sterilized water; after a few minutes, a loopfull of the water containing the bacteria (or spores) is transferred to a drop of sterilized water in a petri dish. Then, serial dilutions are made to several other dishes. Melted agar medium at 37°C is then poured into the dishes. After a day, colonies of bacteria (or fungi) are visible, and transfers of single colonies can be made to an appropriate medium in tubes or in other petri dishes.

INOCULATION TECHNIQUES

Soil Infestation

Bits of mycelium are often used as inoculum when the infection court is a root or seedling stem; mycelium from liquid or solid media can be chopped in water in a commercial food blender.

To test the pathogenicity of damping-off, root-rotting, or wilt-inducing fungi, a small square of PDA infested with the appropriate mycelium can be buried in soil near the center of a clay pot and seeds of the suscept can be planted around the center. Within about a week, damping-off fungi will have induced disease; more time is required by the other pathogens.

Root-invading pathogens (including bacteria, fungi, and nematodes) can be mixed with potting soil into which seeds or transplants are placed. Suspensions of bacterial cells, fungal spores, or nematodes may be used; fungal mycelium may be chopped in water in a blending machine and added to the soil; or the fungus may be grown on sterilized cereal grains, which should then be added to soil in which the suscept is to be planted.

Inoculation of Leaves and Stems with Fungi or Bacteria

Leaves and stems of susceptible plants can be inoculated with suspensions of fungal spores or bacterial cells in the following manner: Culture dishes or tubes are flooded with sterilized water to which has been added a drop of wetting agent, and the spores or bacteria are gently scraped off the substrate into the water. The suspension is then placed in an atomizer, from which it is sprayed on the susceptible plants. When a fungus is the inoculum, the sprayed plants should be kept in a moist atmosphere for 12–48 hours. With leaf-attacking bacteria, however, best results are usually experienced when the plants are inoculated in the greenhouse, preferably between 9:00 AM and 3:00 PM, when their stomata are most likely to be open.

When dry spores, such as those of cereal rusts, can be collected *en masse*, dry inoculum can be prepared by mixing one part of the spores with three or four parts of talc. After plants to be inoculated have been sprayed with water containing a little wetting agent, the spore–talc mixture is dusted onto the plants. Inoculated plants should be kept in a moist chamber for 12–24 hours.

To inoculate plants with fungi or bacteria that cause systemic infections, suspensions of fungal spores or bacterial cells can be injected with a hypodermic syringe. Toothpicks infested with wilt-inducing bacteria or fungi can be inserted into susceptible stems for successful inoculation.

Inoculations with Plant Viruses

Mechanical Inoculation. From a plant affected by a virus disease, select a leaf that shows well-developed symptoms of the disease. Break the leaf from the plant and grind it thoroughly in a sterilized mortar and pestle. The expressed juice, which will contain the infectious principle, may be used as inoculum. When the juice is to be diluted, it should first be filtered to remove the pulp; this is best accomplished by pressing the juice through a sterilized gauze pad.

Only young, rapidly growing plants should be selected for inoculation. Their leaves should be turgid, free from dirt or other extraneous material, and should have dry surfaces. Prior to inoculation, dust the leaves to be inoculated with Carborundum (400–600 mesh). Support the leaf being inoculated with the palm of one hand and apply the liquid inoculum to its upper surface with a moistened forefinger, pestle, or glass spatula, or with a saturated gauze pad. Rub the infective juice over the leaf surface with quick, gentle strokes, and then rinse the inoculated leaf with a fine spray of tap water. Keep inoculated plants in the greenhouse.

Graft Inoculation. Any plant virus may be transmitted by grafting when the donor and receiving tissues, which must be graft-compatible, are both susceptible to the virus. The best grafting method to use is the one best suited to the propagation of the host plant in question. Usually, the best results are obtained when diseased scions are grafted to healthy, vigorously growing stocks.

Inoculation by Dodder. Some plant viruses can be transmitted from diseased to healthy plants by dodder (*Cuscuta* spp.). Establish one or two strands of dodder on a young plant that shows symptoms of viral disease. After the dodder has begun to grow vigorously, trim off all but one or two of its strands, which should be brought into contact with the healthy test plant. After dodder has become established on the test plant (2–4 days), remove the leaves of that plant. If successful transmission has occurred, symptoms of the disease will appear in the new growth of the test plant.

Inoculation by Insects. Inoculations of plants by insects should be accomplished under controlled conditions. The choice of host plants, insect species, and environmental conditions will depend upon the virus to be transmitted. Many of the viruses that cause mosaic diseases are transmitted by aphids, whereas certain leafhoppers are likely to be the vectors of the viruses that cause certain "yellows" diseases. Thrips, white flies, and mealy bugs transmit some viruses. Recently, certain nematodes and one fungus have been shown to transmit particular plant viruses.

The general procedure for the controlled transmission of plant viruses by insects begins with caging healthy individuals of the appropriate insect species on a diseased plant. After a period of feeding, these individuals become viruliferous, and if caged on healthy test plants, they will feed upon them and transmit virus to them in the process.

COLLECTING AND PRESERVING DISEASED PLANT MATERIALS

Diseased plant materials can often be collected and stored for future study. While collecting in the field, care should be exercised to select specimens that include samples of each phase of disease development.

Many specimens can be kept in cold storage just above 0°C; pathogens can be isolated from such specimens after storage periods of several weeks. The material to be stored should be carefully selected in the field and wrapped in waxed paper or placed in plastic bags. These collections may then be packed in large tins or glass jars and put into storage.

Some material is stored best by pressing and drying. The selected material may be placed in the plant press in the field, or it may be kept temporarily between the pages of large magazines (or in a vasculum, or in plastic bags) and transferred to the press at the end of the day. Such material is useful in the study of pathogen structures and pathological histology.

Sometimes it is desirable to wet-preserve diseased plant material. Formalin–acetic acid–alcohol (FAA) is a good killing and fixing agent. FAA is prepared by mixing 5 ml of commercial formalin (37% formaldehyde in water) and 5 ml of glacial acetic acid with 90 ml of 50% or 70% ethanol. After treatment for at least one hour with FAA, the material can be permanently stored in 5% commercial formalin.

Demonstration specimens may be prepared by killing and coloring the appropriate plant material in a hot solution of copper acetate in acetic acid. The stock solution consists of 50% glacial acetic acid saturated with copper acetate. For preparation of demonstration specimens, 300 ml of water are added to 100 ml of the stock solution. This solution is brought to a boil; the plant material is added to the solution and allowed to simmer for a few minutes. First, the plant material will become cleared; then it will begin to take on a green color again. When the desired color has developed, the material is taken out of the solution and rinsed in water. Specimens are then stored in 5% commercial formalin in glass containers with screw tops. Materials prepared in this way may be used for demonstration purposes, or for the study of symptomatology and etiology. Colored material can be blotted and pressed for mounting on herbarium sheets.

Selected References

Altman, J., 1966. *Phytopathological Techniques Laboratory Manual.* Boulder, Colo.: Pruett.

American Phytopathological Society, Sourcebook Committee, 1967. *Source-*

book of Laboratory Exercises in Plant Pathology. San Francisco: W. H. Freeman and Company.

Commonwealth Mycological Institute, 1968. *Plant Pathologist's Pocketbook*. Kew: Commonwealth Mycological Institute.

Filipjev, I. N., and J. H. Schuurmans Stekhoven, Jr., 1941. *A Manual of Agricultural Helminthology*, pp. 168–196 (Technical methods for collection and preparation of nematodes). Leiden: E. G. Brill.

Heald, F. D., 1937. *Introduction to Plant Pathology* (1st ed.), pp. 527–548 (Plant pathology methods). New York: McGraw-Hill.

Johnson, L. F., E. A. Curl, J. H. Bond, and H. A. Fribourg, 1959. *Methods for Studying Soil Microflora–Plant Disease Relationships*. Minneapolis: Burgess.

Rawlins, T. E., 1933. *Phytopathological and Botanical Research Methods*. New York: Wiley.

———, and W. N. Takahashi, 1952. *Technics of Plant Histochemistry and Virology*. Millbrae, Calif.: National Press.

Riker, A. J., and R. S. Riker, 1936. *Introduction to Research on Plant Diseases*. Madison, Wis.: published by the authors.

Spencer Lens Company, 1935. *The Microscope, Its Construction, Use and Care*. Buffalo, N. Y.: Spencer Lens Company.

Tuite, J., 1969. *Plant Pathological Methods, Fungi and Bacteria*. Minneapolis: Burgess.

Yarwood, C. E., 1946. Isolation of *Thielaviopsis basicola* from soil by means of carrot disks. *Mycologia* 38:346–348.

APPENDIX 2

Use of the Literature of Plant Pathology

SOURCES OF LITERATURE ON PLANT DISEASES

Articles on plant diseases appear in hundreds of periodical and occasional publications of scientific societies, experiment stations, colleges, universities, and associations of growers. A thorough search of the extensive literature of plant pathology would be an almost hopeless undertaking, were it not for abstracting journals, which summarize articles and bring together the references to the literature on the subject.

Abstracting Journals

Review of Applied Mycology (renamed *Review of Plant Pathology* in 1970) is a publication of the Commonwealth Mycological Institute, which is located in Kew, Surrey, England. It has abstracted almost every important paper on plant diseases published since 1922. Abstracts in this journal are more complete than those found in other abstracting journals, and it is usually the first such journal one should consult when undertaking a review of the literature on a plant-pathological topic.

Biological Abstracts, an abstracting journal that has been published in the United States since 1922, has been issued in five sections since 1939. Section D covers the literature on plant pathology and other plant sciences. *Experiment Station Record,* published between 1889 and 1948 by the U.S. Department of Agriculture, covers much of the biological science literature published in the United States during that era. Many articles on plant pathology are also reviewed in *Chemical Abstracts. Helminthological Abstracts* reviews those articles dealing with nematodes.

Bibliographies

Bibliographies are lists of references only and do not contain abstracts of original papers. Two useful lists are *Bibliography of Agriculture,* published since 1937 by the U.S. Department of Agriculture, and *Bibliographie der Pflanzenschutzliteratur,* published since 1914.

Books

Lists of references on plant diseases appear in most books on the subject. Some useful books on plant pathology are listed below.

Agrios, George N., 1969. *Plant Pathology.* New York: Academic Press.

Anderson, H. W., 1956. *Diseases of Fruit Crops.* New York: McGraw-Hill.

Baker, K. F., and W. C. Snyder, 1964. *Ecology of Soil-borne Plant Pathogens, Prelude to Biological Control.* Berkeley and Los Angeles: University of California Press.

Barnes, E. H., 1968. *Atlas and Manual of Plant Pathology.* New York: Appleton.

deBary, A., 1884. *Vergleichende Morphologie und Biologie der Pilze, Mycetozoen und Bakterien.* Leipzig. [English transl. by H. E. F. Garnsey and I. B. Balfour, 1887. *Comparative Morphology and Biology of Fungi, Mycetozoa and Bacteria.* Oxford: The Clarendon Press.]

Bawden, F. C., 1948. *Plant Diseases.* London: Nelson.

———, 1964. *Plant Viruses and Virus Diseases* (4th ed.). New York: Ronald Press.

Baxter, D. V., 1952. *Pathology in Forest Practice* (2nd ed.). New York: Wiley.

Boyce, J. S., 1961. *Forest Pathology* (3rd ed.). New York: McGraw-Hill.

Brooks, F. T., 1953. *Plant Diseases* (2nd ed.). London: Oxford University Press.

Butler, Sir Edwin J., and S. G. Jones, 1949. *Plant Pathology.* London: Macmillan.

Carter, W., 1962. *Insects in Relation to Plant Disease.* New York: Interscience.

Chester, K. S., 1947. *Nature and Prevention of Plant Diseases.* (2nd ed.). Philadelphia: Blakiston.

Christie, J. R., 1959. *Plant Nematodes, Their Bionomics and Control.* Gainesville, Fla.: University of Florida Agricultural Experiment Station.

Chupp, C., and A. F. Sherf, 1960. *Vegetable Diseases and Their Control.* New York: Macmillan.

Commonwealth Mycological Institute, 1968. *Plant Pathologist's Pocketbook.* Kew: Commonwealth Mycological Institute.

Cook, M. T., 1947. *Viruses and Virus Diseases of Plants.* Minneapolis: Burgess.

Corbett, M. K., and H. D. Sisler, eds., 1964. *Plant Virology.* Gainesville, Fla.: University of Florida Press.

Couch, H. B., 1962. *Diseases of Turfgrasses.* New York: Reinhold.

Dickson, J. G., 1956. *Diseases of Field Crops* (2nd ed.). New York: McGraw-Hill.

Duggar, B. M., 1909. *Fungous Diseases of Plants.* New York: Ginn & Co.

Elliott, C., 1951. *Manual of Bacterial Plant Pathogens* (2nd ed.). Waltham, Mass.: Chronica Botanica.

Fawcett, H. S., 1936. *Citrus Diseases and Their Control* (2nd ed.). New York: McGraw-Hill.

Felt, E. P., and W. H. Rankin, 1932. *Insects and Diseases of Ornamental Trees and Shrubs.* New York: Macmillan.

Filipjev, J. N., and J. H. Schuurmans Stekhoven, Jr., 1941. *A Manual of Agricultural Helminthology.* Leiden: E. J. Brill.

Fischer, G. W., 1951. *The Smut Fungi, a Guide to the Literature, with Bibliography.* New York: Ronald Press.

Forsberg, J. L., 1963. *Diseases of Ornamental Plants* (Univ. Ill. Coll. Agr. Spec. Pub. No. 3). Urbana, Ill.: University of Illinois College of Agriculture.

Fulton, J. P., D. A. Slack, N. D. Fulton, J. L. Dale, M. J. Goode, and G. E. Templeton, 1962. *Plant Pathology Laboratory Manual.* Minneapolis: Burgess.

Garrett, S. D., 1944. *Root Disease Fungi.* Waltham, Mass.: Chronica Botanica.

———, 1956. *Biology of Root-infecting Fungi.* London: Cambridge University Press.

Gäumann, E. A., 1951. *Pflanzliche Infektionslehre* (2nd ed.). Basel: Birkhauser. [English translation (of 1st ed.) by W. B. Brierley, 1950. *Principles of Plant Infections.* London: Crosby Lockwood and Son.]

Goodey, T., 1933. *Plant Parasitic Nematodes and the Diseases They Cause.* London: Methuen.

Goodman, R. N., Z. Kiraly, and M. Zaitlin, 1967. *The Biochemistry and Physiology of Infectious Plant Disease.* Princeton, N. J.: Van Nostrand.

Heald, F. D., 1933. *Manual of Plant Diseases* (2nd ed.). New York: McGraw-Hill.

———, 1943. *Introduction of Plant Pathology* (2nd ed.). New York: McGraw-Hill.

Hesler, L. R., and H. H. Whetzel, 1917. *Manual of Fruit Diseases.* New York: Macmillan.

Holmes, F. O., 1949. *The Filterable Viruses,* pp. 1127–1286. Baltimore: Williams & Wilkins.

Holton, C. S., et al., eds., 1959. *Plant Pathology—Problems and Progress, 1908–1958.* Madison, Wis.: University of Wisconsin Press.

Horsfall, J. G., 1956. *Principles of Fungicidal Action.* Waltham, Mass.: Chronica Botanica.

———, and A. E. Dimond, eds., 1959–60. *Plant Pathology—An Advanced Treatise,* vol. I (The diseased plant), vol. II (The pathogen), vol. III (The diseased population—epidemics and control). New York: Academic Press.

Hubert, E. E., 1931. *An Outline of Forest Pathology.* New York: Wiley.

Kenaga, C. B. 1970. *Principles of Phytopathology.* Lafayette, Ind: Balt.

Leach, J. G., 1940. *Insect Transmission of Plant Diseases.* New York: McGraw-Hill.

Martin, H., 1964. *The Scientific Principles of Crop Protection* (5th ed.). New York: St. Martin's Press.

Melhus, I. E., and G. C. Kent, 1939. *Elements of Plant Pathology.* New York: Macmillan.

Mirocha, C. J., and I. Uritani, eds., 1967. *The Dynamic Role of Molecular Constituents in Plant Parasite Interaction* (proceedings of a conference held 15–21 May, 1967, at Gamagori, Japan). St. Paul, Minn.: Bruce.

Owens, C. E., 1928. *Principles of Plant Pathology.* New York: Wiley.

Parris, G. K., 1970. *Basic Plant Pathology.* State College, Miss.: published privately by the author.

Peace, T. R., 1962. *Pathology of Trees and Shrubs.* New York: Oxford University Press.

Pirone, P. P., B. O. Dodge, and H. W. Rickett, 1960. *Diseases and Pests of Ornamental Plants* (3rd ed.). New York: Ronald Press.

Sasser, J. N., and W. R. Jenkins, eds., 1960. *Nematology.* Chapel Hill, N. C.: University of North Carolina Press.

Sharvelle, E. G., 1961. *The Nature and Uses of Modern Fungicides.* Minneapolis: Burgess.

Shurtleff, M. C., 1966. *How to Control Plant Diseases in Home and Garden* (2nd ed.). Ames, Iowa: Iowa State University Press.

Smith, E. F., 1905–14. *Bacteria in Relation to Plant Diseases* (3 vols.). Washington, D. C.: Carnegie Institution of Washington.

———, 1920. *An Introduction to Bacterial Diseases of Plants.* Philadelphia: Saunders.

Smith, K. M., 1951. *Recent Advances in the Study of Plant Viruses* (2nd ed.). Philadelphia: Blakiston.

———, 1959. *A Text Book of Plant Virus Diseases* (2nd ed.). Boston: Little, Brown.

———, 1960. *Plant Viruses.* New York: Wiley.

———, 1962. *Viruses.* London: Cambridge University Press.

Sprague, R., 1950. *Diseases of Cereals and Grasses in North America (Fungi, Except Smuts and Rusts).* New York: Ronald Press.

Stakman, E. C., and J. G. Harrar, 1957. *Principles of Plant Pathology.* New York: Ronald Press.

Stapp, C., 1961. *Bacterial Plant Pathogens.* London: Oxford University Press.

Stevens, F. L., 1913. *The Fungi Which Cause Plant Disease.* New York: Macmillan.

———, 1925. *Plant Disease Fungi.* New York: Macmillan.

Stevens, N. E., and R. B. Stevens, 1952. *Disease in Plants.* Waltham, Mass.: Chronica Botanica.

Strobel, G. A., and D. E. Mathre, 1970. *Outlines of Plant Pathology.* New York: Van Nostrand Reinhold.

Thorne, G., 1961. *Principles of Nematology.* New York: McGraw-Hill.

Walker, J. C., 1952. *Diseases of Vegetable Crops.* New York: McGraw-Hill.

———, 1969. *Plant Pathology* (3rd ed.). New York: McGraw-Hill.

Wallace, H. R., 1964. *The Biology of Plant Parasitic Nematodes.* New York: St. Martin's Press.

Wallace, T., 1961. *The Diagnosis of Mineral Deficiencies in Plants* (2nd ed.). New York: Chemical Publishing Co.

Westcott, C., 1950. *Plant Disease Handbook.* New York: Van Nostrand.

Wheeler, B. E. J., 1969. *An Introduction to Plant Diseases.* New York: Wiley.

Wood, R. K. S., 1967. *Physiological Plant Pathology.* Oxford and Edinburgh: Blackwell.

In addition to these books, and numerous others written in English by American and British plant pathologists, there are a great many fine books on the subject in other languages, including German, French, Dutch, Russian, Japanese, and the Scandanavian languages. Often, it is desirable to consult books written in languages other than English in order to obtain complete lists of references to the literature on certain diseases.

The five-volume *Thesaurus of Mycological Literature,* by Lindau and Sydow, is an excellent source of references to mycological and plant-pathological papers by early workers. Oudeman's *Enumeratio Systematica Fungorum,* also in five volumes, is a valuable source for early references on plant-pathogenic fungi. Saccardo's *Sylloge Fungorum* is the classical work in which Latin descriptions are given for all fungi described before 1931. The *Check List of Fungi,* published by the Commonwealth Mycological Institute, cites references containing descriptions of recently described fungi.

Journals in Which Original Articles are Published

Original articles usually have more or less comprehensive bibliographies that serve as ready sources of references; only rarely, however, are such bibliographies complete. The following is a list of some journals in which articles on plant pathology and related subjects are published. The recommended abbreviation is listed after the name of each journal.

American Journal of Botany—*Amer. J. Bot.*

American Potato Journal—*Amer. Potato J.*

Annals of Applied Biology—*Ann. Appl. Biol.*

Annals of Botany—*Ann. Bot.*

Annals of the Missouri Botanical Garden—*Ann. Mo. Bot. Gard.*

Australian Journal of Agricultural Research—*Aust. J. Agr. Res.*

Australian Journal of Biological Sciences—*Aust. J. Biol. Sci.*

Australian Journal of Botany—*Aust. J. Bot.*

Botanical Gazette—*Bot. Gaz.*

Canadian Journal of Botany—*Can. J. Bot.* [continuation of *Can. J. Res.* (C)]

Canadian Journal of Research, Section C—*Can. J. Res.* (C) [continued as *Can. J. Bot.*]

Comptes Rendus Hebdomadaires des Seances de l'Academie des Sciences—*C. R. Hebd. Seances Acad. Sci.*

Contributions of the Boyce Thompson Institute for Plant Research—*Contrib. Boyce Thompson Inst. Plant Res.*

Journal of Agricultural Research—*J. Agr. Res.* [terminated in 1949]
Journal of General Microbiology—*J. Gen. Microbiol.*
Journal of Helminthology—*J. Helminthol.*
Mycologia—not abbreviated
Nature—not abbreviated
Netherlands Journal of Plant Pathology—*Neth. J. Plant Pathol.* [continuation of *Tijdschr. Plantenziekten*]
Phytopathologische Zeitschrift—*Phytopathol. Z.*
Phytopathology—not abbreviated
Plant Disease Reporter—*Plant Dis. Rep.*
Plant Pathology—*Plant Pathol.*
Science—not abbreviated
Tijdschrift over Plantenziekten—*Tijdschr. Plantenziekten* [continued as *Neth. J. Plant Pathol.*]
Transactions of the British Mycological Society—*Trans. Brit. Mycol. Soc.*
Virology—not abbreviated
Zeitschrift für Pflanzenkrankheiten (Pflanzenpathologie) und Pflanzenschutz—*Z. Pflanzenkr. (Pflanzenpathol.) Pflanzenshutz*

Journals in Which Review Articles are Published

Articles in review journals are often useful in determining the scope of a topic, but these articles are not substitutes for original research papers. Review articles on plant pathology appear in the journals listed below.

Advances in Virus Research—*Advan. Virus Res.*
Annual Review of Microbiology—*Annu. Rev. Microbiol.*
Annual Review of Phytopathology—*Annu. Rev. Phytopathol.*
Annual Review of Plant Physiology—*Annu. Rev. Plant Physiol.*
Bacteriological Reviews—*Bacteriol. Rev.*
Biological Reviews (Cambridge)—*Biol. Rev. (Cambridge)*
Botanical Review—*Bot. Rev.*
Quarterly Review of Biology—*Quart. Rev. Biol.*

PREPARING A BIBLIOGRAPHY

A bibliography is a list of references to literature pertinent to a subject. A systematic search of abstracting journals, beginning with the most recent publications, is the best way to begin the preparation of the bibliography.

The Key-Words Sheet

The key-words sheet is a work sheet on which should be written the important words to be looked up in the indices of the abstracting journals that you are searching for references. Additional key words should be added as they are discovered during the preparation of the bibliography.

The Check List

The check list, the second work sheet, constitutes a record of the literature that you have searched for references. A thorough and systematic search of the literature cannot be made unless a complete and accurate record of the sources is kept.

Searching the Abstracting Journals

Although appropriate abstracting journals are the best sources of references, textbooks should also be checked, because they usually contain listings of important references. First, check one book in which there is an account of the subject of the paper you are preparing. Read the account in order to learn something of the subject itself, then check the list of references: select the more important references, particularly those to papers published before 1920, and cite each on a separate card or sheet of paper.

Next, select the appropriate abstracting journals, and proceed to check them for references to the subject at hand. In most instances, the *Review of Applied Mycology* is the best source of references, and it should be checked first. Next, *Experiment Station Record* can be checked for references to literature published in the United States before about 1920. Then, if it is necessary, other abstracting journals and bibliographies can be checked. Begin the search of literature by selecting the latest available volume of the *Review of Applied Mycology*.

After you have checked the index of a volume of the abstracting journal, look up, in that volume, the abstracts on the subject you are studying. Read each abstract, and decide whether or not the information in it will be useful in preparation of your paper. If it is an article that might prove useful, write the citation to the reference on a card or sheet of paper. Repeat this procedure for all volumes of each abstracting journal that you consult. When this has been done, there will be a stack of reference sheets that make up the bibliography.

Citing References on Bibliography Sheets

Various sets of rules for arranging items in citations to references are prescribed by various journals as matters of editorial policy. Consistency and, above all, accuracy in the citation of references make it possible for readers to find such references readily.

Abbreviations used in citing references are those published in *BIOSIS 1970 List of Serials with Title Abbreviations.* This reference is listed at the end of the chapter.

ABSTRACTING ORIGINAL PAPERS

After your references have been compiled into a bibliography, and after you have decided upon the outline of your paper, you are ready to obtain the facts that will be woven together to make up the presentation. Often it is possible to determine from the literature citation which of the papers that make up the bibliography are likely to be most important. These should be consulted and abstracted first.

There are two common abstracting systems: the single-abstract system and the multiple-abstract system. Select one system and use it throughout. In the single-abstract system, abstracts are written on bibliography sheets, beginning just below the literature citation. In the multiple-abstract system, nothing is written on bibliography sheets except the reference citation and the source of the reference; the abstracts themselves are written on sheets or cards with topical headings, and are cross-referenced to the bibliography sheets. When the multiple-abstract system is used, information from a single article may appear on several abstracting sheets.

WRITING A PAPER

After the sources of literature have been checked, and after pertinent information has been abstracted from the original publications that make up the bibliography, the paper itself can be written.

First, prepare the section of literature cited. Arrange the references in alphabetical order by the surnames of the senior authors. If the single-abstract system is used, the abstracting sheets themselves will be so arranged. If the multiple-abstract system is used, only the reference sheets will be so arranged, and the abstracting sheets will be arranged in order according to subject headings.

When you are ready to write the text, first arrange the abstracts so they can be reviewed conveniently. Then, write a clear, concise summary and a concise introductory paragraph. Review carefully the abstracts that contain information on the topics of your paper, and formulate paragraphs in which the information on each topic is presented. Above all, be accurate, clear, and concise throughout the paper. Use good English. Use the technical terms of plant pathology wherever they are applicable, and cite sources of information in the text. Use direct quotations only when the point under discussion demands their use for the sake of clarity. Critically review and revise the first draft of your completed paper, and have it reviewed by a responsible critic before submitting it to an editor.

Selected References

Anderson, J. A., and M. W. Thistle, 1947. On writing scientific papers. *Can. J. Res. Bull.* (1691).

BIOSIS 1970 List of Serials with Title Abbreviations. Philadelphia: Bio-Sciences Information Service of Biological Abstracts.

Conference of Biological Editors, Committee on Form and Style, 1964. *Style Manual for Biological Journals* (2nd ed.). Washington, D.C.: American Institute of Biological Sciences.

Price, W. C., 1954. Preparing manuscripts for *Phytopathology. Phytopathology* 44:667–674.

Riker, A. J., 1946. The preparation of manuscripts for *Phytopathology. Phytopathology* 36:955–977.

Strunk, William, Jr., and E. B. White, 1959. *The Elements of Style.* New York: Macmillan.

Glossary

Most of the definitions herein are based on those that appear in one or more of the references cited at the end of the Glossary.

abscission: The shedding of leaves or fruit as a result of cambial activity of cells in a zone (abscission layer) that extends across the base of the petiole; premature formation of abscission layers (proleptic abscission) is a hyperplastic symptom of some plant diseases.

acervulus (*pl.* acervuli): A fruiting body of certain imperfect fungi: a shallow, saucer-shaped structure with a layer of conidiophores that bear conidia; as the acervulus expands to form a cushionlike mass, it ruptures the cell layers of the substrate, exposing the spore-bearing surface.

acropetal: Describes the development of structures (such as spores) in succession from the base towards the apex.

actinomycetes: Organisms characterized by fine hyphae, usually less than 1.0 μ in diameter, that readily break into fragments resembling bacterial cells. They are classified as bacteria by some, as fungi by others.

adaptation: From an evolutionary standpoint, a characteristic of a living organism that improves its chances for survival in the environment of its habitat; change brought about in a population of an organism as a result of exposure to a particular set of environmental conditions, the change enabling the organism to adjust to the environmental conditions in question.

aecidium (*pl.* aecidia): A cup-shaped aecium that contains a peridium. Aecidia usually are brightly colored (orange) and occur in groups that make up the "cluster-cup" stage of certain rust fungi.

aecium (*pl.* aecia): The fruiting body of certain rust fungi produced immediately after the pycnium: the aecium produces dicaryotic spores in chains; spores may infect the suscept in which they are formed, but usually infect an unrelated alternate suscept. Aecia may be of the aecidium, caeoma, peridermium, or roestelium type.

agar: A polysaccharide obtained from certain seaweeds; agar forms a gel with water, and is used at a concentration of 1.5–2.0 percent to solidify media used for culturing microorganisms.

agent of inoculation: That which transports inoculum from its source to or into the infection court. Examples: wind, splashing rain, running water, insects, man, other animals.

albication: Whiteness; a hypoplastic symptom of plant disease characterized by the failure of color to develop in organs that are normally colored. Albication is one kind of achromatosis or suppression of color development.

alga (*pl.* algae; *adj.* algal): Aquatic plants or plants of damp habitats: unicellular or multicellular plants with filamentous thalli; distinguished from the fungi by the presence of chlorophyll.

alternate host: One of two species of plants required as a host by heteroecious rust fungi for the completion of their developmental cycles.

alternative host: A plant other than the main host that is fed upon by a parasite; alternative hosts are not required for completion of the developmental cycle of the parasite.

amoeboid: Amoebalike; moving by temporary processes (pseudopodia) from the surface of the vegetative body.

amphiospore: Urediniospore with thickened walls and capable of hibernating.

amphispore: Amphiospore.

anastomosis: Union of one organ (for example, hypha) with another.

antagonistic symbiosis: Parasitism; one organism of an association benefits at the expense of the other.

antheridium: The fingerlike male sexual organ of the oomycetes.

anthocyanescence: The condition of being anthocyanescent; reddish-purple color in tissues that are normally green—often a plesionecrotic symptom of plant disease caused by an infectious agent, usually appearing in the margins around holonecrotic spots in green leaves.

anthracnose: Common name of plant diseases usually characterized by development of ulcerous lesions on stems, leaves, and fruit; often caused by certain imperfect fungi that produce conidia in acervuli (examples: *Colletotrichum. Gloeosporium, Kabatiella*).

antibiosis: An association between two organisms that is detrimental to the vital activities of one of them.

antibody: A protein produced in an animal in response to activities of a normally foreign substance (antigen) that gains access to its tissues; it combines chemically with the antigen and renders it harmless.

antigen: A substance capable of stimulating the formation of antibodies.

apothecium (*pl.* apothecia): Sessile or stalked saucer-shaped fruiting structure of ascomycetes of the series Discomycetes; the inner surface is lined with a hymenium consisting of asci and paraphyses (sterile cells).

appressorium (*pl.* appressoria): A bulbous or lobed swelling of a hyphal tip; it is often held to the surface of the substrate by a gelatinous secretion. Suscept tissue is penetrated by a slender peg that forms at the center of the appressorium.

ascigerous: Ascus-bearing (perfect) stage of development of an ascomycete.

ascocarp: The fruiting body (apothecium, cleistothecium, or perithecium) of an ascomycete.

Ascomycetes: A class of fungi characterized by endogenous production of spores (ascospores) in the organ of meiosis (ascus).

ascospore: Sexually produced spore borne within a saclike cell, the ascus.

ascus (*pl.* asci): Organ of meiosis and ascospore-producing cell; usually contains eight ascospores.

asexual reproduction: Reproduction without gametes (for example, reproduction by conidia in the imperfect fungi).

autoecious: Having all different spore forms (rust fungi) produced on only one species of host plant.

autotrophic: Capable of growth independent of outside sources (for example, green plants and certain bacteria).

bactericidal: Lethal to bacteria.

bacteriophage (phage): A virus that replicates in bacteria.

bacteriostatic: Inhibiting the growth of, but not lethal to, bacteria.

bacterium (*pl.* bacteria; *adj;* bacterial): Small (0.5—2.0 μ) motile or nonmotile microorganism that contains no chlorophyll and that reproduces by simple cell division.

Basidiomycetes: A class of fungi characterized by exogenous production of spores (basidiospores) on club-shaped organs of meiosis (basidia).

basidiospore: Sexually produced exogenous spore of a basidiomycete.

basidium (*pl.* basidia): Typically club-shaped organ of meiosis of basidiomycetes; basidiospores (usually four) are borne externally on stalks (sterigmata) that extend from the apex of the basidium.

basipetally: Successively from apex to base.

binomial nomenclature: The method of scientifically naming plants and animals in descriptive Latin terms; the first term identifies the genus, the second the species to which an organism belongs. The first letter of the generic name is capitalized and both names are italicized. The name (often abbreviated) of the author responsible for naming the organism may follow the Latin binomial. When another author transfers a species to another genus, the name of the first author is placed in parentheses and the name of the second author follows. Thus, the scientific name of the fungus that causes brown rot of peach is written *Monilinia fructicola* (Wint.) Honey. (See Stafleu et al., 1972.).

biotype: A population of individuals that have identical genetic but varying physiological characters; a subdivision of a pathologic race.

blasting: A symptom of plant disease characterized by shedding of unopened buds; classically, the failure to produce fruit or seed.

blight: A nonrestricted necrotic symptom characterized by ultimate death of tissues throughout entire organs, such as leaves or flowers.

blotch: A necrotic spot with fibrillose margins of visible mycelial strands.

caeoma: An aecium that is surrounded by fungal filaments but that has no peridium.

callus: Parenchymatous tissue of cambial origin that forms in response to wounding.

canker: a necrotic symptom of disease in woody plant parts: the necrosis is restricted to a definite area that is surrounded by callus.

carrier: An organism that bears an infectious agent internally but that shows no marked symptoms of the disease caused by that agent.

caryogamy (karyogamy): Fusion of two haploid nuclei.

causal agent of disease: That which is capable of causing disease.

chemotherapy: Treatment of disease by chemicals that work internally.

chlamydospore: A thick-walled asexual spore that develops directly from hyphal cells.

chloranemia: The necrotic symptom of yellowing; a loss of chlorophyll.

chlorosis (*pl.* chloroses; *adj.* chlorotic): A hypoplastic symptom of plant disease characterized by a deficiency of chlorophyll that is due to its failure to develop fully.

cleistothecium (*pl.* cleistothecia): Completely enclosed ascigerous fruiting body of certain fungi (for example, powdery mildews).

cluster cup: Aecidium.

coenocytic: Adjective that describes thalli, especially those of archimycetes and phycomycetes, consisting of multinucleate masses of protoplasm.

commensalism: Symbiosis in which neither organism is injured; one or neither may be benefited.

conidium (*pl.* conidia): Asexual fungal spore produced exogenously on a specialized hypha (conidiophore).

control: Reduction of crop losses that are due to plant diseases.

coremium (*pl.* coremia): A fungal fruiting structure consisting of a cluster of erect intertwined hyphae that bear conidia at their apices.

crozier formation: Process of ascus development from coiled tips of ascigerous hyphae.

damping-off: Disease or necrotic symptom of disease in seedlings in which the seedling stem is decayed near the soil line and the seedling topples. Damping-off pathogens may also prevent seed germination and kill the sprout before it emerges from the soil.

dichotomous: Dividing into two equal branches.

dieback: Necrotic symptom of disease in which death of shoot tissues begins at the tip and progresses back toward the main stem.

dicaryon (dikaryon): A pair of haploid nuclei that occur in a cell; the nuclei undergo simultaneous division (conjugate division) upon formation of each new cell. Typified in the diploid phase (dicaryophase) of basidiomycetes.

diploid: State of having homologous chromosomes in pairs in the nucleus so that twice the haploid number is present.

discomycete: An ascomycete whose fruiting body is an apothecium (for example, *Sclerotinia sclerotiorum*).

disease: Harmful physiological processes caused by continuous irritation of a plant by a primary causal agent. It results in morbific cellular activity and is expressed in characteristic pathological responses called symptoms.

disinfest: To kill pathogens that have not yet initiated disease, but that occur in or on such inanimate objects as soil, tools, and so on, or that occur on the surface of such plant parts as seeds.

dispersal (dissemination): Spread of a pathogen within an area of its graphical range. "Effective dissemination" is synonymous with "inoculation."

distribution: Spread of a pathogen to areas outside of its previous geographical range; "geographical distribution" is synonymous with "range."

ectoparasite: Parasite living on the outside of its host (for example, stubby-root and dagger nematodes). *Cf.* "endoparasite."

ectosymbiosis: Symbiosis in which one member (microsymbiote) develops on the outside of the other member. *Cf.* "endosymbiosis."

ectotrophic: Refers to a mycorrhiza in which the mycelium forms an external covering on the root (in pine, for example). *Cf.* "endotrophic."

edema: Oedema.

eelworms: Nemas, nematodes.

encysted: Surrounded by a hard shell (cyst).

endemic: Refers to a disease restricted to a particular area.

endogenous: Arising from within the generating structure (for example, endoconidia of *Thielaviopsis basicola*). *Cf.* "exogenous."

endoparasite: Parasite living within its host. *Cf.* "ectoparasite."

endosymbiosis: Symbiosis in which one member (microsymbiote) lives within the other. *Cf.* "ectosymbiosis."

endotrophic: Refers to a mycorrhiza in which the mycelium grows within the cortical cells of the root (in orchids, for example). *Cf.* "ectotrophic."

enphytotic: Plant disease occurring regularly in a restricted area.

enzootic: Regularly occurring animal disease of restricted range.

epidemic: Outbreak of a disease in a high percentage of a population.

epidemiology: The study of epidemics. *Cf.* "epiphytology."

epinasty: Downward curling of a leaf blade due to cell growth on the upper side of a petiole being more rapid than that on the lower side; often a hyperplastic symptom of plant disease (for example, the disease caused by potato virus Y in *Physalis floridana*).

epiphyte: Nonparasitic plant that is attached to another plant for mechanical support only. Examples: various orchids, lichens, mosses, and the bromeliad known as Spanish moss.

epiphytology: The study of epiphytotics, with special reference to occurrence and severity of a plant disease as influenced by such environmental factors as moisture, temperature, wind, the nature of the soil, and so forth. Analogous to "epidemiology."

epiphytotic: Occurrence of a plant disease in abundant proportions in an extensive area.

epizootic: Occurrence of an animal disease in abundant proportions.

eradication: Principle of plant-disease prevention characterized by destruction or removal of a pathogen already established in a given area.

ergot: Disease of certain grasses and cereals, especially rye, caused by *Claviceps purpurea;* also the spur-shaped sclerotium of *C. purpurea* that replaces the grain in a diseased inflorescence. The sclerotium contains toxic alkaloids, which have been used by physicians for medicinal purposes, including the induction of labor in women. Consumption of ergot-infested grain by animals, or human consumption of bread made from ergot-diseased grain, results in ergotism, a dreaded disease that is either paralytic or gangrenous.

erumpent: Breaking through (for example, growth of sporodochia through the epidermis of susceptible plant organs).

escape: Failure of inherently susceptible plants to become diseased, even though disease is prevalent.

etiolation: A phenomenon exhibited by plants grown in the dark: etiolated plants are pale yellow and have long internodes and small leaves.

etiology: The study of cause; that phase of plant pathology which deals with the causal agent and its relations with the susceptible plant.

exclusion: The principle of plant-disease prevention in which the pathogen is prevented from entering a given region (for example, by seed disinfestation or quarantine).

exogenous: Arising on the outside of the generating structure. *Cf.* "endogenous."

extracellular: Outside the cells. *Cf.* "intracellular."

facultative parasite: An organism that is normally saprophytic but that can live as a parasite. *Cf.* "facultative saprophyte," "obligate."

facultative saprophyte: An organism that is normally parasitic but that can live saprophytically. *Cf.* "facultative parasite."

fasciation: Hyperplastic symptom characterized by a fusing (and flattening) of such plant organs as stems.

fasciculation: Hyperplastic symptom characterized by a clustering of such plant organs as shoots into such structures as "witches'-brooms."

filiform: Theadlike.

flagellate: Having one or more flagella.

flagellum (*pl.* flagella): Fine whiplike filament of a cell that enables the cell to move about in a liquid medium.

forma specialis (*pl.* formae speciales; *abbr.* f. sp.): Special form: a biotype (or group of biotypes) of a species of pathogen that differs from others in the ability to infect selected genera or species of susceptible plants. *Cf.* "physiologic race."

fructification: Spore production by fungi; or, a structure in or on which spores are formed.

fruiting body: A complex structure that bears fungal spores. Examples: sporangia, coremia, sporodochia, acervuli, pycnidia, apothecia, perithecia.

f. sp.: Forma specialis.

fumigant: Liquid or solid chemical that forms vapors that kill organisms; usually, they are used on soils or within closed structures to eradicate pathogens.

Fungi Imperfecti: A form class erected to contain those fungi whose sexual reproduction is unknown.

fungicide: A substance that kills fungi.

fungus (*pl.* fungi; *adj.* fungal or fungous): Nonchlorophyllous plant whose vegetative body (thallus) consists of threadlike filaments (hyphae) aggregated into branched systems (mycelia).

gall: A tumefaction or tumor; hyperplastic symptom of plant disease characterized by localized swelling or outgrowth of tissue composed of unorganized cells. *Cf.* "tumefaction."

gametangium (*pl.* gametangia): A structure in which gametes are produced.

gamete: "Germ-cell"; a reproductive cell whose haploid nucleus is capable of fusion with that of a gamete of the opposite mating type. *Cf.* "zygote."

gametophyte: A phase of the life history of a plant that arises from a haploid spore resulting from meiosis in a diploid sporophyte; plants have haploid nuclei during the gametophyte phase.

genus (*pl.* genera): A taxonomic category ranking above species and below family; the generic name of an organism is the first of the binomial.

germ tube: The hypha produced by a fungal spore when it begins to grow.

germination: The beginning of growth of a spore or seed.

gill: A lamella of a gill fungus.

gill fungi: Mushrooms; members of the family Agaricaceae of the class Basidiomycetes, which produce fruiting bodies consisting of a stipe that supports a pileus whose lower surface has radially arranged lamellae bearing the hymenium.

habitat: A place with a particular kind of environment that is suitable for the growth of an organism.

haploid: Having a single set of unpaired chromosomes in each nucleus.

haustorium (*pl.* haustoria): Absorbing organ of certain parasites. The haustoria of plant-parasitic fungi enter plant cells and invaginate the cytoplast as they enlarge into simple or branched structures.

heteroecious: Of rust fungi, requiring more than one host for a complete developmental cycle. *Cf.* "autoecious."

heterocaryosis (heterokaryosis): A condition in which two or more genetically different nuclei occur in one cell.

heterothallism: Of fungi and algae, a condition in which the fusion of two mycelia of different nuclear components (mating types) is required for sexual reproduction, even though each strain may produce male and female organs. *Cf.* "homothallism."

heterotopy: Hyperplastic symptom in which an organ develops in a position other than its normal one (for example, the development of "ears" in the tassels of corn plants).

heterotrophic: Refers to organisms that are dependent upon outside sources for growth, being incapable of synthesizing required organic materials from inorganic sources. Heterotrophs obtain their food from other organisms, living or dead. *Cf.* "autotrophic."

holocarpic reproduction: In fungi, reproduction in which the entire fungal body (thallus) is segmented into spores.

holonecrosis (*pl.* holonecroses): The condition of being dead. *Cf.* "plesionecrosis."

homothallism: Of fungi and algae, a condition in which nuclei of both mating types occur in the same haploid thallus. *Cf.* "heterothallism."

host: A living organism from which a parasite derives its sustenance.

hyaline: Transparent.

hybridization: Cross-breeding.

hydrosis (*pl.* hydroses; *adj.* hydrotic): Necrotic symptom of disease characterized by water-soaking of tissues.

hymenium (*pl.* hymenia): A layer of regularly arranged asci or basidia in fungal fruiting bodies.

hyperparasite: An organism that parasitizes another parasite.

hyperplasia: An overgrowth resulting from an abnormal increase in the numbers of cells. *Cf.* "hypertrophy," "hypoplasia."

hypertrophy: An overgrowth resulting from an abnormal increase in the size of cells. *Cf.* "hyperplasia."

hypha (*pl.* hyphae; *adj.* hyphal): A filament of fungal mycelium (thallus) that is composed of one or more cylindrical cells and that increases in length by growth at its tip. New hyphae arise as lateral branches.

hyponasty: More rapid growth of the lower side of an organ (a petiole, for example) than of the upper side. *Cf.* "epinasty."

hypoplasia: Underdevelopment resulting from an abnormal paucity of cells. *Cf.* "hyperplasia."

immunity: In plants, the ability to remain free from disease because of inherent structural or functional properties. *Cf.* "resistance."

immunization: In plants, complete resistance to disease. *Cf.* "resistance."

imperfect stage: Of fungi, the period of the life cycle other than that in which spores are formed as a result of a sexual process. *Cf.* "perfect stage."

in vitro: Refers to biological processes when they are allowed to occur in isolation from the whole organism. *Cf. "in vivo."*

in vivo: Refers to biological processes when they occur within the living organism. *Cf. "in vitro."*

incubation: Of a pathogen, its development, growth, and penetration of a plant prior to infection; or its activity within plant tissues subsequent to penetration and up to appearance of symptoms, signs of disease, or both.

indexing: Process of determining the presence of disease in a plant by transferring inoculum from the plant to another in which diagnostic symptoms develop. The second plant is termed a "test plant" or an "indicator plant."

infect (*n.* infection): Of a parasite or pathogen, to enter and grow, or to replicate, within plant tissues. If the association is noninjurious, the relationship is simply parasitic; if it is injurious, the relationship is pathogenic as well as parasitic.

infection court: The place on or in the susceptible plant where infection may be initiated.

infection thread: The specialized hypha of a pathogenic fungus that invades tissue of the susceptible plant.

infectious: Refers to a pathogen that can be transmitted from one suscept to another by an external agent; also refers to any disease whose pathogen can be so transmitted.

infective: Of an agent of inoculation, capable of transmitting inoculum.

infest (n. infestation): To be present in numbers. Infestation does not imply disease and is not to be confused with infection.

ingress: The act, by a plant pathogen, of gaining entrance into the tissues of a susceptible plant.

inoculate (n. inoculation): To transfer inoculum to or into an infection court.

inoculum (*pl.* inocula): That portion of a pathogen that may be transferred to an infection court. Examples: a fungal spore, a bacterial cell.

intercellular: Between the cells.

intracellular: Within a cell.

intumescence: Hyperplastic symptom characterized by blisterlike swelling (due to water excess) on the surfaces of plant organs.

invagination: Retraction, under force of pressure, of an outer surface toward the inside.

invasion: Spread of a pathogen through tissues of a diseased plant.

irritability: Responsiveness to change in the physical or biological environment; a property of all living things.

isogamy: The condition in which gametes are morphologically similar, as in the members of the subclass Zygomycetes of the class Phycomycetes.

juvenillody: Condition in which tissues and organs remain immature.

karyogamy: Caryogamy.

klendusity: A special kind of disease escape, in which a susceptible plant avoids disease because of an intrinsic property of the plant itself that greatly reduces the chances of its being inoculated, even though there may be an abundance of inoculum in the area.

lamella (*pl.* lamellae): In mushrooms, one of the thin, spore-bearing gills attached to the underside of the cap.

lesion: A localized area of diseased tissue. Examples: a spot, a scab.

life cycle (life history): The complete succession of changes undergone by an organism during its life. A new cycle occurs when an identical succession of changes is initiated.

local infection: An infection affecting a limited part of a plant.

mechanical inoculation: Of plant viruses, a method of experimentally transmitting the pathogen from plant to plant: juice from diseased plants is rubbed on test-plant leaves that usually have been dusted with Carborundum or some other abrasive material.

medulla: Central part of an organ.

meiosis: Reduction division of chromosomes, through which haploid nuclei arise from diploid ones.

metaplasia. Changed condition of a structure or organ; hyperplastic class of symptoms characterized by overdevelopment other than that due to hypertrophy or hyperplasia. Examples: abnormal starch accumulation, virescence.

microbe: Microscopic organism.

micron: (μ): One-thousandth of a millimeter. A millimicron ($m\mu$) is one-thousandth of a micron and is equivalent to 10 Ångstroms (Å).

microorganism: Microscopic organism. *Cf.* "microbe."

mildew: Downy or powdery fungal growth on the surface of diseased plant or other substrate.

mitosis: Nuclear division in which the chromosome number remains unchanged.

mold: Superficial fungal growth on various substrates.

mosaic: A symptom of certain viral diseases of plants: a patchy pattern of green and light green or yellow in leaves or fruit that are normally green.

mummy: A dried, shriveled fruit (for example, a peach fruit infected with *Monilinia fructicola*).

mushroom (toadstool): A fleshy fruiting body of a fungus, especially of a basidiomycete of the family Agaricaceae.

mutation: A sudden, permanent genetic change.

mutualistic symbiosis (mutualism): Symbiosis beneficial to both members of the association.

mycelium (*pl.* mycelia): The aggregate of the hyphae that constitute the thallus or vegetative body of a fungus.

mycology: The study of fungi.

mycophagous: Feeding on fungi.

mycoplasma: Microscopic organisms intermediate in size between bacteria and viruses.

mycorrhiza (*pl.* mycorrhizae): Literally, "fungus root": an association of fungal mycelium with the roots of a higher plant, from which association the higher plant may derive benefit. The relation may be one of mutualistic symbiosis.

mycosis (*pl.* mycoses): Disease caused by a fungus.

mycotrophic: Refers to green plants having mycorrhizae.

myxamoeba: A naked cell capable of amoeboid movement; characteristic of the vegetative phase of myxomycetes and such plasmodiophoromycetes as *Plasmodiophora brassicae.*

Myxomycetes (Mycetozoa): The slime molds, a class of fungi characterized by amoeboid vegetative protoplasts, plasmodia, and by brightly colored spore-bearing capillitia.

Myxomycota: A division of fungi, to which belong the myxomycetes and the plasmodiophoromycetes, the "primitive fungi."

necrosis (*pl.* necroses; *adj.* necrotic): Death.

nema: Nematode, eelworm.

nematicide (nematocide): A chemical lethal to nematodes.

nematode: Nema, eelworm; a round- or thread-worm of the phylum Nematoda (in some classifications, a class of the phylum Aschelminthes). They are triploblastic, bilaterally symmetrical, unsegmented, invertebrate animals. Many are free-living, whereas others parasitize animals or plants.

nucleic acid: A compound of high molecular weight that consist of pentose (ribose or deoxyribose), phosphoric acid, and nitrogen bases (purines and pyrimidines). Nucleic acids are present in all living things. (The infectious parts of plant viruses consist of nucleic acids.)

nucleoprotein: A compound of nucleic acid and protein. (Most plant viruses thus far chemically characterized are nucleoproteins.)

obligate: Refers to an organism that is restricted to a particular set of environmental conditions, without which it cannot survive. An obligate parasite, for example, is an organism that can function only by parasitizing another organism. *Cf.* "facultative saprophyte," "facultative parasite."

oedema (edema): Intumescence or blister formation due to an increase in intercellular water, as in leaves.

oogonium (*pl.* oogonia): In fungi, the female reproductive structure of such oomycetes as *Pythium* spp. and *Phytophthora* spp.

oospore: A spore formed after the fertilization of a large, passive female cell (oogonium) by a small, active male cell.

ostiole: A pore; an opening in a fruiting body (a perithecium or a pycnidium) through which spores are discharged.

paraphysis (*pl.* paraphyses): In fungi, one of the sterile filaments or cells interspersed between the asci or basidia in a hymenium.

parasexualism: Caryogamy in a heterocaryotic hypha.

parasite: An organism that obtains nourishment from cells of another living organism, the host, while contributing nothing to the host's survival.

parenchyma: Plant tissue consisting of closely packed cells whose walls are predominantly cellulosic (for example, pith and mesophyll).

pathogen: A causal agent of disease in a plant, the suscept.

pathogenesis: That portion of the life cycle of a pathogen during which it becomes, and continues to be, associated with its suscept.

pathogenicity: The ability of a pathogen to cause disease.

pathology: The study of disease.

perfect stage: Of a fungus, that part of the life cycle during which spores are produced sexually. *Cf.* "imperfect stage."

peridermium (*pl.* peridermia): An aecium, of the sort found in *Cronartium, Melampsora,* and *Coleosporium,* with a peridium that extends prominently beyond the chains of aeciospores. *Cf.* "roestelium."

peridium (*pl.* peridia): A membrane of sterile cells that forms around the aecia produced in aecidia or in peridermia or roestelia.

perithecium (*pl.* perithecia): A spherical or flask-shaped, thick-walled fruiting body of ascomycetes, consisting of an internal hymenium of asci and paraphyses and an ostiole through which ascospores are discharged.

phage: Bacteriophage.

phycomycetes: A class of fungi (the "algal fungi") characterized by coenocytic mycelium and sexual reproduction by the union of two sex cells.

phyllody: A change of floral leaves (petals) to foliage leaves.

physiologic race: A biotype, or group of closely related biotypes, that differ from other biotypes of that species in the ability to infect particular varieties of the susceptible plant species. *Cf.* "forma specialis."

physiologic specialization: The existence of a number of races or forms of one species of pathogen.

phytoalexin: In hypersensitive hosts, a metabolic by-product, toxic both to parasite and host, that is formed only in response to attack by the parasite.

pileus (*pl.* pilei): The cap of a mushroom, bearing the hymenium on its lower surface.

plasmogamy: The initiation of the diploid phase in the life cycles of certain fungi: nuclei or different mating types come together and divide conjugately as dicaryons, but do not fuse until later. *Cf.* "caryogamy."

pleomorphism (polymorphism): The occurrence of several forms in the life cycle (for example, many rust fungi are pleomorphic in that they produce as many as five different spore forms in their complete life cycles).

plesionecrosis (*pl.* plesionecroses): A symptom exhibited by tissues not yet dead but in the process of dying (wilting, for example).

polymorphism: Pleomorphism.

predisposition: An increase in susceptibility resulting from the influence of environment upon the suscept.

primary cycle: The first cycle of a pathogen initiated after a period of pathological inactivity, usually after rest or seasonal inactivity (dormancy). *Cf.* "life cycle."

proliferation: A rapid and repeated production of new cells, tissues, or organs; specifically, a hyperplastic symptom of plant disease in which organs continue to develop after they have reached the point beyond which they normally do not grow.

prolepsis (*pl.* prolepses): A hyperplastic symptom of disease in which organs appear before the natural time (for example, the sprouting of shoots from adventitious buds after disease has impaired the metabolism of the organ in question).

promycelium (*pl.* promycelia): An organ of meiosis of smuts and rusts that bears spores exogenously; analagous to, and often termed, a basidium.

protection: As a principle of plant-disease control, the placing of a barrier between suscept and pathogen (for example, the use of protective chemical dusts or sprays).

protoplast: The unit of living substance; a cell with or without a cell wall, although modern usage connotes a wall-less cell, consisting of plasma membrane, cytoplasm, nucleus, and other organelles.

pustule: A pimplelike, eruptive fruiting structure, such as a uredinium of a rust fungus.

pycnidium (*pl.* pycnidia): A hard-walled, flask-shaped, fungal fruiting body that contains conidia.

pycnium (*pl.* pycnia): A spermagonium, a flask-shaped fruiting body of a rust fungus: it contains haploid pycniospores (spermatia) and filaments (receptive hyphae) that extend through the ostiole. A pycnium and its contents are of one mating type.

pygmism: The state of being dwarfed or reduced in size.

quiescence: That period of the prepenetration stage during which a pathogen may be inactive because environmental conditions are unfavorable for its growth.

race: Physiologic race.

range: Of a plant pathogen, the geographical region or regions in which it is known to occur.

resistance: The ability of a plant to remain relatively unaffected by a disease because of its inherent genetic and physiological or structural characteristics.

response: The change produced in an organism by a stimulus.

resting spore: A fungal spore, usually thick-walled, that can remain viable in a dormant condition for an extended period. Examples: chlamydospores, oospores, amphiospores, teliospores.

rhizomorph: A compact strand of hyphae formed by the longitudinal joining of hyphal strands into a bundle; it has a hard outer covering, grows from the apex, and serves as a survival organ and in the transport of food materials within the thallus.

roestelium (*pl.* roestelia): In *Gymnosporangium* spp., an aecium with a strongly developed peridium extending beyond the chains of aeciospores. *Cf.* "peridermium."

rogue: A variation from the standard varietal type; also, to remove such undesirable plants (especially those infected by viruses) from the growing crop.

roundworms: Nematodes.

rugose: Rough; said of leaves roughened or crinkled by viral diseases (for example, rugose mosaic of potatoes).

russetting: A hyperplastic symptom of disease in which brownish, roughened areas form on the skin of fruit or tubers because of excessive cork-cell production.

rust: A disease caused by a rust fungus (order Uredinales); also, the fungus itself.

saltation: A mutation occurring in the asexual stage of fungal growth, especially one occurring in culture.

saprogenesis: Survival; that phase of the life cycle of a pathogen during which it is not actively causing disease in a living suscept.

saprophyte: An organism that obtains its nourishment from nonliving organic matter. *Cf.* "parasite."

sarcody: A hyperplastic symptom in which swellings occur above and below portions of organs that are tightly encircled, as a stem might be "choked" by a twining vine.

scab: A hyperplastic symptom characterized by rough, crusty lesions formed by excessive cork production.

sclerotium (*pl.* sclerotia): A hard, compact mass of fungal tissue consisting of an outer sclerotized rind and an inner parenchymatous medulla; capable of surviving long periods of adverse environmental conditions.

secondary cycle: A cycle initiated by inoculum from a primary (or from another secondary) cycle without an interposed resting or dormant period for the pathogen. *Cf.* "primary cycle."

septate: Of hyphae, divided into compartments ("cells") by crosswalls.

septum (*pl.* septa): A partition or crosswall, as of a hypha or a spore. Hyphal septa are discoid with a central pore.

seta (*pl.* setae): Of fungi, bristlelike or hairlike structure occurring in certain fruiting bodies (in acervuli, for example) and on certain spores.

shot hole: A symptom characterized by the dropping out of roundish fragments of leaf tissue attacked by certain leaf-spotting pathogens.

signs: Of pathogens, visible structures produced in or on diseased tissues.

smut: A disease caused by smut fungi (order Ustilaginales); also, the fungus itself.

smut spore: A dark, thick-walled resting spore of a smut fungus; may germinate to produce a promycelium, the organ of meiosis; often improperly termed chlamydospore. *Cf.* "teliospore."

sorus (*pl.* sori): Of smut and rust fungi, fruiting bodies in which masses of spores are produced.

species: The fundamental taxonomic category in the classification of living things, ranking below genus. The individuals of a species are, and remain, morphologically distinct from these of other species in their genus. The specific name of an organism (its binomial) is formed from the generic name and a second name known as the specific epithet.

spermagonium (*pl.* spermagonia): A pycnium (also called spermogonium, spermagone).

sporangiophore: A hypha on which one or more sporangia are borne.

sporangium (*pl.* sporangia): A more or less spherical body in which asexual spores are produced.

spore: Of fungi, a one- to many-celled reproductive unit that becomes detached from the parent and that can germinate to give rise to a new individual.

sporidium (*pl.* sporidia): A small spore produced on a promycelium, as in the smuts and rusts.

sporodochium (*pl.* sporodochia): Of fungi, an asexual fruiting body in which conidiophores develop over the surface of an erumpent, cushionlike fungal structure.

sporophore: A structure that bears spores.

sporophyte: A phase of the life cycle during which the plant has diploid nuclei and in which spores are produced after meiosis.

sporulation: The process of producing spores.

spot: A symptom of disease characterized by a limited necrotic area, as on leaves, flowers, and stems.

sterigma (*pl.* sterigmata): Of fungi, a minute stalk that bears spores; basidiospores are borne on sterigmata that protrude as parts of walls of basidia.

sterile: Unable to reproduce sexually; also taken by many to mean free from living microorganisms. *Cf.* "sterilized."

sterilized: Free from living microorganisms.

stimulus (*pl.* stimuli): An environmental change capable of inciting a change in the activities of an organism without itself providing energy for the new activities. *Cf.* "response."

stipe: A stalk, as of mushrooms.

strain: A pure line (genetically homogeneous group) of an organism.

stroma (*pl.* stromata): A mass of fungal tissue, often sclerotized, in which fruiting bodies are produced, as the perithecial stromata of the ergot fungus, *Claviceps purpurea.*

suberized: Refers to cell walls hardened by their conversion to cork (suberin).

summer spore: A fungal spore that germinates without resting and that is usually the inoculum for secondary infections during the growing season of the suscept.

suppression: A hypoplastic symptom characterized by the failure of plant organs or substances to develop.

suscept: Any plant or species of plant that is susceptible to disease; an abbreviated term denoting "susceptible plant" or "susceptible species."

susceptibility: The condition of being susceptible; the inability of a plant to resist or avoid disease.

swarmspore: Zoospore.

symbiont: One member of a symbiotic relationship.

symbiosis: A vital association of two dissimilar organisms; often used in the restricted sense to mean mutualistic symbiosis.

symbiote: Symbiont.

symptom: A visible expression by a suscept of a pathologic condition.

symptomatology: The study of symptoms of disease and signs of pathogens for the purpose of diagnosis.

systemic: Of disease, one in which there is general spread of the pathogen throughout the plant.

teleutosorus (*pl.* teleutorsori): Telium.

teleutospore: Teliospore.

teliospore: A teleutospore: a spore of the rust fungi, usually a resting spore, that germinates to produce a promycelium in which meiosis occurs.

telium (*pl.* telia): A sorus of a rust fungus in which teliospores are produced.

thermal death point: That high temperature at which death of an organism occurs after a specified length of time, usually 10 minutes.

threadworm: Nematode.

toadstool: Mushroom.

tolerance: Of diseased plants, the ability to endure disease without serious injury or crop loss.

transmission: The dissemination of pathogens and the inoculation of suscepts.

trichogyne: In some algae, lichens, and fungi, a projection from the female sex organ that receives the male gamete or nuclei before fertilization (caryogamy).

tumefaction: A plant tumor or gall.

tylosis (*pl.* tyloses): A balloonlike outgrowth of a membrane of a vessel or a tracheid pit that protrudes into and blocks the vessel or tracheid cavity.

type specimen: The original collection of an organism on which the description of a new species is based.

urediniospore: Urediospore, uredospore. *Cf.* "uredinium."

uredinium (*pl.* uredinium): Uredium, uredosorus: in a rust fungus, a group of spore-bearing filaments crowded together on which masses of red "summer spores" (urediniospores) are formed.

urediospore: Urediniospore, uredospore.

uredium (*pl.* uredia): Uredinium.

uredosorus (*pl.* uredosori): Uredinium, uredium.

uredospore: Urediniospore, urediospore.

vector: An agent of dissemination or inoculation (for example, an insect that transmits a virus).

vegetative reproduction: Asexual reproduction.

vein banding: A symptom of virus-infected leaves in which tissues along the veins are darker green than other laminar tissue.

vein clearing: A symptom of virus-infected leaves in which veinal tissue is lighter green that that of healthy plants.

viable: Able to live; of spores, able to germinate.

virescence: A symptom in which green pigmentation occurs in plant tissues not normally green.

virulence: Relative ability to cause disease; a measure of pathogenicity.

viruliferous: Containing a virus; of an insect vector, containing virus and being capable of introducing it into a suscept.

virus: An ultramicroscopic (one dimension less than 200 $m\mu$), obligately parasitic, and infectious pathogen of disease. A virus consists of a core of nucleic acid in a protein coat.

wild type: The phenotype characteristic of the majority of individuals of a species under natural conditions.

wilting: Of plant disease, a plesionecrotic symptom characterized by loss of turgor, which results in drooping of leaves, stems, and flowers.

witches'-broom: A hyperplastic symptom due to proleptic development of many weak shoots from adventitious buds in a stem.

yeast: A unicellular ascomycete that multiplies typically by a budding process.

yellowing: A plesionecrotic symptom characterized by the turning yellow of plant tissues that were once green. *Cf.* "chloranemia."

zoosporangium (*pl.* zoosporangia): A sporangium that bears zoospores.

zoospore: A swarm-spore: a naked spore (motile by means of one to many flagella) produced within a sporangium.

zygospore: A thick-walled resting spore formed by the union of similar sex cells (gametes).

zygote: An individual formed by the union of two gametes.

References

Abercrombie, M., C. L. Hickman, and M. L. Johnson, 1951. *A Dictionary of Biology*. Baltimore, Md.: Penguin.

Commonwealth Mycological Institute, 1968. *Plant Pathologist's Pocketbook*. Kew: Commonwealth Mycological Institute.

George, F., 1953. Glossary. *In* U. S. Department of Agriculture, *Plant Diseases* (The Yearbook of Agriculture, 1953), pp. 897–907. Washington, D. C.: U. S. Government Printing Office.

Leach, J. G., 1940. *Insect Transmission of Plant Diseases,* pp. 585–593 (Glossary). New York: McGraw-Hill.

Stafleu, F. A., et al., eds., 1972. International Code of Botanical Nomenclature, adopted by the Eleventh Botanical Congress, Seattle, August 1969. *Regnum Veg.* 82:1–426.

Whetzel, H. H., ca. 1925. Glossary of phytopathological terms. Mimeographed paper, Plant Pathology Department, Cornell University.

Whetzel, H. H., L. R. Hesler, C. T. Gregory, and W. H. Rankin, 1925. *Laboratory Outlines in Plant Pathology* (2nd ed.), pp. 223–226 (Glossary). Philadelphia: Saunders.

Index